T0185746

Graduate Texts in Physics

Series Editors

Kurt H. Becker, NYU Polytechnic School of Engineering, Brooklyn, NY, USA

Jean-Marc Di Meglio, Matière et Systèmes Complexes, Bâtiment Condorcet, Université Paris Diderot, Paris, France

Sadri Hassani, Department of Physics, Illinois State University, Normal, IL, USA

Morten Hjorth-Jensen, Department of Physics, Blindern, University of Oslo, Oslo, Norway

Bill Munro, NTT Basic Research Laboratories, Atsugi, Japan

Richard Needs, Cavendish Laboratory, University of Cambridge, Cambridge, UK

William T. Rhodes, Department of Computer and Electrical Engineering and Computer Science, Florida Atlantic University, Boca Raton, FL, USA

Susan Scott, Australian National University, Acton, Australia

H. Eugene Stanley, Center for Polymer Studies, Physics Department, Boston University, Boston, MA, USA

Martin Stutzmann, Walter Schottky Institute, Technical University of Munich, Garching, Germany

Andreas Wipf, Institute of Theoretical Physics, Friedrich-Schiller-University Jena, Jena, Germany

Graduate Texts in Physics publishes core learning/teaching material for graduate- and advanced-level undergraduate courses on topics of current and emerging fields within physics, both pure and applied. These textbooks serve students at the MS- or PhD-level and their instructors as comprehensive sources of principles, definitions, derivations, experiments and applications (as relevant) for their mastery and teaching, respectively. International in scope and relevance, the textbooks correspond to course syllabi sufficiently to serve as required reading. Their didactic style, comprehensiveness and coverage of fundamental material also make them suitable as introductions or references for scientists entering, or requiring timely knowledge of, a research field.

More information about this series at http://www.springer.com/series/8431

János A. Bergou • Mark Hillery • Mark Saffman

Quantum Information Processing

Theory and Implementation

Second Edition

 Springer

János A. Bergou
Physics and Astronomy
Hunter College
New York, NY, USA

Mark Hillery
Physics and Astronomy
Hunter College
New York, NY, USA

Mark Saffman
Department of Physics
University of Wisconsin–Madison
Madison, WI, USA

ColdQuanta, Inc.
Madison, WI, USA

ISSN 1868-4513   ISSN 1868-4521   (electronic)
Graduate Texts in Physics
ISBN 978-3-030-75438-9   ISBN 978-3-030-75436-5   (eBook)
https://doi.org/10.1007/978-3-030-75436-5

1st edition: © Springer Science+Business Media New York 2013
2nd edition: © Springer Nature Switzerland AG 2021
This work is subject to copyright. All rights are reserved by the Publisher, whether the whole or part of the material is concerned, specifically the rights of translation, reprinting, reuse of illustrations, recitation, broadcasting, reproduction on microfilms or in any other physical way, and transmission or information storage and retrieval, electronic adaptation, computer software, or by similar or dissimilar methodology now known or hereafter developed.
The use of general descriptive names, registered names, trademarks, service marks, etc. in this publication does not imply, even in the absence of a specific statement, that such names are exempt from the relevant protective laws and regulations and therefore free for general use.
The publisher, the authors, and the editors are safe to assume that the advice and information in this book are believed to be true and accurate at the date of publication. Neither the publisher nor the authors or the editors give a warranty, expressed or implied, with respect to the material contained herein or for any errors or omissions that may have been made. The publisher remains neutral with regard to jurisdictional claims in published maps and institutional affiliations.

This Springer imprint is published by the registered company Springer Nature Switzerland AG
The registered company address is: Gewerbestrasse 11, 6330 Cham, Switzerland

To Valéria, Attila, Miklós and Katalin
JB

To Carol
MH

To Dasha, Alexandra, Andrey, and Nadine
MS

Preface to the Second Edition

While we have made many changes in putting together the second edition, the largest is the addition of a third author, Mark Saffman of the University of Wisconsin-Madison. Mark, being an experimentalist, brings a different perspective to the subject, and he has contributed several chapters on implementations of quantum information processing. These are based on classes he has taught in Wisconsin. Consequently, the book is no longer just about the theory of quantum information processing, but about quantum information processing, theory, and experiment.

We have also made additions to the theory parts. There are new chapters on the stabilizer formalism and on quantum information theory as applied to quantum communication. We have also made additions to the existing chapters, including, for example, discussions of different models of quantum computation, entanglement swapping, and repeated measurements. We hope that these additions will improve the usefulness of the book for our readers.

New York, NY, USA

Madison, WI, USA

János A. Bergou
Mark Hillery
Mark Saffman

Preface

These notes are the result of a one-semester graduate course that was first taught during the spring semester in 2003 at the CUNY Graduate Center and has been offered several times since. The students in the courses were all physicists, so a familiarity with quantum mechanics at the first-year graduate level was assumed. The hope was that after taking the course, students could explore the original literature in the subject on their own.

The course covers a range of topics in quantum information, but, given the limited amount of time, it is not by any means exhaustive. We begin with the density matrix and its representations. Next, we study entanglement, starting with Bell's inequalities and continuing with tests for entanglement, in particular the Peres partial transposition test. It is also possible to quantify entanglement, and we show how this can be done for both pure and mixed states, finishing with a discussion of concurrence as a measure of entanglement for states of two qubits. Entanglement is a resource that can be used for quantum communication. Teleportation and dense coding are examples of this. Next, we consider quantum dynamics. In particular, we study generalized quantum dynamics, which generalizes the standard unitary evolution of quantum states. The Kraus representation of quantum maps is derived and applied to examples, such as the depolarizing channel. There are also certain kinds of maps that are impossible, such as the cloning map, a map that produces a perfect copy of an arbitrary input state.

We then move on to the study of quantum measurements. Just as quantum maps generalize the standard unitary evolution, positive operator valued measures (POVMs) generalize the standard projective measurements. Here, we develop an extensive theory of generalized measurements that are described by POVMs. The problem of discriminating between two, nonorthogonal quantum states provides a useful illustration of this type of measurement, and the two commonly employed strategies, the minimum-error strategy and the unambiguous state discrimination strategy, are discussed. These POVMs lead to a discussion of quantum cryptography. In particular, the B92 proposal and the original BB84 proposal are studied from this perspective. Many of the fascinating applications of quantum information theory in the area of quantum communication, such as secret sharing, rely on the impossibility of certain maps.

In quantum computation, the other major area of quantum information processing, consequences of the superposition principle are exploited. In the area of quantum algorithms, we focus primarily on the Deutsch–Jozsa algorithm, the Bernstein–Vazirani algorithm, the Grover search algorithm, and period finding. We also explore a technique that has been useful in finding new algorithms, the quantum walk. In a real quantum computation, it is necessary to protect against errors, and for this, quantum error-detecting codes are necessary. We develop the general theory of such codes, and discuss some examples such as the Shor code and CSS codes.

We also have a chapter on quantum machines, devices that perform certain operations on quantum systems. These may be single purpose or programmable, and we discuss the limits on programmable machines. We conclude with an example of a programmable state discriminator, in which the states to be discriminated are provided as a program rather than hard wired into the machine.

This covers a lot of material, but it also leaves out a lot. In a single semester, we cannot touch on subjects such as the applications of information theory to quantum information or the physical implementations of quantum information protocols, both important subjects. We also do not treat the Shor algorithm for finding the prime factors of a number, not because it is not important, but because it requires some background in number theory. When teaching a one-semester course, time constraints are a very real consideration, and we felt that an adequate presentation of the Shor algorithm and its background would take too much time. Our choice of subjects has been guided by the requirement of providing a firm foundation for further study and by our own interests as we have explored the field.

The chapters are completed with problems and a cursory list of the most relevant literature. The references are not meant to be exhaustive, but to serve as a guide to further reading.

We should also mention two standard sources that we found useful in preparing the notes from which this book originated. One is *Quantum Computation and Quantum Information* by Michael Nielsen and Isaac Chuang. The second is the set of lecture notes by John Preskill for Physics 219 at Caltech, which can be found at http://www.theory.caltech.edu/people/preskill/ph229/. These cover some of the topics we discuss in more depth and also treat many topics that we do not. A more recent book, which can also supplement what we present here, is *Quantum Information* by Stephen Barnett.

Over the years, we benefitted from numerous discussions and close collaborations with many colleagues and friends. Among them we want to particularly thank Erika Andersson, Emilio Bagan, Stephen Barnett, Sam Braunstein, Vladimir Bužek, Luiz Davidovich, Berge Englert, Edgar Feldman, Ulrike Herzog, Igor Jex, Miguel Orszag, Daniel Reitzner, Wolfgang Schleich, Aephraim Steinberg, Mario Ziman, and M. Suhail Zubairy.

Finally, we are most grateful for the love and support of our families to whom this book is dedicated.

New York, NY, USA   János A. Bergou
                                                                                    Mark Hillery

Contents

Chapter 1
Introduction

The field of quantum information encompasses the study of the representation, storing, processing and accessing of information by quantum mechanical systems. The field grew from the investigations of the physical limits to computation initiated by Bennett and Landauer. One of the first questions studied was whether quantum mechanics imposes any limits on what a computer can do, and it was shown by Feynman that it does not. Earlier work by Paul Benioff had explored the possibilities of quantum Turing machines. Shortly after Feynman's work, David Deutsch realized that not only is quantum mechanics not a problem for computation, it can also be an advantage. The major breakthrough in the field was Peter Shor's factoring algorithm, which showed that a quantum computer can find the prime factors of integers in a time that scales as a polynomial of the size (number of digits) of the integer.

1.1 The Qubit

The basic unit of classical information is the bit, which can be 0 or 1. The corresponding object in quantum information is the qubit, which is a two-level quantum system. The two levels are often denoted by $|0\rangle$ and $|1\rangle$, which correspond to logical 0 and 1, respectively. Natural physical systems that can be used to represent a qubit are electronic or nuclear spins, and the polarization of a photon. The key difference between a bit and a qubit, is that the latter can exist in a superposition state

$$|\psi\rangle = \alpha|0\rangle + \beta|1\rangle, \tag{1.1}$$

while the former is definitely either 0 or 1. This leads to significant differences in what can be done with information represented by bits and that represented by qubits.

© Springer Nature Switzerland AG 2021
J. A. Bergou et al., *Quantum Information Processing*, Graduate Texts in Physics,
https://doi.org/10.1007/978-3-030-75436-5_1

Fig. 1.1 Bloch sphere representation of the qubit in Eq. (1.2)

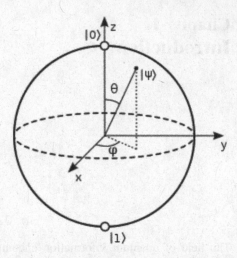

A convenient representation of the state of a qubit is given by the Bloch sphere. The state is parameterized by two angles, $0 \leq \theta \leq \pi$ and $0 \leq \phi < 2\pi$,

$$|\psi\rangle = \cos(\theta/2)|0\rangle + e^{i\phi}\sin(\theta/2)|1\rangle. \tag{1.2}$$

It is represented by a point on the unit sphere whose polar angle is θ and whose azimuthal angle is ϕ. That is, the vector from the origin to the point representing the state makes an angle of θ with the z axis, and its component in the x-y plane makes an angle of ϕ with the x axis. The state $|0\rangle$ is the North Pole of the sphere, and the state $|1\rangle$ is the South Pole. The Bloch sphere of a qubit is shown in Fig. 1.1. The Bloch sphere is often useful in illustrating how the state of a qubit is changed by a quantum map.

The state of N qubits is spanned by the tensor product basis

$$|0\rangle \otimes \ldots |0\rangle \otimes |0\rangle = |0\ldots00\rangle$$
$$|0\rangle \otimes \ldots |0\rangle \otimes |1\rangle = |0\ldots01\rangle$$
$$\vdots$$
$$|1\rangle \otimes \ldots |1\rangle \otimes |1\rangle = |1\ldots11\rangle \tag{1.3}$$

Note that we have expressed these states as $|x\rangle$, where x is an N-digit binary number. The most general N-qubit state can be expressed as

$$|\Psi\rangle = \sum_{x=0}^{2^N-1} c_x|x\rangle. \tag{1.4}$$

1.2 Quantum Gates

Quantum gates are unitary operators that act on one or more qubits. They are unitary, because they represent the effect of some kind of time evolution on the state of the qubit, and the time development transformation is a unitary operator. Because of this fact, quantum gates must be reversible, that is, if we know the output state of the gate, we can infer what the input state was. This rules out quantum versions of certain classical gates. For example, the AND gate is a gate with a two-bit input and a one bit output. The output is given by the product of the inputs, which implies that the output 0 can be produced by the inputs, 00, 01, or 10. Thus, this gate is not reversible and, therefore, has no quantum version.

On the other hand, the NOT gate, which simply flips a bit, $0 \rightarrow 1$ and $1 \rightarrow 0$, is reversible, so a quantum version, which performs the operations, $|0\rangle \rightarrow |1\rangle$ and $|1\rangle \rightarrow |0\rangle$, exists. It has the following action on a general qubit state

$$\alpha|0\rangle + \beta|1\rangle \rightarrow \alpha|1\rangle + \beta|0\rangle. \tag{1.5}$$

If we represent the qubit's state as a two-component column vector,

$$\alpha|0\rangle + \beta|1\rangle = \begin{pmatrix} \alpha \\ \beta \end{pmatrix}, \tag{1.6}$$

then the quantum NOT gate can be represented as the Pauli matrix, σ_x,

$$\begin{pmatrix} 0 & 1 \\ 1 & 0 \end{pmatrix} \begin{pmatrix} \alpha \\ \beta \end{pmatrix} = \begin{pmatrix} \beta \\ \alpha \end{pmatrix}. \tag{1.7}$$

The quantum NOT gate, also known as the X gate, is represented in Fig. 1.2. Here, and in the following, the left line represents the input qubit and the right line the output qubit.

There are also quantum gates that have no classical analogue. One particularly useful one is the Hadamard gate. It is a single-qubit gate, and is represented by the circuit symbol in Fig. 1.3.

Fig. 1.2 Circuit symbol for the NOT or X gate

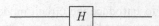

Fig. 1.3 Circuit symbol for the Hadamard gate

Fig. 1.4 Circuit symbol for
the C-NOT gate

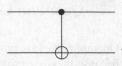

The Hadamard gate performs the following transformation

$$H|0\rangle = \frac{1}{\sqrt{2}}(|0\rangle + |1\rangle)$$

$$H|1\rangle = \frac{1}{\sqrt{2}}(|0\rangle - |1\rangle). \qquad (1.8)$$

There is no classical analogue of this gate, because it takes the computational basis states, $\{|0\rangle, |1\rangle\}$ and maps them into superposition states. Note that $H^2 = I$, where I is the identity operator.

A third important gate, which does have a classical analogue, is the Controlled-NOT gate (or C-NOT gate for short), which is also known as the exclusive OR gate (or XOR gate for short). It is a two-qubit gate and its circuit symbol is given in Fig. 1.4.

The inputs are again on the left and the outputs on the right. The upper qubit is called the control qubit, and the lower one the target qubit. The state of the control qubit is not changed by the gate, and the change in the state of the target qubit depends on what the state of the control qubit is. In particular, if the control bit is $|0\rangle$, nothing happens to the target qubit, but if the control bit is $|1\rangle$, then the target qubit is flipped. In more detail, if the first qubit is the control and the second target we have

$$|0\rangle|0\rangle \rightarrow |0\rangle|0\rangle \qquad |0\rangle|1\rangle \rightarrow |0\rangle|1\rangle$$

$$|1\rangle|0\rangle \rightarrow |1\rangle|1\rangle \qquad |1\rangle|1\rangle \rightarrow |1\rangle|0\rangle. \qquad (1.9)$$

From these relations the matrix elements of the transformation corresponding to the C-NOT gate can be read out. This and the verification that the matrix is unitary are left as a problem at the end of the chapter.

1.3   Quantum Circuits

At this stage, we are ready to introduce the circuit model of quantum computing. In this model, qubits are represented by lines and quantum gates, i.e., unitary operators, by their symbols. In particular, single qubit gates are denoted by their symbol on the line representing the qubit, two-qubit gates are denoted by their symbol connecting two lines corresponding to two qubits, and so on. What makes the circuit representation extremely useful is that the set of gates consisting of the C-NOT and all single qubit rotations is a universal set, which means that any unitary transformation on any number of qubits can be constructed from them. We shall not give the proof of this statement here. It is the reason that many schemes for physically implementing quantum information protocols concentrate on the construction of C-NOT gates. Instead of a more formal discussion, in the next section we will give an example illustrating how this type of description works.

1.4 The Deutsch Algorithm

The standard introductory example that is used to illustrate the fact that quantum information processing can be more powerful than classical information processing is Deutsch's problem. Consider a function, f, that maps the set $\{0, 1\}$ to $\{0, 1\}$. If $f(0) = f(1)$, then the function is constant, and if $f(0) \neq f(1)$, then we call f balanced. The problem is, given an unknown function, we want to determine whether it is constant or balanced. Classically we have to evaluate the function twice to determine this but, using a quantum circuit, it is only necessary to evaluate it once. The quantum circuit that solves Deutsch's problem is shown in Fig. 1.5.

Let us see how this circuit works. The lines represent qubits, and the action proceeds from left to right. The gate labeled U_f is a two-qubit gate called an f-Controlled-NOT. Like the C-NOT gate it has both a control (upper) qubit and a target (lower) qubit. The control qubit is not changed by the action of the gate, but the target bit has $f(x)$ added to it, modulo 2, where x is the value of the control bit. That is, if the input to the gate is $|x\rangle|y\rangle$, where x and y are the values of the control and target qubits, respectively, and are either 0 or 1, then the output is given by $|x\rangle|y + f(x)\rangle$. We have been given this gate, but we do not know what f is.

We shall now follow the qubits through the circuit. We start them in the state

$$|\Psi_0\rangle = |0\rangle_1 \frac{1}{\sqrt{2}}(|0\rangle_2 - |1\rangle_2), \tag{1.10}$$

where qubit 1 is the upper qubit and qubit 2 is the lower qubit. After the first gate the state is

$$|\Psi_1\rangle = \frac{1}{2}(|0\rangle_1 + |1\rangle_1)(|0\rangle_2 - |1\rangle_2), \tag{1.11}$$

Fig. 1.5 Quantum circuit for Deutsch's problem

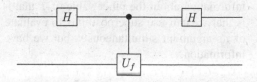

and after the f-Controlled-NOT it is

$$|\Psi_2\rangle = \frac{1}{2}[|0\rangle_1(|0 + f(0)\rangle_2 - |1 + f(0)\rangle_2)$$
$$+ |1\rangle_1(|0 + f(1)\rangle_2 - |1 + f(1)\rangle_2). \tag{1.12}$$

Noting that

$$|0 + f(x)\rangle_2 - |1 + f(x)\rangle_2 = (-1)^{f(x)}(|0\rangle_2 - |1\rangle_2), \tag{1.13}$$

we have that

$$|\Psi_2\rangle = \frac{1}{2}[(-1)^{f(0)}|0\rangle_1 + (-1)^{f(1)}|1\rangle_1](|0\rangle_2 - |1\rangle_2). \tag{1.14}$$

Finally, after passing through the second Hadamard gate, the state is

$$|\Psi_3\rangle = \frac{1}{2\sqrt{2}}\{|0\rangle_1[(-1)^{f(0)} + (-1)^{f(1)}]$$
$$+ |1\rangle_1[(-1)^{f(0)} - (-1)^{f(1)}]\}(|0\rangle_2 - |1\rangle_2). \tag{1.15}$$

Examining this expression, we see that if the function is constant, the first qubit is in the state $|0\rangle$, and if the function is balanced the first qubit is in the state $|1\rangle$. Therefore, by measuring the first qubit in the computational basis, we can determine whether f is constant or balanced. Note that the f-Controlled-NOT was used only once, so that the function was only evaluated once. The reason this procedure works is that what goes into the f-Controlled-NOT gate is a superposition of two input values, $|0\rangle$ and $|1\rangle$, and the function is evaluated on both of them at once (in the expression for $|\Psi_2\rangle$, both $f(0)$ and $f(1)$ appear). By carefully manipulating these values we can obtain information about the global properties of the function.

What we cannot do is to obtain more than one value of the function if we evaluate it only once. Suppose we send the state $(1/\sqrt{2})(|0\rangle_1 + |1\rangle_1)|0\rangle_2$ into the f-Controlled-NOT gate. We have

$$\frac{1}{\sqrt{2}}(|0\rangle_1 + |1\rangle_1)|0\rangle_2 \rightarrow \frac{1}{\sqrt{2}}(|0\rangle_1|f(0)\rangle_2 + |1\rangle_1|f(1)\rangle_2). \tag{1.16}$$

If we measure this state in the computational basis, we will obtain one of the values of f, and which one we obtain will be random. The measurement destroys the information about the other value of f that is present in the state. The lesson here is that we can use superpositions to evaluate a function on many different values of its argument simultaneously, but we have to be clever about how we use this information.

1.5 An Interferometer Version

The operation of the Deutsch algorithm can seem rather mysterious, and one way to gain a better understanding is to look at an analogy based on a Mach-Zehnder interferometer. This interferometer is composed of two beam splitters, two mirrors and phase shifters (see Fig. 1.7). What this analogy will show is that what makes the Deutsch algorithm work is quantum interference.

Let us begin by discussing how the components of the interferometer work. Mirrors just reflect, so their action is straightforward. A phase shifter is an optical element that induces a phase shift in a mode. It could be, for example, a piece of glass, or anything that alters the optical path length. If the creation operator corresponding to the optical mode is $a^\dagger$, and U_ϕ is the unitary operator that describes the action of the phase shifter, then $U_\phi a^\dagger U_\phi^\dagger = e^{i\phi}a^\dagger$, and $U_\phi^\dagger|0\rangle = |0\rangle$, where $|0\rangle$ is the vacuum state. The beam splitter is more complicated; it couples two modes (see Fig. 1.6). Assuming we have a 50–50 beam splitter one possible action is to transform input mode $a^\dagger$ to the output mode $(a^\dagger + b^\dagger)/\sqrt{2}$ and the input mode $b^\dagger$ to $(b^\dagger - a^\dagger)/\sqrt{2}$. That is, if U_{BS} is the unitary operator corresponding to the beam splitter, we have

$$U_{BS}a^\dagger U_{BS}^\dagger = \frac{1}{\sqrt{2}}(a^\dagger + b^\dagger)$$

$$U_{BS}b^\dagger U_{BS}^\dagger = \frac{1}{\sqrt{2}}(b^\dagger - a^\dagger), \tag{1.17}$$

with $U_{BS}^\dagger|0\rangle = |0\rangle$. What this means is that if a photon in the state $a^\dagger|0\rangle$ enters the beam splitter, it has an amplitude of $1/\sqrt{2}$ to be transmitted and an amplitude of $1/\sqrt{2}$ to be reflected. If it enters in the state $b^\dagger|0\rangle$, it has an amplitude of $1/\sqrt{2}$ to be transmitted and an amplitude of $-1/\sqrt{2}$ to be reflected.

Now that we know how the individual components act, we can see how a state is transformed by our interferometer. Let us send a photon in the $a^\dagger$ mode into the

Fig. 1.6 A beam splitter that couples two modes

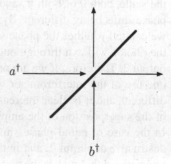

Fig. 1.7 A Mach-Zehnder
interferometer with two phase
shifters. If the allowed phase
shifts are 0 and π, its
behavior is analogous to that
of the Deutsch algorithm

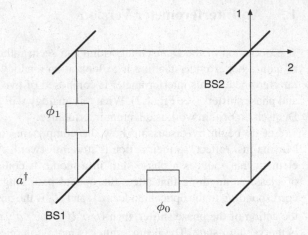

interferometer, so our input state is $a^\dagger|0\rangle$. The output state is given by

$$|\psi_{out}\rangle = U_{BS2}U_{\phi_1}U_{\phi_0}U_{BS1}a^\dagger|0\rangle$$

$$= \frac{1}{2}(e^{i\phi_0} + e^{i\phi_1})a^\dagger|0\rangle + \frac{1}{2}(-e^{i\phi_0} + e^{i\phi_1})b^\dagger|0\rangle, \qquad (1.18)$$

where U_{BS1} and U_{BS2} are the beam-splitter transformations for the first and second beam splitters, respectively, and U_{ϕ_0} and U_{ϕ_1} are the phase shifter transformations for the output modes of the first beam splitter (see Fig. 1.7). Note that U_{ϕ_0} acts only on the a mode and U_{ϕ_1} acts only on the b mode. In this expression, $a^\dagger$ corresponds to output 1, and $b^\dagger$ corresponds to output 2.

In order to make a connection with the Deutsch algorithm, we will look at the cases in which the phase shifts are either 0 or π. We can view the phase shifters as corresponding to the outputs of the Boolean function, $f(x)$, with $f(0)$ corresponding to ϕ_0 and $f(1)$ corresponding to ϕ_1. A phase shift of 0 will correspond to a value of 0 for the function and a phase shift of π will correspond to a value of 1. The function being constant will correspond to ϕ_0 and ϕ_1 being the same, both 0 or both π, and the function being balanced will correspond to the phase shifts being different. By sending a single photon through the interferometer, we can tell whether the phase shifts are the same or different. If they are the same, the photon will exit through output 1, while if they are different, it will exit through output 2. Therefore, if we are presented with the interferometer in a black box, with one use of the interferometer we can tell whether its phase shifts are the same or different, and it is clear the reason for this is interference. Each output is the result of the superposition of the amplitudes for the two paths through the interferometer. In the case of equal phase shifts, the interference is constructive at output 1 and destructive at output 2, and in the case of unequal phase shifts, vice versa. It is also interference that makes the circuit for the Deutsch algorithm work.

1.6 Other Models of Quantum Computation

In this chapter so far, we have presented the circuit model of quantum computation, and this is the model we will use throughout the book. However, there are two additional models, which we will now briefly discuss, adiabatic quantum computing and measurement-based-one-way quantum computing.

In adiabatic quantum computing, one starts with a simple Hamiltonian, H_0, with a simple ground state, and the system in that ground state. One then slowly varies the Hamiltonian until a new one, H_1, is reached, whose ground state encodes the solution to a computational problem. For example, we could choose $H(s) = (1 - s)H_0 + sH_1$ and slowly vary s from 0 to 1. The evolution of the system from $t = 0$ to $t = t_f$ is then described by the Schrödinger equation

$$i\hbar \frac{d}{dt}|\psi(t)\rangle = H(t)|\psi(t)\rangle, \tag{1.19}$$

where we have set $s = t/t_f$. The adiabatic theorem of quantum mechanics guarantees that if the change is made sufficiently slowly, then the system will remain in the ground state, and at the end of the process be in the ground state of H_1. This ground state can then be measured to obtain the solution to the problem. The key parameter that determines how fast the change in the Hamiltonian can be made is the minimum energy gap between the ground state and the lowest excited state, where the minimum is taken over the parameter s. If this is too small, the run time of the algorithm will be large, and no quantum speedup will be obtained. It has been shown, however, that a quantum speedup can be obtained for search problems in the adiabatic model.

A second model is that of one-way-measurement-based computing. In that case, one prepares a multi-qubit state, known as a cluster state. One then performs the computation by making measurements on this state, where the measurements chosen can depend on the results of previous ones. How this works is best illustrated by some examples.

We have a qubit in the state $|\psi\rangle$ and we would like to apply the operator $W(\theta)$ to it, where

$$W(\theta)|0\rangle = |+x\rangle = \frac{1}{\sqrt{2}}(|0\rangle + |1\rangle)$$

$$W(\theta)|1\rangle = \frac{1}{\sqrt{2}}e^{i\theta}(|0\rangle - |1\rangle). \tag{1.20}$$

We append a second qubit, which we will refer to as qubit 2, in the state $|+x\rangle$ to the first, which we shall refer to as qubit 1, and apply a controlled-PHASE (CPHASE) gate to the two qubits. The CPHASE gate acts as $\text{CPHASE}|j\rangle|k\rangle = (-1)^{jk}|j\rangle|k\rangle$, where $j, k = 0, 1$. The CPHASE gate is symmetric in its two inputs, so it does not matter which qubit is the control and which is the target. Defining the basis

$|\pm\theta\rangle = (1/\sqrt{2})(|0\rangle \pm e^{-i\theta}|1\rangle)$, we can express the resulting state as

$$\frac{1}{\sqrt{2}}(|+\theta\rangle_1 W(\theta)|\psi\rangle_2 + |-\theta\rangle_1 \sigma_x W(\theta)|\psi\rangle_2). \qquad (1.21)$$

We now measure qubit 1 in the $|\pm\theta\rangle$ basis. If we obtain $|+\theta\rangle$, qubit 2 is in the desired state. If we obtain $|-\theta\rangle$, then qubit 2 is in the state $\sigma_x W(\theta)|\psi\rangle_2$. Now if this is the end of our calculation, and the next step is to measure the state in the computational basis, the presence of the extra σ_x does not matter; one simply switches the results. For example, the probability to obtain $|0\rangle$ in the state $W(\theta)|\psi\rangle_2$ is the same as the probability to obtain $|1\rangle$ in the state $\sigma_x W(\theta)|\psi\rangle_2$.

If this is not the end of our calculation, then the extra σ_x still does not present a problem. Suppose we would like to obtain the state $W(\theta')W(\theta)|\psi\rangle$, and we obtained $|+\theta\rangle$ for the first measurement. We now append a third qubit to the system in the state $|+x\rangle$, so the total state is $|-\theta\rangle_1|W(\theta)|\psi\rangle_2|+x\rangle_3$. We now apply the CPHASE gate to qubits 2 and 3, and measure qubit 2 in the $|\pm\theta'\rangle$ basis. If we obtained $|+\theta'\rangle$, then the state of qubit 3 is $W(\theta')W(\theta)|\psi\rangle$, and if it is $|-\theta'\rangle$, then the state of qubit 3 is $\sigma_x W(\theta')W(\theta)|\psi\rangle$. If this is now the end of the calculation, then, as we have argued, the extra σ_x doesn't matter. What happens if the result of our first measurement was $|-\theta\rangle$? In that case, we measure the second qubit in the basis $|\pm\theta'_*\rangle = (1/\sqrt{2})(|0\rangle \pm e^{i\theta'}|1\rangle)$. If we obtain $|+\theta'_*\rangle$, then the state of qubit 3 is $W(-\theta')\sigma_x W(\theta)|\psi\rangle$. We can now use the identity $W(\theta)\sigma_x = \sigma_z W(-\theta)$, to express it as $\sigma_z W(\theta')W(\theta)|\psi\rangle$. If this is the end of our calculation, and we now measure in the computational basis, the presence of the extra σ_z will not affect the probabilities of the outcomes. If we obtained $|-\theta'_*\rangle$, then our state, after making use of the identity, is $\sigma_z\sigma_x W(\theta')W(\theta)|\psi\rangle$ and again the probabilities of the measurement results in computational basis for the desired state, the state without the $\sigma_z\sigma_x$, are obtained simply by switching the results obtained from the actual output state. Summarizing, we can obtain the results we want simply by performing a series of measurements.

These examples give the flavor of measurement-based quantum computing, and it can be shown that any quantum computation that can be performed in the circuit model, can also be performed in the measurement-based model.

1.7 Discussion

The example of Deutsch's problem tells us several things. First that there are gains to be had by representing information by quantum systems. The second, is that finding how to produce these gains is far from straightforward. And last but not least, it shows that the final step in the algorithm is a measurement to read out the final state of the system. These are general features of all protocols and in the following chapters we will take an in-depth look at all of these ingredients.

First, in Chap. 2, we take a look at pure and mixed quantum states, and how they can be used to represent information. In Chap. 3 we study their intrinsic quantum features, entanglement as a resource for quantum information and quantum computing, in particular. Generalized dynamics, operations and the Kraus representation, is the subject of Chap. 4. The theory of quantum measurements, including generalized measurements, is presented in Chap. 5. Thus, the first five chapters contain what we can call the toolbox of quantum information theory. We then put these tools to good use in the rest of the book. In Chap. 6 we take a look at quantum communication, Chap. 7 deals with quantum computing (quantum algorithms, in particular). We included the field of quantum machines in Chap. 8. Chapter 9 deals with the main enemy: the inevitable influence of the environment, leading to decoherence and the protecting of quantum information (quantum error correction, in particular). Chapter 10 presents the stabilizer formalism, which provides another method of constructing error-correcting codes, and the Gottesman–Knill theorem. Chapter 11 is a very brief introduction to quantum information theory and, in particular, its application to quantum communication. The last four chapters of the book are devoted to physical implementations of quantum computations. In Chap. 12 we introduce the general principles behind implementations. Chapter 13 deals with atomic qubits, while Chap. 14 deals with optical qubits. Finally, Chap. 15 deals with solid state qubits, including also superconducting qubits, which at present, is perhaps the most promising platform for the implementation of quantum computational schemes.

1.8 Problems

1. (a) Find the matrix corresponding to the Hadamard gate in the computational basis. (b) Find the matrix corresponding to the C-NOT gate in the two-qubits computational basis. (c) Check to see if the matrices in (a) and (b) are unitary.

2. Let us denote the unitary operator that implements a Controlled-NOT gate by D_{ab}, where a is the control bit and b is the target bit. Let $|\psi\rangle = \alpha|0\rangle + \beta|1\rangle$ be a general qubit state, and let $|\pm x\rangle$ be the states given by $|\pm x\rangle = (|0\rangle \pm |1\rangle)/\sqrt{2}$.

   (a) We want to see what happens if the states $|\pm x\rangle$ are the input target bit states. Find the states $D_{ab}|\psi\rangle_a |\pm x\rangle_b$.
   (b) We can use the C-NOT to implement a one-parameter group of operations on a qubit probabilistically. Start by calculating

   $$D_{ab}|\psi\rangle_a (\cos\theta| + x\rangle_b + i \sin\theta| - x\rangle_b).$$

   Now measure the target qubit to see if it is in $|0\rangle_b$, or $|1\rangle_b$. Find the probability that it is in $|0\rangle_b$, and show that if it is, the control qubit is in the state $\exp(i\theta\sigma_z)|\psi\rangle_a$, where $\sigma_z|0\rangle = |0\rangle$ and $\sigma_z|1\rangle = -|1\rangle$.

3. The function that we considered in the class in connection with Deutsch's algorithm is a special case of the so-called Boolean functions. A Boolean function $f^{(n)}$ maps the set of binary numbers $\{0, 1, \cdots, 2^n - 1\}$ to $\{0, 1\}$. We have shown in the class that there are four different Boolean functions for $n = 1$, $f_1^{(1)}, f_2^{(1)}, f_3^{(1)}, f_4^{(1)}$. Two of them are constant and two of them are balanced.

   (a) Work out the truth table for each of these functions.
   (b) Find the corresponding f-Controlled-NOT (f-CNOT) gate for each.
   (c) Show that they are unitary.

4. (a) A SWAP gate is a two-qubit gate that has the action $|\psi\rangle_a|\phi\rangle_b \rightarrow |\phi\rangle_a|\psi\rangle_b$. Show that a SWAP gate can be constructed from three CNOT gates.
   (b) A Controlled-PHASE gate is a two-qubit gate, with qubit a as the control qubit and qubit b as the target qubit. The control qubit does not change, and if the control qubit is in the state $|0\rangle_a$ neither does the target qubit. If the control bit is in the state $|1\rangle_a$, then the gate acts as $|1\rangle_a|0\rangle_b \rightarrow |1\rangle_a|0\rangle_b$ and $|1\rangle_a|1\rangle_b \rightarrow -|1\rangle_a|1\rangle_b$. Show that a Controlled-PHASE gate can be constructed from a CNOT gate and two single-qubit gates.

5. A Controlled-SWAP gate is a three qubit gate. It has a control qubit and two other qubit inputs. As usual, the control qubit is not changed by the gate, and if it is in the state $|0\rangle_c$, the other two inputs are unchanged as well. If it is in the state $|1\rangle_c$, then the other two inputs are swapped. Summarizing, $|0\rangle_c|\psi\rangle_a|\phi\rangle_b \rightarrow |0\rangle_c|\psi\rangle_a|\phi\rangle_b$ and $|1\rangle_c|\psi\rangle_a|\phi\rangle_b \rightarrow |1\rangle_c|\phi\rangle_a|\psi\rangle_b$. Combine the Controlled-SWAP gate with two Hadamard gates, where the Hadamard gates are in the control qubit line, one before and one after the Controlled-SWAP gate. Show that if we measure the control qubit at the output to determine whether it is in the state $|0\rangle$, this enables us to gain information about the overlap between the states $|\psi\rangle$ and $|\phi\rangle$, that is about $\langle\psi|\phi\rangle$. This procedure is known as the SWAP test.

6. We can get a better estimate of the overlap between vectors if we have more information. Let's consider a controlled-permutation gate, S. This gate has a qutrit for a control, and three qubits, which we shall label as 1, 2, and 3 as targets. The control qutrit states $|0\rangle$, $|1\rangle$, and $|2\rangle$ are not changed by the gate. If the control is in the state $|0\rangle$, the gate does nothing to the qubits. If it is in the state $|1\rangle$ it performs the permutation $1 \rightarrow 2 \rightarrow 3 \rightarrow 1$, and if it is in the state $|2\rangle$, it performs the permutation $1 \rightarrow 3 \rightarrow 2 \rightarrow 1$, Now set $|\alpha\rangle = (1/\sqrt{3})(|0\rangle + |1\rangle + |2\rangle)$ and start with the state $|\alpha\rangle_c|\psi\rangle_1|\phi\rangle_2|\phi\rangle_3$, where the subscript c denotes the control qutrit. Run the state through the controlled-permutation gate and find the probability that the control qutrit is still in the state $|\alpha\rangle$. This should give you an estimate of the overlap between $|\psi\rangle$ and $|\phi\rangle$.

Further Reading

1. D. Deutsch, Quantum theory, the Church-Turing Principle, and the universal quantum computer. Proc. R. Soc. Lond. A **400**, 97 (1985)
2. D.P. DiVincenzo, Two qubit gates are universal for quantum computation. Phys. Rev. A **51**, 1015 (1995)
3. A. Barenco, C.H. Bennett, R. Cleve, D.P. DiVincenzo, N. Margolus, P. Shor, T. Sleator, J. Smolin, H. Weinfurter, Elementary gates for quantum computation. Phys. Rev. A **52**, 3457 (1995)
4. R. Feynman, A. Hey, *Feynman Lectures on Computation* (CRC Press, Boca Raton, 2000)
5. E. Farhi, J. Goldstone, S. Gutman, M. Sipser, arXiv:quant-ph/0001106 (2000)
6. T. Albash, D. Lidar, Rev. Mod. Phys. **90**, 015002 (2018)
7. R. Raussendorf, H. Briegel, Phys. Rev. Lett. **86**, 5188 (2001)
8. R. Jozsa, arXiv:quant-ph/0508124

Chapter 2
The Density Matrix

We are going to require a more general description of a quantum state than that given by a state vector. The density matrix provides such a description. Its use is required when we are discussing an ensemble of pure states, or when we are describing a subsystem of a larger system.

2.1 Ensembles and Subsystems

Let us look at ensembles first. Suppose that we have a collection of objects some of which are in the quantum state $|\psi_1\rangle$, some of which are in $|\psi_2\rangle$, and so on. In particular, if we choose an object from the ensemble, the probability that it is in state $|\psi_j\rangle$ is p_j. We want to find the expectation value of some observable, Q, in this ensemble. We pick one of the objects in the ensemble and measure Q, pick another and do the same. We repeat this process many times. If all of the objects were in the state $|\psi_j\rangle$, the expectation value of Q would be $\langle\psi_j|Q|\psi_j\rangle$, but in reality, objects with this state appear only with a probability p_j. Therefore, the expectation of Q in the ensemble is given by

$$\langle Q \rangle = \sum_j p_j \langle\psi_j|Q|\psi_j\rangle = \mathrm{Tr}(Q\rho), \tag{2.1}$$

where we have defined the operator ρ, which is the density matrix corresponding to the ensemble, to be

$$\rho = \sum_j p_j |\psi_j\rangle\langle\psi_j|. \tag{2.2}$$

© Springer Nature Switzerland AG 2021
J. A. Bergou et al., *Quantum Information Processing*, Graduate Texts in Physics,
https://doi.org/10.1007/978-3-030-75436-5_2

Now let us look at subsystems. Suppose we have a large system composed of two subsystems, A and B. The Hilbert spaces for the quantum states of the subsystems are $\mathcal{H}_A$ and $\mathcal{H}_B$, so that the Hilbert space for the entire system is $\mathcal{H}_A \otimes \mathcal{H}_B$. Let $\{|m\rangle_A\}$ be an orthonormal basis for $\mathcal{H}_A$ and $\{|n\rangle_B\}$ be an orthonormal basis for $\mathcal{H}_B$. Now if X_A is an observable on subsystem A, then the operator corresponding to it in the total Hilbert space is $X_A \otimes I_B$, where I_B is the identity on $\mathcal{H}_B$. If $|\Psi\rangle$ is the state of the entire system, then the expectation value of X_A is given by

$$
\begin{aligned}
\langle X_A \rangle &= \langle \Psi | X_A \otimes I_B | \Psi \rangle \\
&= \sum_m \sum_n \langle \Psi | X_A \otimes I_B (|m\rangle_A |n\rangle_B)(_A\langle m| \, _B\langle n|)|\Psi\rangle \\
&= \sum_m {}_A\langle m|(\sum_n {}_B\langle n|\Psi\rangle\langle\Psi|n\rangle_B) X_A |m\rangle_A.
\end{aligned} \tag{2.3}
$$

If we now define

$$
\rho_A = \sum_n {}_B\langle n|\Psi\rangle\langle\Psi|n\rangle_B = \mathrm{Tr}_B(|\Psi\rangle\langle\Psi|), \tag{2.4}
$$

then we have that

$$
\langle X_A \rangle = Tr_A(\rho_A X_A). \tag{2.5}
$$

The operator ρ_A is known as the reduced density operator for subsystem A, and it can be used to evaluate the expectation value of any observable that pertains only to subsystem A.

Now let us look at two examples. First, we have an ensemble in which half of the qubits are in the state $|0\rangle$ and the other half are in the state $|1\rangle$. The density matrix for this ensemble is

$$
\rho = \frac{1}{2}|0\rangle\langle 0| + \frac{1}{2}|1\rangle\langle 1|. \tag{2.6}
$$

Suppose we want to find the expectation value of σ_z in this ensemble, where $\sigma_z|0\rangle = |0\rangle$, and $\sigma_z|1\rangle = -|1\rangle$. We have that

$$
\langle \sigma_z \rangle = \mathrm{Tr}(\sigma_z \rho) = 0. \tag{2.7}
$$

Next, we have a two-qubit state

$$
|\Psi\rangle = \frac{1}{\sqrt{2}}(|0\rangle_A|1\rangle_B + |1\rangle_A|0\rangle_B), \tag{2.8}
$$

and we would like to find the reduced density matrix for subsystem A. We have that

$$\rho_A = \text{Tr}(|\Psi\rangle\langle\Psi|) = \frac{1}{2}I_A. \tag{2.9}$$

Defining $\sigma_{zA} = \sigma_z \otimes I_B$, we have that

$$\langle\sigma_{Az}\rangle = \text{Tr}(\sigma_z\rho_A) = 0. \tag{2.10}$$

Note that if ρ is a one-dimensional projection, i.e. $\rho = |\psi\rangle\langle\psi|$, then for any observable Q,

$$\langle Q\rangle = \text{Tr}(\rho Q) = \langle\psi|Q|\psi\rangle, \tag{2.11}$$

so that the density matrix corresponds to the system being in the state $|\psi\rangle$. If ρ is of this form we call it a pure state. If not, it is called a mixed state.

2.2 Properties

In order for an operator to be a density matrix, it must satisfy several properties. In fact, any operator satisfying these properties is a valid density matrix.

1. $\text{Tr}(\rho) = 1$.
   This follows from the fact that

$$\text{Tr}(\rho) = \text{Tr}\left(\sum_j p_j|\psi_j\rangle\langle\psi_j|\right) = \sum_j p_j = 1. \tag{2.12}$$

2. A density matrix is hermitian, $\rho = \rho^\dagger$.
3. A density matrix is positive, $\langle\psi|\rho|\psi\rangle \geq 0$ for all $|\psi\rangle$.
   This follows from

$$\langle\psi|\rho|\psi\rangle = \sum_j p_j|\langle\psi|\psi_j\rangle|^2 \geq 0. \tag{2.13}$$

We also note that an operator is positive if and only if all of its eigenvalues are greater than or equal to zero, which implies that the eigenvalues of any density matrix must satisfy this property. In addition, because the trace of a density matrix is one, and the trace is just the sum of the eigenvalues, we have that if λ_j is an eigenvalue of a density matrix, then $0 \leq \lambda_j \leq 1$.

We now want to use these requirements to find additional properties of the set of density matrices. The first is a simple way of identifying pure states.

Theorem *The density matrix ρ is a pure state if and only if $\mathrm{Tr}(\rho^2) = 1$.*

Proof If ρ is pure, then $\rho = |\psi\rangle\langle\psi|$, and $\rho^2 = \rho$. This immediately implies that $\mathrm{Tr}(\rho^2) = 1$. Now assume that $\mathrm{Tr}(\rho^2) = 1$. Because ρ is hermitian, we can express it as

$$\rho = \sum_j \lambda_j P_j, \tag{2.14}$$

where λ_j are the nonzero eigenvalues of ρ, and P_j are the corresponding spectral projections. This immediately implies that

$$\rho^2 = \sum_j \lambda_j^2 P_j. \tag{2.15}$$

Denoting the rank of P_j by n_j we have that

$$\mathrm{Tr}(\rho) = 1 \Rightarrow \sum_j \lambda_j n_j = 1$$

$$\mathrm{Tr}(\rho^2) = 1 \Rightarrow \sum_j \lambda_j^2 n_j = 1, \tag{2.16}$$

and subtracting these two equations gives us

$$\sum_j (\lambda_j - \lambda_j^2) n_j = 0. \tag{2.17}$$

Because each eigenvalue is between 0 and 1, each term in the above sum is greater than or equal to zero, which further implies that each term must be equal to zero. The only way this can happen is if each λ_j is equal to zero or one, and we have assumed that $\lambda_j > 0$, so that $\lambda_j = 1$. The only way that this can be consistent with the fact that the sum of the eigenvalues times their multiciplicities is one, is if only one of them is nonzero, and this eigenvalue has a multiplicity of one. Therefore, ρ is equal to a rank-one projection, which means that it is a pure state.

2.3   Pure States and Mixed States of a Qubit

In the previous chapter we introduced the Bloch sphere as a convenient representation for state vectors of qubits, that is qubit pure states. If we extend this representation to include the interior of the sphere, it can be used to represent mixed states of qubits as well. In order to see this we expand a general qubit density matrix, which is a 2×2 matrix, in terms of the identity matrix and the Pauli matrices, which form a complete basis for the space of a set of 2×2 matrices

$$\rho = \frac{1}{2}(I + n_x \sigma_x + n_y \sigma_y + n_z \sigma_z). \tag{2.18}$$

This satisfies the condition $\mathrm{Tr}(\rho) = 1$, and the fact that ρ is hermitian implies that n_x, n_y, and n_z are real. This equation implies that

$$\rho = \frac{1}{2} \begin{pmatrix} 1 + n_z & n_x - i n_y \\ n_x + i n_y & 1 - n_z \end{pmatrix}, \tag{2.19}$$

which further implies that $\det \rho = (1 - |\mathbf{n}|^2)/4$. The fact that ρ is positive means that its determinant must be greater than or equal to zero, and, therefore, $1 \geq |\mathbf{n}|$. We represent the density matrix ρ by the vector $\mathbf{n}$, which lies in the unit ball.

We know that if ρ is a pure state, its corresponding vector will have its endpoint on the surface of the Bloch sphere. Let us show the converse. If $|\mathbf{n}| = 1$ then $\mathrm{Tr}(\rho) = 1$ and $\det \rho = 0$. This implies that one of the eigenvalues of ρ is zero and the other is one. If $|u\rangle$ is the eigenvector with eigenvalue one, where $\|u\| = 1$, then $\rho = |u\rangle \langle u|$, and ρ is a pure state.

Given a qubit density matrix, ρ, we can easily find the vector corresponding to it. The identity $\mathrm{Tr}(\sigma_j \sigma_k) = 2\delta_{jk}$, where $j, k \in \{x, y, z\}$, gives us that

$$n_j = \mathrm{Tr}(\rho \sigma_j). \tag{2.20}$$

Most density matrices can correspond to many different ensembles. We give some examples. In the first example, we define the states

$$|\pm x\rangle = \frac{1}{\sqrt{2}}(|0\rangle \pm |1\rangle), \tag{2.21}$$

which are eigenstates of σ_x. Then we can write the density matrix of a maximally mixed state in two ways,

$$\rho = \frac{1}{2}I = \frac{1}{2}(|0\rangle \langle 0| + |1\rangle \langle 1|) = \frac{1}{2}(|+x\rangle \langle +x| + |-x\rangle \langle -x|). \tag{2.22}$$

The first decomposition corresponds to an ensemble in which half of the elements are in the state $|0\rangle$ and half in the state $|1\rangle$, and the second corresponds to an ensemble in which half of the elements are in the state $|+x\rangle$ and half in the state $|-x\rangle$. These ensembles are different, but they are described by the same density matrix.

In the second example, we define the states

$$|u_\pm\rangle = \frac{1}{\sqrt{4 \pm 2\sqrt{2}}}((\sqrt{2} \pm 1)|0\rangle \pm |1\rangle), \tag{2.23}$$

which are the eigenstates of the Hadamard gate, with eigenvalues ± 1, respectively. Then

$$\rho = \frac{1}{2}(|0\rangle\langle 0| + |+x\rangle\langle +x|) = (\frac{1}{2} + \frac{\sqrt{2}}{4})(|u_+\rangle\langle u_+| + (\frac{1}{2} - \frac{\sqrt{2}}{4})|u_-\rangle\langle u_-|)\,,$$

(2.24)

again describing two different ensembles by the same density matrix.

In general, if ρ_1 and ρ_2 are density matrices, so is

$$\rho(\theta) = \theta\rho_1 + (1 - \theta)\rho_2,$$

(2.25)

where $0 \leq \theta \leq 1$. This implies that the set of density matrices is convex. Most density matrices can be expressed as a sum of other density matrices in many different ways, and each of these decompositions will, in general, correspond to a different ensemble. The two examples above were just special cases of this general statement. This, however, is not true for pure sates; they have a unique decomposition. To see this suppose that $\rho = |\psi\rangle\langle\psi|$ is a pure state density matrix and that it can also be expressed as a convex sum of two other density matrices, $\rho(\theta) = \theta\rho_1 + (1 - \theta)\rho_2$. If $|\psi_\perp\rangle$ satisfies $\langle\psi_\perp|\psi\rangle = 0$, then

$$0 = \langle\psi_\perp|\rho(\theta)|\psi_\perp\rangle = \theta\langle\psi_\perp|\rho_1|\psi_\perp\rangle + (1 - \theta)\langle\psi_\perp|\rho_2|\psi_\perp\rangle.$$

(2.26)

Since both terms on the right-hand side are ≥ 0, it follows that

$$\langle\psi_\perp|\rho_1|\psi_\perp\rangle = \langle\psi_\perp|\rho_2|\psi_\perp\rangle = 0.$$

(2.27)

This equation is true for any vector orthogonal to $|\psi\rangle$. Therefore, $\rho_1 = \rho_2 = |\psi\rangle\langle\psi|$ and the representation of any pure state is unique. Pure states cannot be expressed as a sum of other density matrices. These are the only states with this property, because if ρ is mixed, it is given by $\rho = \sum_j p_j|\psi_j\rangle\langle\psi_j|$, which is just a convex sum of pure states.

2.4   Pure State Decompositions and the Ensemble Interpretation

Next we turn our attention to the ways in which a density matrix can be decomposed into pure states. The main result is summarized in the following

Theorem *ρ can be expressed as $\sum_i p_i|\psi_i\rangle\langle\psi_i|$ and $\sum_i q_i|\phi_i\rangle\langle\phi_i|$ iff*

$$\sqrt{p_i}|\psi_i\rangle = \sum_j U_{ij}\sqrt{q_j}|\phi_j\rangle\,,$$

where U_{ij} is a unitary matrix and we "pad" whichever set of vectors is smaller with additional 0 vectors so that the sets have the same number of elements.

Proof Let $|\tilde{\psi}_i\rangle = \sqrt{p_i}|\psi_i\rangle$ and $|\tilde{\phi}_i\rangle = \sqrt{q_i}|\phi_i\rangle$. We first prove the if part, meaning that the condition is sufficient. To this end we suppose $|\tilde{\psi}_i\rangle = \sum_j U_{ij}|\tilde{\phi}_j\rangle$, for U_{ij} unitary. Then

$$\sum_i |\tilde{\psi}_i\rangle\langle\tilde{\psi}_i| = \sum_{i,j,k} U_{ij} U_{ik}^* |\tilde{\phi}_j\rangle\langle\tilde{\phi}_k| = \sum_{j,k} \left(\sum_i U_{ki}^\dagger U_{ij}\right)|\tilde{\phi}_j\rangle\langle\tilde{\phi}_k|$$

$$= \sum_{j,k} \delta_{jk}|\tilde{\phi}_j\rangle\langle\tilde{\phi}_k| = \sum_j |\tilde{\phi}_j\rangle\langle\tilde{\phi}_j| \,. \tag{2.28}$$

To prove the only if part, that the condition is also necessary, is considerably more work. We now suppose

$$\rho = \sum_i^{N_1} |\tilde{\psi}_i\rangle\langle\tilde{\psi}_i| = \sum_i^{N_2} |\tilde{\phi}_i\rangle\langle\tilde{\phi}_i| \,, \tag{2.29}$$

and assume $N_1 \geq N_2$. Since ρ is a positive operator, it has the spectral representation

$$\rho = \sum_{k=1}^{N_k} \lambda_k |k\rangle\langle k| = \sum_{k=1}^{N_k} |\tilde{k}\rangle\langle\tilde{k}| \,, \tag{2.30}$$

where $\langle k|k'\rangle = \delta_{k,k'}$ and $|\tilde{k}\rangle = \sqrt{\lambda_k}|k\rangle$. First, we want to show that $|\tilde{\psi}_i\rangle$ lies in the subspace spanned by $\{|k\rangle\}$. To do this, let $\mathcal{H}_k$ denote the space spanned by $\{|k\rangle\}$. Suppose $|\psi\rangle \in \mathcal{H}_k^\perp$, then

$$\langle\psi|\rho|\psi\rangle = 0 = \sum_{i=1}^{N_1} |\langle\tilde{\psi}_i|\psi\rangle|^2 \,. \tag{2.31}$$

From here $\langle\tilde{\psi}_i|\psi\rangle = 0$ follows and so $|\tilde{\psi}_i\rangle \in (\mathcal{H}_k^\perp)^\perp = \mathcal{H}_k$. Therefore, we can express $|\tilde{\psi}_i\rangle$ as

$$|\tilde{\psi}_i\rangle = \sum_{k=1}^{N_k} c_{ik}|\tilde{k}\rangle \,. \tag{2.32}$$

We can use this representation in (2.29) to obtain

$$\rho = \sum_{i=1}^{N_1} |\tilde{\psi}_i\rangle\langle\tilde{\psi}_i| = \sum_{k,k'=1}^{N_k} \left(\sum_{i=1}^{N_1} c_{ik} c_{ik'}^*\right)|\tilde{k}\rangle\langle\tilde{k}'| = \sum_{k=1}^{N_k} |\tilde{k}\rangle\langle\tilde{k}| \,. \tag{2.33}$$

Since the operators $|\tilde{k}\rangle\langle\tilde{k}'|$ are linearly independent we have that $\sum_{i=1}^{N_1} c_{ik}c_{ik'}^* = \delta_{kk'}$. Thus c_{ik} form N_k orthogonal vectors of dimension N_1 and $N_1 \geq N_k$. In other words, c_{ik} ($k = 1, \ldots, N_k$) form the first N_k columns of a matrix containing N_1 rows that can be extended to an $N_1 \times N_1$ unitary matrix in the following way. Find $N_1 - N_k$ orthonormal vectors of dimension N_1 that are orthogonal to the vectors c_{ik} and call them c_{ik}', where $i = 1, \ldots, N_1$ and $k = N_k + 1, \ldots, N_1$. Obviously, the matrix

$$C_{ik} \equiv \begin{cases} c_{ik} & for \quad k = 1, \ldots, N_k \\ c_{ik}' & for \quad k = N_k + 1, \ldots, N_1 \end{cases} \tag{2.34}$$

for $i = 1, \ldots, N_1$ is an $N_1 \times N_1$ unitary matrix. If we introduce the vectors

$$\left(|\tilde{\psi}\rangle \right) = \begin{pmatrix} |\tilde{\psi}_1\rangle \\ \vdots \\ |\tilde{\psi}_{N_1}\rangle \end{pmatrix}, \tag{2.35}$$

and

$$\left(|\tilde{k}_{N_1}\rangle \right) = \begin{pmatrix} |\tilde{k}_1\rangle \\ \vdots \\ |\tilde{k}_{N_k}\rangle \\ 0 \\ \vdots \\ 0 \end{pmatrix}, \tag{2.36}$$

where the last $N_1 - N_k$ elements of $\left(|\tilde{k}_{N_1}\rangle \right)$ are 0, we can write

$$\left(|\tilde{\psi}\rangle \right) = \begin{pmatrix} \ddots & & \\ & C_{ik} & \\ & & \ddots \end{pmatrix} \left(|\tilde{k}_{N_1}\rangle \right), \tag{2.37}$$

or, formally,

$$|\tilde{\psi}\rangle = C|\tilde{k}_{N_1}\rangle. \tag{2.38}$$

In an entirely similar way, we can show that $|\tilde{\phi}_i\rangle$ also lies in the subspace spanned by $\{|k\rangle\}$. Therefore, we can express $|\tilde{\phi}_i\rangle$ as

$$|\tilde{\phi}_i\rangle = \sum_{k=1}^{N_k} d_{ik}|\tilde{k}\rangle . \tag{2.39}$$

We can use this representation in (2.29) to obtain

$$
\rho = \sum_{i=1}^{N_2} |\tilde{\phi}_i\rangle\langle\tilde{\phi}_i| = \sum_{k,k'=1}^{N_k} (\sum_{i=1}^{N_2} d_{ik}d_{ik'}^*)|\tilde{k}\rangle\langle\tilde{k}'| = \sum_{k=1}^{N_k} |\tilde{k}\rangle\langle\tilde{k}| . \tag{2.40}
$$

Since the operators $|\tilde{k}\rangle\langle\tilde{k}'|$ are linearly independent we have that $\sum_{i=1}^{N_2} d_{ik}d_{ik'}^* = \delta_{kk'}$. Thus d_{ik} form N_k orthogonal vectors of dimension N_2 and $N_1 \geq N_2 \geq N_k$. In other words, d_{ik} $(k = 1, \ldots, N_k)$ form the first N_k columns of a matrix containing N_2 rows that can be extended to an $N_2 \times N_2$ unitary matrix in the following way. Find $N_2 - N_k$ orthonormal vectors of dimension N_2 that are orthogonal to the vectors d_{ik} and call them d'_{ik}, where $i = 1, \ldots, N_2$ and $k = N_k + 1, \ldots, N_1$. Obviously, the matrix

$$
D'_{ik} \equiv \begin{cases} d_{ik} & for \quad k = 1, \ldots, N_k \\ d'_{ik} & for \quad k = N_k + 1, \ldots, N_2 \end{cases} \tag{2.41}
$$

for $i = 1, \ldots, N_2$ is an $N_2 \times N_2$ unitary matrix. Then we introduce the vector

$$
\left(|\tilde{\phi}\rangle \right) = \begin{pmatrix} |\tilde{\phi}_1\rangle \\ \vdots \\ |\tilde{\phi}_{N_2}\rangle \\ 0 \\ \vdots \\ 0 \end{pmatrix}, \tag{2.42}
$$

where the last $N_1 - N_2$ elements are 0 and unitarily extend D'_{ik} into an $N_1 \times N_1$ matrix D_{ik} by the following definition

$$
D = \begin{pmatrix} \ddots & & \\ & D'_{ik} & 0 \\ & & \ddots \\ 0 & & I \end{pmatrix}, \tag{2.43}
$$

so that it is (2.41) for the first N_2 dimensions and identity for the remaining $N_1 - N_2$ dimensions. With these definitions we can now write

$$
\left(|\tilde{\phi}\rangle \right) = \begin{pmatrix} \ddots & & \\ & D'_{ik} & 0 \\ & & \ddots \\ 0 & & I \end{pmatrix} \left(|\tilde{k}_{N_1}\rangle \right), \tag{2.44}
$$

or, formally,

$$|\tilde{\phi}\rangle = D|\tilde{k}_{N_1}\rangle. \tag{2.45}$$

Comparing Eqs. (2.38) and (2.45), we finally obtain

$$|\tilde{\psi}\rangle = CD^{\dagger}|\tilde{\phi}\rangle. \tag{2.46}$$

Since the matrix $U = CD^{\dagger}$ is unitary by construction, this completes the proof.

2.5 A Mathematical Aside: The Schmidt Decomposition of a Bipartite State

In the previous section we have looked at the possible decompositions of the density matrix in terms of convex sums of pure state density matrices. The decomposition is not unique, but the possible decompositions of the same density matrix are connected via the theorem proved in the previous section. Namely, the renormalized pure states, with appropriate zero vectors included if their numbers are different, are connected via a unitary transformation. Each of these decompositions gives rise to a different ensemble interpretation. The ensembles are not unique but the various decompositions cannot be discriminated.

In this section we want to take a look at the other possible interpretation in which the mixed state density matrix emerges as the state of the subsystem of a larger system that itself is in a pure state. Therefore, we now examine the different ways in which a given density matrix ρ can be represented as the reduced density matrix for part of a pure bipartite state. To do this, we first need to derive the Schmidt decomposition of a bipartite state.

Let $|\psi\rangle_{AB} \in \mathcal{H}_A \otimes \mathcal{H}_B$, and $\{|u_i\rangle_A\}$ be an orthonormal basis for $\mathcal{H}_A$ and $\{|v_j\rangle_B\}$ be an orthonormal basis for $\mathcal{H}_B$. Then an arbitrary bipartite state can be expanded as a double sum over the product basis $\{|u_i\rangle|v_j\rangle\}$, as

$$|\psi\rangle_{AB} = \Sigma_{i,j} c_{ij} |u_i\rangle_A |v_j\rangle_B. \tag{2.47}$$

It is easy to see that this double sum expression can be written as a single sum,

$$|\psi\rangle_{AB} = \Sigma_i |u_i\rangle |\tilde{v}_i\rangle_B , \tag{2.48}$$

where we introduced $|\tilde{v}_i\rangle = \Sigma_j c_{ij} |v_j\rangle_B$. The price to pay is that $\{|\tilde{v}_i\rangle_B\}$ are not, in general, orthonormal. Therefore, it is somewhat surprising that for bipartite states there exists a single sum expansion where only diagonal elements of a product basis, $\{|u_i\rangle|w_i\rangle\}$ enter.

In order to show this, suppose that $\{|u_i\rangle\}$ is the basis in which $\rho_A = Tr_B(|\psi\rangle_{AB\,AB}\langle\psi|)$ is diagonal,

$$\rho_A = \Sigma_i \lambda_i |u_i\rangle\langle u_i|, \tag{2.49}$$

where $0 \leq \lambda_i \leq 1$. But we also have

$$\rho_A = Tr_B[\Sigma_{(i,j)}(|u_i\rangle_A|\tilde{v}_i\rangle_B)({}_A\langle u_j|_B\langle\tilde{v}_j|)] = \Sigma_{(i,j)\,B}\langle\tilde{v}_j|\tilde{v}_i\rangle_B|u_i\rangle_{AA}\langle u_j|. \tag{2.50}$$

Therefore, we must have ${}_B\langle\tilde{v}_j|\tilde{v}_i\rangle_B = \delta_{ij}\lambda_i$ and hence $\{|\tilde{v}_i\rangle\}$ are orthogonal.

Let $\{|u_i\rangle_A \mid i = 1, \ldots, N,$ where $N \leq dim\mathcal{H}_A\}$ correspond to nonzero values of λ_i and set $|w_i\rangle_B = \frac{1}{\sqrt{\lambda_i}}|\tilde{v}_i\rangle_B$. Hence $\{|w_i\rangle_B\}$ are orthonormal. Then

$$|\psi\rangle_{AB} = \Sigma_{i=1}^N \sqrt{\lambda_i}|u_i\rangle_A|w_i\rangle_B, \tag{2.51}$$

where $N \leq dim\mathcal{H}_A$ and by a similar argument $N \leq dim\mathcal{H}_B$. Note that

$$\rho_B = Tr_A(|\psi\rangle_{AB\,AB}\langle\psi|) = \Sigma_{i=1}^N \lambda_i |w_i\rangle_{B\,B}\langle w_i|, \tag{2.52}$$

so that $\{|w_i\rangle\}$ are eigenstates of ρ_B having non-zero eigenvalues and ρ_A and ρ_B have the same nonzero eigenvalues. The double sum expansion in Eq. (2.47) always exists. It is somewhat surprising that the single sum expansion of Eq. (2.51), in terms of orthonormal basis vectors, also exists for bipartite systems. This latter is called the Schmidt decomposition.

2.6 Purfication, Reduced Density Matrices and the Subsystem Interpretation

Equipped with the Schmidt decomposition, we now look at purifications. Suppose, we have a density matrix

$$\rho_A = \Sigma_{i=1}^N p_i |\psi_i\rangle_{AA}\langle\psi_i|, \tag{2.53}$$

where $|\psi_i\rangle \in \mathcal{H}_A$. We want to find a state $|\Phi\rangle_{AB} \in \mathcal{H}_A \otimes \mathcal{H}_B$ on a larger space so that

$$\rho_A = Tr_B(|\Phi\rangle_{AB\,AB}\langle\Phi|). \tag{2.54}$$

$|\Phi\rangle_{AB}$ is called a purification of ρ_A.

One way to do this is to choose $dim(\mathcal{H}_B) \geq N$ and let $\{|u_i\rangle\}$ be an orthonormal basis for $\mathcal{H}_B$. Then

$$|\Phi\rangle_{AB} = \Sigma_i \sqrt{p_i} |\psi_i\rangle_A |u_i\rangle_B \tag{2.55}$$

is a purification of ρ_A.

Purifications are not unique. But if two purifications are in the same Hilbert space, we can still say something about their mutual relationship. Suppose we have two different states, $|\Phi_1\rangle_{AB}$ and $|\Phi_2\rangle_{AB}$, both of which are in $\mathcal{H}_A \otimes \mathcal{H}_B$ and both of which are purifications of ρ_A. How are they related? To answer this question, we use the Schmidt decomposition, $|\Phi_1\rangle_{AB} = \Sigma_k \sqrt{\lambda_k} |u_k\rangle_A |v_k\rangle_B$ and $|\Phi_2\rangle_{AB} = \Sigma_k \sqrt{\lambda_k} |u_k\rangle_A |w_k\rangle_B$. A part of both states, eigenvalues and eigenvectors of ρ_A, is the same. $\{|v_k\rangle_B\}$ and $\{|w_k\rangle_B\}$ form orthonormal sets, so there is at least one unitary operator on $\mathcal{H}_B$, which we call U_B, such that

$$|w_k\rangle_B = U_B |u_k\rangle_B . \tag{2.56}$$

Then $|\Phi_2\rangle_{AB} = (I_A \otimes U_B)|\Phi_1\rangle_{AB}$.

2.7 Comparing Density Matices

In many applications, we need to compare two density matrices. We might wish to know how far apart, in some sense, they are or how large a component one has in the direction of the other. There are a number of ways to do this, but in quantum information there are two measures for comparing density matrices, the trace norm and the fidelity, that have proven to be particularly useful.

The trace norm of an operator A, which is denoted by $\|A\|_1$, is given by $\mathrm{Tr}(|A|)$ where $|A| = \sqrt{(A^\dagger A)}$. In practice, to find the trace norm one finds the eigenvalues, λ_j, of $A^\dagger A$, and then $\|A\|_1 = \sum_j |\lambda_j|$. Since the trace norm is a norm, which we will not prove here, it satisfies the triangle inequality, which states that for two operators A and B,

$$\|A + B\|_1 \leq \|A\|_1 + \|B\|_1. \tag{2.57}$$

For an individual density matrix, we clearly have $\|\rho\|_1 = 1$, since the eigenvalues of ρ are non-negative and add to one. We will be more interested in the trace norm as a measure of the distance between two density matrices. One way of expressing that distance is given by the following theorem.

Theorem *If ρ_1 and ρ_2 are density matrices, then*

$$\frac{1}{2}\|\rho_1 - \rho_2\|_1 = \max_P \mathrm{Tr}(P(\rho_1 - \rho_2)), \tag{2.58}$$

where the maximum is taken over all projection operators P.

Proof We can express $\rho_1 - \rho_2$ as the difference of two positive operators, Q_+ and Q_-, $\rho_1 - \rho_2 = Q_+ - Q_-$. This is done by diagonalizing $\rho_1 - \rho_2 = \sum_j \lambda_j |u_j\rangle\langle u_j|$, and setting

$$Q_+ = \sum_{\lambda_j \geq 0} \lambda_j |u_j\rangle\langle u_j| \qquad Q_- = \sum_{\lambda_j < 0} |\lambda_j| |u_j\rangle\langle u_j|. \tag{2.59}$$

Note that Q_+ and Q_- have orthogonal supports. We then have that $|\rho_1 - \rho_2| = Q_+ + Q_-$ so that $\|\rho_1 - \rho_2\|_1 = \text{Tr}(Q_+) + \text{Tr}(Q_-)$, and $\text{Tr}(\rho_1 - \rho_2) = \text{Tr}(Q_+ - Q_-) = 0$. Therefore, $\|\rho_1 - \rho_2\|_1 = 2\text{Tr}(Q_+)$. Now, for P a projection

$$\text{Tr}(P(\rho_1 - \rho_2)) = \text{Tr}(P(Q_+ - Q_-)) \leq \text{Tr}(PQ_+)$$

$$\leq \text{Tr}(Q_+) = \frac{1}{2}\|\rho_1 - \rho_2\|_1. \tag{2.60}$$

Finally, to see that this bound can be achieved, suppose P is the projection onto the support of Q_+. We then have that $\text{Tr}(P(\rho_1 - \rho_2)) = \text{Tr}(P(Q_+ - Q_-)) = \text{Tr}(Q_+)$, which completes the proof.

This result immediately implies that $|\text{Tr}(P\rho_1) - \text{Tr}(P\rho_2)| \leq (1/2)\|\rho_1 - \rho_2\|_1$. This tells us that if the trace norm of two density matrices is small, then the results of projective measurements made on them will be close to each other. In more detail, $\text{Tr}(P\rho)$ is the probability of obtaining the result 1 when measuring the projection operator P. If two density matrices are close in the sense of the trace norm, then the probabilities of obtaining 1 when measuring P in each of them will be close. That implies that the density matrices will be difficult to distinguish by making projective measurements.

The fidelity between two density matrices is based on the following idea. Suppose we want to compare two pure states, $|\psi_1\rangle$ and $|\psi_2\rangle$. One way to do so is to look at their overlap, $|\langle\psi_1|\psi_2\rangle|$. If the overlap is close to one, the states are similar, if it is close to zero, they are almost orthogonal. In order to generalize this to density matrices, we define the fidelity between ρ_1 and ρ_2 to be

$$F(\rho_1, \rho_2) = \text{Tr}\sqrt{\rho_1^{1/2}\rho_2\rho_1^{1/2}}). \tag{2.61}$$

That this is the definition we want is not immediately obvious, so let's look at some special cases. If we set $\rho_1 = |\psi_1\rangle\langle\psi_1|$, we find that $F(\rho_1, \rho_2) = \sqrt{\langle\psi_1|\rho_2|\psi_1\rangle}$, and then if $\rho_2 = |\psi_2\rangle\langle\psi_2|$ the fidelity just becomes the overlap, $|\langle\psi_1|\psi_2\rangle|$. It is also not obvious that the fidelity is symmetric in its two entries, that is that $F(\rho_1, \rho_2) = F(\rho_2, \rho_1)$. This is, however, the case, a result that we will not prove. It is clear that if $\rho_1 = \rho_2$, then their fidelity is one, and that if the two density matrices have orthogonal support, their fidelity is zero. For the general case, the fidelity is between 0 and 1.

Finally, one would suspect that there is a relation between the trace norm distance between two density matrices and their fidelity. If the trace distance between two density matrices is small, their fidelity should be close to one, and if it is close to 2, its maximum value, then the fidelity should be small. We, in fact, have the inequality,

$$1 - F(\rho_1, \rho_2) \leq \frac{1}{2} \|\rho_1 - \rho_2\|_1 \leq \sqrt{1 - F(\rho_1, \rho_2)^2}, \tag{2.62}$$

which we state without proof.

2.8   Problems

1. This problem combines elements from Chaps. 1 and 2 as it uses circuits with mixed states. We shall consider two quantum circuits, each consisting of 3 qubits and two C-NOT gates.

   (a) The operator for the first circuit is given by $U_1 = D_{ac} D_{ab}$. If the input state is $(\alpha|0\rangle_a + \beta|1\rangle_a)|0\rangle_b|0\rangle_c$ find the output state, and find the reduced density matrices for qubit b and qubit c at the output. Note that the effect of this circuit is to transfer information about $|\alpha|$ and $|\beta|$ to qubits b and c.

   (b) The operator for the second circuit is $U_2 = D_{ca} D_{ba}$. If the input state is $(\alpha|0\rangle_a + \beta|1\rangle_a)|+x\rangle_b|+x\rangle_c$, find the output state and the reduced density matrices for qubit b and qubit c at the output. Note that in this case phase information about α and β is transferred to qubits b, and c.

2. The Schmidt representation for states of a bipartite system is extremely convenient, and so it is natural to ask if such a representation exists for tripartite systems. Unfortunately, the answer is no. Show that there exist three-qubit states that cannot be written in the form

$$|\Psi\rangle_{abc} = \sum_{j=0}^{1} \sqrt{\lambda_j} |u_j\rangle_a |v_j\rangle_b |w_j\rangle_c,$$

   where $\{u_j | j = 0, 1\}$, $\{v_j | j = 0, 1\}$, and $\{w_j | j = 0, 1\}$ are orthonormal bases.

3. Suppose that Alice can prepare a density matrix only in the computational basis. She prepares a bipartite state of the form

$$\rho = \sum_{j,k=0}^{1} p_{jk} |j\rangle\langle j| \otimes |k\rangle\langle k|.$$

She sends one qubit to Bob and one qubit to Charlie. If Bob and Charlie do not measure in the computational basis, the correlations they can obtain are limited. Show that if they measure in the $|\pm x\rangle$ basis their results will be uncorrelated, that is, they are equally likely to get the same result as opposite results.

4. The W state is the three-qubit state

$$|\Phi\rangle_{abc} = \frac{1}{\sqrt{3}}(|0\rangle_a|0\rangle_b|1\rangle_c + |0\rangle_a|1\rangle_b|0\rangle_c + |1\rangle_a|0\rangle_b|0\rangle_c).$$

(a) Find the reduced density matrix of qubits b and c together.
(b) Find the reduced density matrix of qubit a.
(c) Considering qubit a as the first system and qubits b and c as the second system, find the Schmidt representation of the W state.

5. Let $\rho_1 = |\psi_1\rangle\langle\psi_1|$ and $\rho_2 = |\psi_2\rangle\langle\psi_2|$. Find $\|\rho_1 - \rho_2\|_1$ in terms of $\langle\psi_1|\psi_2\rangle$.

Further Reading

1. Asher Peres, *Quantum Theory: Concepts and Methods* (Kluwer Academic Publishers, Dordrecht, 1995)
2. M. Nielsen, I. Chuang, *Quantum Computation and Quantum Information* (Cambridge University Press, Cambridge, 2010)

Chapter 3
Entanglement

Entanglement is an essential feature of quantum mechanics. If two systems are entangled, there are not individual quantum states describing each system, only a global state describing both. Systems that are entangled can exhibit strong correlations, and it is possible to make use of those correlations in quantum communication protocols. In this chapter we will describe how entangled states can be used to violate Bell inequalities, and thereby show that the world is not described by local hidden-variable theories. We will then show how entanglement can be detected and measured and then describe two communication protocols, teleportation and dense coding, in which it is essential.

3.1 Definition of Entanglement

We begin with some definitions. Consider a quantum state in a tensor product Hilbert space, $\mathcal{H} = \mathcal{H}_A \otimes \mathcal{H}_B$. A pure state is not entangled if it is of product form

$$|\psi\rangle_{AB} = |\phi_1\rangle_A \otimes |\phi_2\rangle_B, \tag{3.1}$$

otherwise, it is entangled. A density matrix, ρ_{AB}, is separable if it is a mixture of product states, i.e. if it is of the form

$$\rho_{AB} = \sum_i p_i \rho_{Ai} \otimes \rho_{Bi}, \tag{3.2}$$

where $0 \leq p_i \leq 1$, and $\sum_i p_i = 1$. If ρ_{AB} is not separable, it is entangled. For a pure state that is not entangled, measurements on systems A and B are not correlated. For a separable density matrix there are only classical correlations between measurements conducted on the two systems. As we shall see, entangled

© Springer Nature Switzerland AG 2021

J. A. Bergou et al., *Quantum Information Processing*, Graduate Texts in Physics, https://doi.org/10.1007/978-3-030-75436-5_3

states can lead to much stronger correlations than are possible classically. Finally, we call a state maximally entangled if it is entangled and its reduced density matrices are proportional to the identity.

Entanglement can be thought of in terms of state preparation. Suppose Alice is in her laboratory and Bob is in his. If they want to prepare a separable pure state, each party can prepare their part of the state in their own laboratory, and no communication is required. In order to prepare a general separable state, all that is required is that each performs operations in their own laboratory, and they communicate classically. In order to create entanglement, they have to exchange quantum systems.

Let us briefly look at some two-qubit examples of separable and entangled states. The pure state $|0\rangle_A |0\rangle_B$ is not entangled, and neither is the density matrix

$$\rho_{AB} = \frac{1}{3}|0\rangle_A {}_A\langle 0| \otimes |0\rangle_B {}_B\langle 0| + \frac{2}{3}|1\rangle_A {}_A\langle 1| \otimes |1\rangle_B {}_B\langle 1|. \tag{3.3}$$

On the other hand, the so-called Bell states

$$|\Phi_\pm\rangle_{AB} = \frac{1}{\sqrt{2}}(|01\rangle_{AB} \pm |10\rangle_{AB})$$

$$|\Psi_\pm\rangle_{AB} = \frac{1}{\sqrt{2}}(|00\rangle_{AB} \pm |11\rangle_{AB}), \tag{3.4}$$

are maximally entangled.

3.2   Faster Than Light Communication 1

The strong correlations present in entangled states suggest that it might be possible to use them for instantaneous communication. For example, if Alice and Bob share the state $(1/\sqrt{2})(|00\rangle + |11\rangle)$, and Alice measures her particle in the computational basis and gets $|0\rangle$, then Bob's state becomes $|0\rangle$ as well. Alice and Bob could be separated by a large distance, and the fact that Bob's state collapses to $|0\rangle$ when Alice measures her state seems to indicate that they can communicate faster than the speed of light. Needless to say, this doesn't happen, but let us look at a specific scheme to see why.

We assume that Alice and Bob each have one qubit of the state $(1/\sqrt{2})(|0\rangle_a |0\rangle_b + |1\rangle_a |1\rangle_b)$. Bob sends his particle through a one-qubit gate, which we shall call a phase gate, that has the action $|0\rangle \rightarrow |0\rangle$ and $|1\rangle \rightarrow e^{i\phi}|1\rangle$. The idea is to see whether Alice can influence whether Bob sees an interference pattern or not, that is, whether some probabilities for Bob's measurement depend on ϕ or not. For example, if Bob has the state $(1/\sqrt{2})(|0\rangle_b + e^{i\phi}|1\rangle_b)$, sends it through a Hadamard gate, and then measures in the computational basis, the probability of measuring $|0\rangle_b$, p_{0b}, is $(1/4)|1 + e^{i\phi}|^2$, and the probability of measuring $|1\rangle_b$,

p_{1b}, is $(1/4)|1 - e^{i\phi}|^2$. If he does the same with his particle from the state $(1/\sqrt{2})(|0\rangle_a|0\rangle_b + e^{i\phi}|1\rangle_a|1\rangle_b)$, we find $p_{0b} = p_{1b} = 1/2$. The fact that Bob's particle is correlated with Alice's destroys the interference. However, if Alice were to measure her particle in the $|\pm x\rangle$ basis, she will destroy the correlation. If Alice were to get $|+x\rangle$ for her result, then Bob would have the state $(1/\sqrt{2})(|0\rangle_b + e^{i\phi}|1\rangle_b)$, which suggests he would seen an interference pattern. Therefore, the thought is that Bob can tell whether Alice measured or not by determining whether he sees an interference pattern. Variants of this scheme keep being rediscovered and submitted to physics journals.

So what is wrong? The short answer is that the fact that Alice cannot control what measurement result she gets washes out the interference pattern when she does make a measurement. Let us look more carefully. First, consider the case in which Alice makes no measurement. After Bob sends his state through the phase gate and a Hadamard gate, the two-particle state is

$$\frac{1}{2}(|0\rangle_a + e^{i\phi}|1\rangle_a)|0\rangle_b + \frac{1}{2}(|0\rangle_a - e^{i\phi}|1\rangle_a)|1\rangle_b, \tag{3.5}$$

and noting that the two Alice states in the above equation are orthogonal, we see that Bob's reduced density matrix is just $\rho_b = (1/2)I_b$. Now consider what happens if Alice does make a measurement. She will measure her qubit by sending it through a Hadamard gate and then measuring it in the computational basis. If she obtains $|0\rangle_a$, which she does with a probability of $1/2$, then Bob's state is

$$\frac{1}{2}(1 + e^{i\phi})|0\rangle_b + \frac{1}{2}(1 - e^{i\phi})|1\rangle_b, \tag{3.6}$$

and if she obtains $|1\rangle_a$, which she does with a probability of $1/2$, then Bob's state is

$$\frac{1}{2}(1 - e^{i\phi})|0\rangle_b + \frac{1}{2}(1 + e^{i\phi})|1\rangle_b. \tag{3.7}$$

Noting that the two Bob states are orthogonal, we find that Bob's reduced density matrix is again $\rho_b = (1/2)I_b$. Therefore, in both cases, Bob's reduced density matrix is the same, so that he cannot determine by any measurement he makes whether Alice measured or not, and in neither case does he see an interference pattern.

3.3 Bell Inequalities

Because entangled states can have correlations that go beyond what is possible classically, they are a valuable resource in quantum communication protocols, for example, as we shall see, in teleportation and dense coding. Before getting to these, however, let us see what is meant by nonclassical correlations, which means having a look at Bell inequalities.

These inequalities arose from a consideration of alternatives to quantum mechanics known as local hidden-variable theories. The idea behind them is that, unlike in quantum mechanics, observables have actual values but we do not know what they are, because they depend on some "hidden variables" about which we know nothing. In quantum mechanics, observables do not have values until we measure them. Bell inequalities show that under very general assumptions, hidden variables produce predictions that conflict with quantum mechanics. These can then be tested experimentally, and the experiments support quantum mechanics.

The basic setup for Bell inequalities consists of two observers, Alice and Bob, and a source that produces two-particle states. One particle is sent to Alice and the other to Bob. Alice can measure one of two observables for her particle, a_1 and a_2. These observables can each be either 1 or -1. Similarly, Bob can measure either b_1 or b_2, and these can also be either 1 or -1. The idea is to run this Gedanken experiment many times, and use the results to compute the quantities $\langle a_i b_j \rangle$.

Let us first see how a hidden-variable theory would describe this situation. The source produces, along with the particles, instruction sets that go with them. For example, one instruction set might say, if Alice measures a_1 she will get 1, if she measures a_2, she gets -1, and if Bob measures b_1 he gets -1, and if he measures b_2 he gets -1, or more briefly, $(a_1 = 1, a_2 = -1, b_1 = -1, b_2 = -1)$. We do not know which instruction set the source will produce, and so this, the instruction set, is our hidden variable. The adjective local is applied to this kind of a hidden-variable theory, because the instructions to Alice's particle do not depend on what Bob decides to measure. That is, the instruction set does not say something like, if Alice measures a_1, then she gets 1 if Bob measures b_1 and -1 if Bob measures b_2. We shall consider only local theories. We assume that each instruction set occurs with some probability. This is equivalent to assuming that we have a joint probability distribution for the variables, a_1, a_2, b_1, and b_2, which we shall denote as $P(a_1, a_2, b_1, b_2)$. We would then compute the expectation value $\langle a_1 b_1 \rangle$, as

$$\langle a_1 b_1 \rangle = \sum_{a_1=-1}^{1} \cdots \sum_{b_2=-1}^{1} a_1 b_1 P(a_1, a_2, b_1, b_2). \tag{3.8}$$

We now want to consider the quantity

$$S = \langle a_1 b_1 \rangle + \langle a_1 b_2 \rangle + \langle a_2 b_1 \rangle - \langle a_2 b_2 \rangle$$

$$= \sum_{a_1=-1}^{1} \cdots \sum_{b_2=-1}^{1} [a_1(b_1 + b_2) + a_2(b_1 - b_2)] P(a_1, a_2, b_1, b_2). \tag{3.9}$$

Call the term in brackets multiplying the probability distribution X. We see that $X = a_1(b_1 + b_2)$ if $b_1 = b_2$, and $X = a_2(b_1 - b_2)$ if $b_1 = -b_2$. In both cases, $|X| = 2$, so that

$$|S| \leq 2 \sum_{a_1=-1}^{1} \cdots \sum_{b_2=-1}^{1} P(a_1, a_2, b_1, b_2) = 2. \tag{3.10}$$

This is a Bell inequality. Note that we can derive similar inequalities simply by interchanging a_1 and a_2, b_1 and b_2, or both.

There are an infinite number of Bell inequalities. The one we have just proved is known as the CHSH (Clauser, Horne, Shimony, Holt) inequality. Bell inequalities are characterized by three numbers, the number of parties, the number of measurements per party, and the number of outcomes for each measurement. In the case of the CHSH inequality, each of these numbers is 2.

Now let us describe the same experiment using quantum mechanics, and assume that we are measuring the spins of two spin-1/2 particles. Assume

$$a_1 = \sigma_{xa} \quad a_2 = \sigma_{ya}$$
$$b_1 = \sigma_{xb} \quad b_2 = \sigma_{yb}, \tag{3.11}$$

and that the source puts out particles in the state

$$|\Psi\rangle = \frac{1}{\sqrt{2}}(|00\rangle + e^{i\pi/4}|11\rangle), \tag{3.12}$$

where

$$\sigma_x|0\rangle = |1\rangle \quad \sigma_y|0\rangle = i|1\rangle$$
$$\sigma_x|1\rangle = |0\rangle \quad \sigma_y|1\rangle = -i|0\rangle. \tag{3.13}$$

Note that $|\Psi\rangle$ is an entangled state. We have that $\langle a_1 b_1\rangle$, $\langle a_1 b_2\rangle$, and $\langle a_2 b_1\rangle$ are all equal to $\sqrt{2}/2$, and $\langle a_2 b_2\rangle$ is equal to $-\sqrt{2}/2$. This gives us $S = 2\sqrt{2}$, which violates our Bell inequality.

From this we can conclude two things. First, that quantum mechanics cannot be described by a local hidden-variable theory. Second, in the hidden-variable theory, the correlations came from a classical joint distribution function. Therefore, quantum mechanics can produce stronger correlations than classical systems can.

Next we want to study the connection of the CHSH Bell inequality to entanglement. We shall do this by showing that if $|\Psi\rangle$ is not entangled, then the Bell inequality will be satisfied. If $|\Psi\rangle$ is a product state then the expectation values appearing in the Bell inequality factorize, i.e. $\langle a_i b_j\rangle = \langle a_i\rangle\langle b_j\rangle$. Define $x_i = \langle a_i\rangle$ and $y_j = \langle b_j\rangle$, for $i, j = 1, 2$, where $-1 \leq x_i \leq 1$ and $-1 \leq y_j \leq 1$. Let us denote by R the region in the y_1, y_2 plane given by $\{-1 \leq y_j \leq 1 | j = 1, 2\}$. We then have that

$$S = x_1(y_1 + y_2) + x_2(y_1 - y_2). \tag{3.14}$$

Now suppose that $y_1 - y_2 = c > 0$, where $c \leq 2$. This line intersects the boundary of R on the line $y_1 = 1$ at the point $y_1 = 1$, $y_2 = 1 - c$ and on the line $y_2 = -1$ at the point $y_1 = c - 1$, $y_2 = -1$. This implies that $c - 2 \leq y_1 + y_2 \leq 2 - c$. Similarly, if $y_1 - y_2 = c < 0$, where $c > -2$, then this line intersects the boundary of R on the line $y_1 = -1$ at the point $y_1 = -1$, $y_2 = -1 - c$ and on the line $y_2 = 1$ at the point $y_1 = c + 1$, $y_2 = 1$. This implies that $-c - 2 \leq y_1 + y_2 \leq 2 + c$. We can summarize both of these cases by the inequality, for $|c| \leq 2$,

$$|c| - 2 \leq y_1 + y_2 \leq 2 - |c|. \tag{3.15}$$

We, therefore, have that if $y_1 - y_2 = c$, then $S = x_1(y_1 + y_2) + x_2 c$, and

$$-2 \leq |x_1|(|c| - 2) - |x_2||c| \leq S \leq |x_1|(2 - |c|) + |x_2||c| \leq 2. \tag{3.16}$$

Hence, we can conclude that for a pure state that is not entangled, the Bell inequality will be satisfied. This conclusion can be easily extended to separable states, because a separable state is just an incoherent superposition of product states and for each of the product states the Bell inequality is satisfied.

Now, that we have seen what kind of states satisfy our Bell inequality, let us address the opposite end and find the maximum violation of this inequality that quantum mechanics can provide. This is given by Tsirelson's inequality. In order to derive the Tsirelson bound, we note the a_j and b_j are Hermitian operators with eigenvalues ± 1 and hence $a_j^2 = b_j^2 = I$. Let us further define the operator $C = a_1 b_1 + a_1 b_2 + a_2 b_1 - a_2 b_2$. We then have

$$2\sqrt{2} - C = \frac{1}{\sqrt{2}}(a_1^2 + a_2^2 + b_1^2 + b_2^2) - C = \frac{1}{\sqrt{2}}(a_1 - \frac{b_1 + b_2}{\sqrt{2}})^2 + \frac{1}{\sqrt{2}}(a_2 - \frac{b_1 + b_2}{\sqrt{2}})^2 \geq 0. \tag{3.17}$$

Therefore, $\langle C \rangle \leq 2\sqrt{2}$. Similarly, by changing all the negative signs to positive ones in the above equation, one can show that $\langle C \rangle \geq -2\sqrt{2}$, and, since $S = \langle C \rangle$, we have that

$$|S| \leq 2\sqrt{2}, \tag{3.18}$$

which is Tsirelson's inequality.

Bell inequalities are closely related to a problem in classical probability theory. If we have a probability distribution of several variables, such as $P(a_1, a_2, b_1, b_2)$, a probability distribution for a subset of the variables, such as one for a_1 and b_1, is called a marginal distribution. Given a full joint distribution, it is easy to find the marginal distribution for a subset of the variables by summing over the variables in the complementary subset. The reverse problem is not so easy. We are given some distributions of some sets of variables, and the question is whether there exists a full joint distribution for which they are the marginals. In the case of the CHSH inequality, we are given four distributions, $P(a_j, b_k)$ for $j, k = 1, 2$, and we want to know whether there is a joint distribution, $P(a_1, a_2, b_1, b_2)$, for which they are marginals. If so, then the experiment can be described by a local hidden-variable theory. As we have seen, quantum mechanics can generate distributions for which there is no joint distribution.

One might wonder whether there are sets of four distributions, $P(a_j, b_k)$ for $j, k = 1, 2$, that violate the CHSH inequality more strongly than quantum mechanics does. These distributions should satisfy the no-signaling condition, so as to respect relativity. This states that Alice's results should be independent of Bob's choice of measurement, and vice versa. Mathematically this means that

$$\sum_{b_1=\pm 1} P(a_j, b_1) = \sum_{b_2=\pm 1} P(a_j, b_2), \tag{3.19}$$

for $j = 1, 2$ and similarly for b_j. Popescu and Rohrlich found four distributions satisfying these conditions for which the quantity S is equal to 4, its maximum possible value. Details can be found in reference 1. Mathematically, then, there are correlations stronger than those allowed by quantum mechanics, but they have not been found in nature.

3.4 Representative Applications of Entanglement: Dense Coding, Teleportation and Entanglement Swapping

In this section we look at some interesting applications of entanglement that reveal its power as a resource for quantum information related tasks. The first, dense coding, shows how entanglement allows us to pack more classical information into a qubit. If Alice sends a single unentangled qubit to Bob, she can send at most one bit of classical information. Using entanglement, she can send two. The second, teleportation, allows us to transfer a quantum state from one qubit to another. This is very useful for moving around quantum information. The third, entanglement swapping, allows us to change which parties are entangled, and thereby enables us to move around entanglement.

3.4.1 Dense Coding

In this protocol two parties, traditionally called Alice and Bob, can communicate two bits of classical information by exchanging only one qubit. The key is

Table 3.1 Bob's possible operations and the resulting two-qubit state at Alice's site

Bob's operation	I	σ_x	σ_y	σ_z
Alice's state	$|\Phi_-\rangle$	$|\Psi_-\rangle$	$-i|\Psi_+\rangle$	$-|\Phi_+\rangle$

entanglement, of course. We assume that Alice and Bob share an entangled pair of qubits in the state $|\Phi_-\rangle = \frac{1}{\sqrt{2}}(|01\rangle_{AB} - |10\rangle_{AB})$. Out of a pair of qubits in this state, Alice has one of the qubits, labeled A, in her possession while the other, labeled B, is in Bob's possession. Bob then performs one of four operations on his qubit and sends it back to Alice. The four operations and the resulting two-qubit state, now entirely at Alice's site, are listed in Table 3.1.

The point is that Alice now has one of four orthogonal states and she can distinguish them perfectly. After performing a measurement in the Bell basis, Alice will know with certainty which of the four operations Bob performed. Bob sent only one particle, a single qubit, to Alice, but Alice can perfectly distinguish among four classical alternatives, i.e. one (entangled) qubit carried two classical bits of information.

3.4.2 Teleportation

Alice has a qubit, say A_1 in some quantum state $|\psi\rangle$, in her possession. She wants to transfer the quantum state of her qubit A_1 onto Bob's qubit B. Alice may not even know what $|\psi\rangle$ is. Measuring $|\psi\rangle$ and transmitting the classical information that is the result will not work; it is not enough information to reconstruct the state.

In the teleportation procedure Alice and Bob share an entangled pair A_2, B in the state $|\Phi_-\rangle_{A_2B} = \frac{1}{\sqrt{2}}(|01\rangle_{A_2B} - |10\rangle_{A_2B})$. The total state of the three qubits, the one whose state is to be teleported and the entangled pair, is then

$$|\psi\rangle_{A_1}|\Phi_-\rangle_{A_2B} = \frac{1}{\sqrt{2}}(\alpha|0\rangle_{A_1} + \beta|1\rangle_{A_1})(|01\rangle_{A_2B} - |10\rangle_{A_2B})$$

$$= \frac{1}{\sqrt{2}}(\alpha|00\rangle_{A_1A_2}|1\rangle_B - \alpha|01\rangle_{A_1A_2}|0\rangle_B$$

$$+ \beta|10\rangle_{A_1A_2}|1\rangle_B - \beta|11\rangle_{A_1A_2}|0\rangle_B)$$

$$= \frac{1}{2}\{|\Phi_+\rangle_{A_1A_2}(-\sigma_z|\psi\rangle_B) + |\Phi_-\rangle_{A_1A_2}(-|\psi\rangle_B)$$

$$+ |\Psi_+\rangle_{A_1A_2}(\sigma_x\sigma_z|\psi\rangle_B) + |\Psi_-\rangle_{A_1A_2}(\sigma_x|\psi\rangle_B)\}. \quad (3.20)$$

The key is in the last line. When the total three-qubit state is decomposed in terms of the four Bell basis states of the two qubits of Alice, the state of Bob's qubit associated with each of these terms is related in a simple way to the state to be teleported. When Alice measures her state in the Bell basis, she tells Bob

Table 3.2 Alice's measurement outcomes and Bob's subsequent operations

Alice's measurement yields	$|\Phi_+\rangle$	$|\Phi_-\rangle$	$|\Psi_+\rangle$	$|\Psi_-\rangle$
Bob performs	σ_z	I (nothing)	$\sigma_z\sigma_x$	σ_x

over a classical channel what she got, and then Bob can apply the appropriate operator to his qubit to recover Alice's state. The four possible outcomes of Alice's measurement and the operations Bob performs corresponding to each of the measurement results, are listed in the Table 3.2.

All information about $|\psi\rangle$ is transferred to Bob, none is left with Alice. After teleportation Alice is left in possession of a Bell state. If someone else prepared the state of the original A_1 qubit for Alice, she will never learn its state in the process. Nevertheless, the state will be faithfully teleported to Bob's qubit B.

3.4.3 Entanglement Swapping

Alice and Bob share an entangled pair, and Bob and Charlie share an entangled pair. By making a measurement on his two qubits and communicating the result to Alice and Charlie, Bob can create entanglement between Alice and Charlie. Alice and Charlie were not initially entangled, but after the entanglement swapping procedure, they are, and Bob is entangled with neither of them.

This protocol is useful in the transmission of quantum information. Suppose Alice wants to teleport a state to Charlie. They have to share an entangled pair of qubits. These qubits are often polarization states of photons, and, in that case, Alice and Charlie would often be connected by an optical fiber. Alice would create two entangled photons, keep one and send the other to Charlie through the fiber. A photon traveling through a fiber is subject to losses and decoherence that will degrade the quantum information it carries. In a classical system, in which one only cares about losses, the answer is to place amplifiers along the fiber to periodically boost the signal. Amplifiers, however, add noise as well as boosting signals, and if one is sending quantum information, the added noise will ruin the quantum information. This limits the direct transmission of entanglement in a fiber to about 100 km. Entanglement swapping provides a solution. Suppose Alice is separated from Bob by 100 km and Bob is separated from Charlie by 100 km. Alice and Charlie can each send Bob one photon from an entangled pair, Bob can perform the entanglement swapping protocol, and then Alice and Charlie, who are separated by 200 km, share an entangled pair. This procedure could be repeated over more links in order to transmit entanglement over longer distances. The devices that use entanglement swapping to enable the transmission of entanglement, and thereby quantum information, are called quantum repeaters.

Now let us go to the procedure. Alice, Bob, and Charlie start in the state $|\Phi_-\rangle_{AB_1}|\Phi_-\rangle_{B_2C}$, where Alice has qubit A, Bob has qubits B_1 and B_2, and Charlie has qubit C. This state can be rewritten in the form

$$|\Phi_-\rangle_{AB_1}|\Phi_-\rangle_{B_2C} = \frac{1}{2}(|\Phi_+\rangle_{B_1B_2}|\Phi_+\rangle_{AC} - |\Phi_-\rangle_{B_1B_2}|\Phi_-\rangle_{AC}$$

$$-|\Psi_+\rangle_{B_1B_2}|\Psi_+\rangle_{AC} + |\Psi_-\rangle_{B_1B_2}|\Psi_-\rangle_{AC}). \quad (3.21)$$

From this we can see that if Bob measures his two qubits in the Bell basis, Alice and Charlie will share one of the Bell states. Bob tells them the result of his measurement, and then they know which Bell state they share. After the procedure, Bob is left with a Bell state, which is unentangled with both Alice and Charlie.

3.5   Conditions of Separability

How can we tell if a given density matrix is separable? Necessary and sufficient conditions are known to exist for the simplest cases only. In general, there are no known necessary and sufficient conditions to determine whether the state is separable or entangled. There are, however, some sufficient conditions.

One of them is a Bell inequality. For two qubits, choose $a_1 = \vec{n}_1 \cdot \vec{\sigma}$, $a_2 = \vec{n}_2 \cdot \vec{\sigma}$, $b_1 = \vec{n}_3 \cdot \vec{\sigma}$, $b_2 = \vec{n}_4 \cdot \vec{\sigma}$, with $\vec{n}_j$ being unit vectors. If ρ_{AB} violates a Bell inequality for some choice of the unit vectors $\{\vec{n}_j | j = 1, \ldots, 4\}$, it is entangled. This is not a particularly strong criterion as there is a large class of entangled states that satisfies Bell inequalities.

A stronger and more general test was found by Peres, which is known as the positive partial transpose (PPT) criterion. Consider a density matrix on $\mathcal{H}_A \otimes \mathcal{H}_B$ of arbitrary dimensions. We have the density matrix elements in some product basis $\rho_{m\mu;n\nu} = {}_A\langle m| \otimes {}_B\langle \mu|\rho|n\rangle_A \otimes |\nu\rangle_B$.

The partial transposition of ρ is the density matrix with the matrix elements

$$\rho^{T_B}_{m\mu;n\nu} = \rho_{m\nu;n\mu}. \quad (3.22)$$

The operator ρ^{T_B} depends on the basis in which the transpose is defined, but its eigenvalues do not. We say a state is PPT if $\rho^{T_B} \geq 0$. A separable state is always PPT. This is because if ρ_{AB} is separable then $\rho^{T_B}_{AB} = \Sigma_i p_i \rho_{Ai} \otimes \rho^T_{Bi}$, and if $\rho_{Bi} \geq 0$, then $\rho^T_{Bi} \geq 0$.

Therefore, if a partial transpose is not positive the state is entangled. Thus, the PPT condition is sufficient. For $2 \otimes 2$ (two-qubit) and $2 \otimes 3$ (qubit-qutrit) systems the converse is also true: if a state is entangled, the partial transpose is not positive. Thus, for these systems, the PPT condition is also necessary.

As an example, consider the two-qubit state

$$\rho_{AB} = p|\Phi_-\rangle_{AB\,AB}\langle\Phi_-| + (1-p)|00\rangle_{AB\,AB}\langle00|. \tag{3.23}$$

It can be shown that if $p \leq \frac{1}{\sqrt{2}}$ all Bell inequalities will be satisfied by this state.

Let us, however, apply the PPT condition to the same state. In the computational basis $\{|00\rangle, |01\rangle, |10\rangle, |11\rangle\}$, the above density matrix can be written as

$$\rho = \begin{pmatrix} 1-p & 0 & 0 & 0 \\ 0 & \frac{p}{2} & -\frac{p}{2} & 0 \\ 0 & -\frac{p}{2} & \frac{p}{2} & 0 \\ 0 & 0 & 0 & 0 \end{pmatrix}. \tag{3.24}$$

Its partial transpose with respect to B is

$$\rho = \begin{pmatrix} 1-p & 0 & 0 & -\frac{p}{2} \\ 0 & \frac{p}{2} & 0 & 0 \\ 0 & 0 & \frac{p}{2} & 0 \\ -\frac{p}{2} & 0 & 0 & 0 \end{pmatrix}. \tag{3.25}$$

The eigenvalues can be determined from the secular equation $det(\rho^{T_B} - \lambda I) = 0$, which yields

$$(\frac{p}{2} - \lambda)^2(\lambda^2 - (1-p)\lambda - \frac{p^2}{4}) = 0, \tag{3.26}$$

so that the eigenvalues are $\lambda_{1,2} = \frac{p}{2}$ and $\lambda_{3,4} = \frac{1}{2}[(1-p) \pm (1 - 2p + 2p^2)^{1/2}]$. Three of them are obviously positive. The fourth one, $\lambda_4 = \frac{1}{2}\{(1-p) - [(1-p)^2 + p^2]^{1/2}\} < 0$ for $p > 0$. Therefore, for $p > 0$ the partial transpose is not positive and the state is entangled. Note, that the Bell inequalities are not violated for $p \leq \frac{1}{\sqrt{2}}$, so the PPT condition is stronger than the condition of violating the Bell inequality.

Another way of detecting entangled states is by means of entanglement witnesses. An entanglement witness, W, is an Hermitian operator that satisfies two properties. The first is that $Tr(\rho_s W) \geq 0$ for all separable density matrices, ρ_s. The second is that there is at least one entangled density matrix, ρ_e, such that $Tr(\rho_e W) < 0$. Since W is an Hermitian operator, it is, at least in principle, an observable and can be measured. Entanglement witnesses provide a method of experimentally determining whether a state is entangled.

Constructing an entanglement witness for a state whose partial transpose is negative is straightforward. Suppose that ρ^{T_B} has a negative eigenvalue, λ_- with a corresponding eigenvector $|\eta\rangle$. Making use of the fact that for any two operators, X and Y on $\mathcal{H}_A \otimes \mathcal{H}_B$, $Tr(X^{T_B} Y) = Tr(XY^{T_B})$, we have that

$$Tr\left(\rho(|\eta\rangle\langle\eta|)^{T_B}\right) = Tr\left(\rho^{T_B}(|\eta\rangle\langle\eta|)\right) = \lambda_- < 0. \tag{3.27}$$

On the other hand, for ρ_s separable,

$$\mathrm{Tr}\left(\rho_s(|\eta\rangle\langle\eta|)^{T_B}\right) = \mathrm{Tr}\left(\rho_s^{T_B}(|\eta\rangle\langle\eta|)\right) > 0, \tag{3.28}$$

because $\rho_s^{T_B}$ is a positive operator. Therefore, $(|\eta\rangle\langle\eta|)^{T_B}$ is an entanglement witness for the state ρ.

There are many other separability conditions that have been developed during the last few years, so our discussion of this subject will be far from complete. What we will do is cover a few conditions that can be used with continuous-variable systems. With these systems, because they are infinite dimensional, applying the partial transpose condition can be difficult. Hence, having simpler conditions can be useful. All of these conditions can be derived from the partial transpose condition, but our derivations will not explicitly make use of this condition.

Let us consider two particles on a line, or, alternatively, two modes of the electromagnetic field. Each particle has a position operator, x_j, and a momentum operator, p_j, where $j = 1, 2$ and $[x_j, p_j] = i$. In the case of field modes, these would be the quadrature operators $x_j = (a_j^\dagger + a_j)/\sqrt{2}$ and $p_j = i(a_j^\dagger - a_j)/\sqrt{2}$, where a_j and $a_j^\dagger$, $j = 1, 2$, are the annihilation and creation operators for the modes. The commutation relations obeyed by x_j and p_j imply that $(\Delta x_j)(\Delta p_j) \geq 1/2$, where $(\Delta x_j)^2 = \langle x_j^2\rangle - \langle x_j\rangle^2$, and similarly for $(\Delta p_j)^2$. Now define the two operators

$$u = |\alpha|x_1 + \frac{1}{\alpha}x_2$$

$$v = |\alpha|p_1 - \frac{1}{\alpha}p_2, \tag{3.29}$$

where α is a real number. What we will show is that for all separable states

$$(\Delta u)^2 + (\Delta v)^2 \geq \alpha^2 + \frac{1}{\alpha^2}. \tag{3.30}$$

That means that if this condition is violated for a particular state, that state is entangled. However, if the condition is satisfied, we can conclude nothing about the entanglement of the state. Violation of this inequality, then, is a sufficient condition for entanglement, but not a necessary one.

We now want to prove this statement. We assume that the density matrix is separable, so it can be expressed as

$$\rho = \sum_k p_k \rho_{1k} \otimes \rho_{2k}. \tag{3.31}$$

We than have that

$$(\Delta u)^2 + (\Delta v)^2 = \sum_k p_k(\langle u^2 \rangle_k + \langle v^2 \rangle_k) - \langle u \rangle^2 - \langle v \rangle^2$$

$$= \sum_k p_k \left(\alpha^2 \langle x_1^2 \rangle_k + \frac{1}{\alpha^2} \langle x_2^2 \rangle_k + \alpha^2 \langle p_1^2 \rangle_k + \frac{1}{\alpha^2} \langle p_2^2 \rangle_k \right)$$

$$+ 2 \frac{\alpha}{|\alpha|} \sum_k p_k(\langle x_1 \rangle_k \langle x_2 \rangle_k - \langle p_1 \rangle_k \langle p_2 \rangle_k) - \langle u \rangle^2 - \langle v \rangle^2,$$

$$(3.32)$$

where expectation values with respect to $\rho_{1k} \otimes \rho_{2k}$ are denoted by a subscript k and expectation values with respect to the entire density matrix, ρ, do not have a subscript. Continuing

$$(\Delta u)^2 + (\Delta v)^2 = \sum_k p_k \left(\alpha^2 (\Delta x_1)_k^2 + \frac{1}{\alpha^2} (\Delta x_2)_k^2 + \Delta(p_1)_k^2 + \frac{1}{\alpha^2} (\Delta p_2)_k^2 \right)$$

$$+ \sum_k p_k \langle u \rangle_k^2 - \left(\sum_k p_k \langle u \rangle_k \right)^2$$

$$+ \sum_k p_k \langle v \rangle_k^2 - \left(\sum_k p_k \langle v \rangle_k \right)^2. \qquad (3.33)$$

The Schwarz inequality implies that

$$\left(\sum_k p_k \langle u \rangle_k \right)^2 \le \sum_k p_k \langle u \rangle_k^2, \qquad (3.34)$$

and similarly for v. Therefore, we have that

$$(\Delta u)^2 + (\Delta v)^2 \ge \sum_k p_k \left(\alpha^2 (\Delta x_1)_k^2 + \frac{1}{\alpha^2} (\Delta x_2)_k^2 \right.$$

$$\left. + \Delta(p_1)_k^2 + \frac{1}{\alpha^2} (\Delta p_2)_k^2 \right). \qquad (3.35)$$

Now the uncertainty relation between x_1 and p_1 implies that

$$(\Delta x_1)_k^2 + (\Delta p_1)_k^2 \ge (\Delta x_1)_k^2 + \frac{1}{4(\Delta x_1)_k^2} > 1, \qquad (3.36)$$

and similarly for x_2 and p_2. Inserting these inequalities into Eq. (3.35) gives us the desired result.

The case $\alpha = 1$ gives a particularly simple result. In that case, we find that a state is entangled if

$$(\Delta(x_1 + x_2))^2 + (\Delta(p_1 - p_2))^2 < 2. \tag{3.37}$$

Noting that $[x_1 + x_2, p_1 - p_2] = 0$, both uncertainties can be made as small as we wish. What we see from the above inequality is that if they are sufficiently small, the state must be entangled.

Only a subset of entangled states will result in a violation of the inequality in Eq. (3.30). For example, the two-mode state $(|0\rangle_1|1\rangle_2 + |1\rangle_1|0\rangle_2)/\sqrt{2}$, that is, one photon in mode 1 and no photons in mode 2 plus no photons in mode 1 and one photon in mode 2 is an entangled state, but its entanglement will not be detected by Eq. (3.30). Consequently, there is room for more entanglement conditions. We will discuss one final condition, which will, in fact, show that the two-mode state we just mentioned is entangled.

We will prove this condition for an arbitrary system. Let A be an operator on $\mathcal{H}_A$ and B be an operator on $\mathcal{H}_B$. For a product state on $\mathcal{H}_A \otimes \mathcal{H}_B$, we have that

$$|\langle AB^{\dagger}\rangle| = |\langle A\rangle\langle B^{\dagger}\rangle| = |\langle AB\rangle| \le \langle A^{\dagger}AB^{\dagger}B\rangle^{1/2}. \tag{3.38}$$

Now consider the density matrix for a general separable state given by $\rho = \sum_k p_k \rho_k$, where ρ_k is a density matrix corresponding to a pure product state, and p_k is the probability of ρ_k. The probabilities satisfy the condition $\sum_k p_k = 1$. We then have that

$$|\langle AB^{\dagger}\rangle| \le \sum_k p_k |\mathrm{Tr}(\rho_k AB^{\dagger})|$$

$$\le \sum_k p_k (\langle A^{\dagger}AB^{\dagger}B\rangle_k)^{1/2}, \tag{3.39}$$

where $\langle A^{\dagger}AB^{\dagger}B\rangle_k = \mathrm{Tr}(\rho_k A^{\dagger}AB^{\dagger}B)$. We can now apply the Schwarz inequality to obtain

$$|\langle AB^{\dagger}\rangle| \le \left(\sum_k p_k\right)^{1/2} \left(\sum_k p_k \langle A^{\dagger}AB^{\dagger}B\rangle_k\right)^{1/2}$$

$$\le (\langle A^{\dagger}AB^{\dagger}B\rangle)^{1/2}. \tag{3.40}$$

If a state violates this inequality, it is entangled. Note that this condition is very general, because we have not specified what A and B have to be. This condition can apply to finite dimensional spaces, infinite dimensional spaces, or a mixture of the two.

If we now go back to our two mode state and choose $A = a_1$ and $B = a_2$, we find that for this state $|\langle a_1 a_2^\dagger \rangle| = 1/\sqrt{2}$ and $\langle a_1^\dagger a_1 a_2^\dagger a_2 \rangle = 0$. This clearly violates the above inequality, and thus proves that the state is entangled.

3.6   Entanglement Distillation and Formation

As we just saw in the examples of the previous sections, maximally entangled states of a pair of qubits are useful resources for several basic tasks in quantum communication, including dense coding and teleportation. In fact, they are so useful that they deserve their own name. If Alice and Bob share one maximally entangled two-qubit state, e.g. a singlet, then we say they share 1 ebit. Ebits, that is shared entanglement, are important resources and we now want to consider two other processes where they prove to be useful. These are

- *Entanglement distillation*. Alice and Bob share n non-maximally entangled states. How many maximally entangled pairs (for example, singlets or ebits) can they produce from them using only local operations and classical communication (LOCC)?
- *Entanglement formation*. Alice and Bob share n ebits and want to produce copies of some non-maximally entangled state $|\psi\rangle_{AB}$. How many copies of $|\psi\rangle_{AB}$ can they produce from them using only LOCC? Note that this is essentially the inverse of entanglement distillation, so we might as well just call it *entanglement dilution*.

3.6.1   *Local Operations and Classical Communication [LOCC]*

Of course, the first question we have to answer is: What is meant by LOCC? The meaning of classical communication is intuitively obvious and does not require further clarification. Local operations, on the other hand, need some explanation. They are operations performed by one party (either Alice or Bob) alone. The possibilities include:

 (i) appending ancillary systems, not entangled with other party;
 (ii) unitary operations;
 (iii) orthogonal measurements;
 (iv) throwing away part of the system.

Note that the possibilities do not include the exchange of qubits.

Now, equipped with the concept of LOCC, we shall look at two simple examples that make use of these possibilities.

3.6.2 Entanglement Distillation: Procrustean Method

In this protocol, which is not optimal, Alice and Bob initially share the non-maximally entangled state $|\psi\rangle_{AB} = \cos\theta|00\rangle_{AB} + \sin\theta|11\rangle_{AB}$, where $\cos\theta > \sin\theta$, and want to extract the maximally entangled Bell state $|\Psi_+\rangle = \frac{1}{\sqrt{2}}(|00\rangle_{AB} + |11\rangle_{AB})$. Note that $0 \leq \theta \leq \frac{\pi}{4}$. The protocol has three steps.

Step 1. Alice appends ancilla qubit A' in state $|0\rangle_{A'}$, so that the total state becomes
$|\psi\rangle_{AB} \otimes |0\rangle_{A'} = \cos\theta|00\rangle_{AA'} \otimes |0\rangle_B + \sin\theta|10\rangle_{AA'} \otimes |1\rangle_B$.

Step 2. Alice applies a unitary transformation U_A that performs the mapping

$$U_A|10\rangle_{AA'} = |10\rangle_{AA'},$$
$$U_A|00\rangle_{AA'} = \tan\theta|00\rangle_{AA'} + (1 - \tan^2\theta)^{1/2}|01\rangle_{AA'},$$

so that the total state becomes

$$(U_A \otimes I_B)(|\psi\rangle_{AB} \otimes |0\rangle_{A'}) = [\sin\theta|00\rangle_{AA'} + (1 - 2\sin^2\theta)^{1/2}|01\rangle_{AA'}] \otimes |0\rangle_B$$
$$+ \sin\theta|10\rangle_{AA'} \otimes |1\rangle_B$$
$$= \sqrt{2}\sin\theta|0\rangle_{A'} \otimes \frac{1}{\sqrt{2}}(|00\rangle_{AB} + |11\rangle_{AB})$$
$$+ (1 - 2\sin^2\theta)^{1/2}|1\rangle_{A'} \otimes |10\rangle_{AB}.$$

Step 3. Alice measures the state of qubit A' and if she finds $|0\rangle_{A'}$, Alice and Bob keep the result because they just generated the ebit state $|\Psi_+\rangle_{AB}$. Otherwise they throw away the result and repeat the steps.

The probability of success is $p_s = 2\sin^2\theta = 1 - \cos(2\theta)$. Thus, if Alice and Bob initially share n copies of $|\psi\rangle_{AB}$, the expected number of ebits resulting from this procedure is $n[1 - \cos(2\theta)]$.

3.6.3 Entanglement Formation

Again, the method is not optimal but demonstrates the power of entanglement. Initially, Alice and Bob share 1 ebit in the $|\Phi_-\rangle_{AB}$ state and they want to generate the state $|\psi\rangle_{AB} = \cos\theta|00\rangle_{AB} + \sin\theta|11\rangle_{AB}$. The protocol has two steps.

Step 1. Alice prepares the state $|\psi\rangle_{AA'} = \cos\theta|00\rangle_{AA'} + \sin\theta|11\rangle_{AA'}$ in her laboratory.

Step 2. Alice uses the ebit (the singlet state shared with Bob) to teleport the state of particle A' to Bob.

The net result is that after the teleportation Alice and Bob share the state $|\bar{\psi}\rangle_{AB} = \cos\theta|00\rangle_{AB} + \sin\theta|11\rangle_{AB}$.

3.7 Entanglement Measures

Up until now we have spoken of entanglement in qualitative terms only. In this section we want to introduce entanglement measures that will tell us how entangled a state is. We will begin with pure states and gradually extend our measures to mixed states.

3.7.1 The von Neumann Entropy as an Entanglement Measure for Pure Bipartite States: A First Set of Properties

For a pure bipartite state $|\psi\rangle_{AB}$ we use the von Neumann entropy of one of the reduced density matrices as a measure of entanglement, E,

$$E(|\psi\rangle_{AB}) = S(\rho_A) = S(\rho_B). \tag{3.41}$$

Here $S(\rho) = -Tr(\rho \log_2 \rho) = -\Sigma_i \lambda_i \log_2 \lambda_i$ is the von Neumann entropy. Note that if $|\psi\rangle_{AB} = |\psi\rangle_A \otimes |\psi\rangle_B$ then $E(|\psi\rangle_{AB}) = 0$ since the von Neumann entropy of a pure state is 0.

The properties of E that make it a good entanglement measure can be listed as follows.

- **The entanglement of independent systems is additive.**

  Proof If we have two independent pure state bi-partite systems then tracing out one member of each system will leave the remaining two particles in independent mixed states, $\text{Tr}_{BB'}\{|\psi\rangle_{AB} \otimes |\psi'\rangle_{A'B'\,AB}\langle\psi| \otimes {}_{A'B'}\langle\psi'|\} = \rho_A \otimes \rho_{A'}$, where $\rho_A = \text{Tr}_B(|\psi\rangle_{AB\,AB}\langle\psi|)$ and $\rho_{A'} = \text{Tr}_{B'}(|\psi'\rangle_{A'B'\,A'B'}\langle\psi'|)$. So, now we need to show that $S(\rho_A \otimes \rho_{A'}) = S(\rho_A) + S(\rho_{A'})$. To this end we employ the diagonal representations, $\rho_A = \Sigma_n \lambda_n |n\rangle\langle n|$ and $\rho_{A'} = \Sigma_{n'} \lambda_{n'} |n'\rangle\langle n'|$, yielding $\rho_A \otimes \rho_{A'} = \Sigma_n \lambda_n \lambda_{n'} (|n\rangle\langle n|) \otimes (|n'\rangle\langle n'|)$. This finally gives $S(\rho_A \otimes \rho_{A'}) = -Tr\{\Sigma_{n,n'} \lambda_n \lambda_{n'} \log_2(\lambda_n \lambda_{n'})(|n\rangle\langle n|) \otimes (|n'\rangle\langle n'|)\} = -\Sigma_{n,n'} \lambda_n \lambda_{n'} (\log_2(\lambda_n) + \log_2 \lambda_{n'}) = S(\rho_A) + S(\rho_{A'})$. □

- **E is conserved under local unitary operations.**

  Proof This follows from the cyclic property of the trace and can be shown in a straightforward manner. The most general local unitary operation can be written as $|\psi'\rangle_{AB} = U_A \otimes U_B |\psi\rangle_{AB}$ from which it immediately follows that $\rho' = U_A \rho_A U_A^{-1}$ and using the cyclic property of the trace operation we have that $S(\rho'_A) = S(\rho_A)$. □

- **E or, rather, the average value of E cannot be increased by LOCC.**

  The proof will be presented later (see Sections 3.7.3-3.7.4).

- **Entanglement can be concentrated and distilled with asymptotic efficiency E using LOCC only** [4]. Note that this is the best that can be done if E does not increase under LOCC.

  What this means is the following.

  (i) Alice and Bob share k copies of $|\psi\rangle_{AB}$ and from these produce n singlet pairs. Then, as $k \to \infty$, $\frac{n}{k} \to E(|\psi\rangle_{AB})$, using LOCC only.

  (ii) Alice and Bob share k copies of $|\Phi_-\rangle_{AB}$ (singlets) and from these produce n copies of $|\psi\rangle_{AB}$. Then, as $k \to \infty$, $\frac{k}{n} \to E(|\psi\rangle_{AB})$, using LOCC only.

3.7.2 A Useful Auxiliary Quantity: Relative Entropy and Klein's Inequality

To find further properties of the von Neumann entropy it is useful if we first introduce an auxiliary quantity, the so-called relative entropy. We define the relative quantum entropy of a state ρ relative to another state σ by

$$S(\rho \parallel \sigma) = Tr(\rho \log \rho) - Tr(\rho \log \sigma). \tag{3.42}$$

It can be thought of as a kind of distance between two density matrices though it does not obey one important property of a real distance, the triangle inequality. Note that it does obey the requirement of a distance that the distance between ρ and itself is zero, as $S(\rho \parallel \rho) = 0$. It can also be equal to $+\infty$, if there are states, $|\psi\rangle$, for which $\sigma|\psi\rangle = 0$ but $\langle\psi|\rho|\psi\rangle$ does not.

An important feature of the relative entropy is that it is non-negative, i. e. satisfies the Klein inequality,

$$S(\rho \parallel \sigma) \geq 0. \tag{3.43}$$

Proof Let $\rho = \Sigma_i p_i |u_i\rangle\langle u_i|$ and $\sigma = \Sigma_i q_i |v_i\rangle\langle v_i|$ be the diagonal representations of ρ and σ. Then $S(\rho \parallel \sigma) = \Sigma_i p_i(\log p_i - \langle u_i| \log \sigma |u_i\rangle)$ and since $\langle u_i| \log \sigma |u_i\rangle = \Sigma_j \log q_j \cdot |\langle u_i|v_j\rangle|^2$ we have that

$$S(\rho \parallel \sigma) = \sum_i p_i(\log p_i - \sum_j \log q_j \cdot |\langle u_i|v_j\rangle|^2). \tag{3.44}$$

Now, we want to use the fact that $\log x$ is a concave function of x. This means that any straight line connecting two points $\log x_1$ and $\log x_2$ on the curve $\log x$ lies below $\log x$. Mathematically, the line $y = \log x_1 + \frac{\log x_2 - \log x_1}{x_2 - x_1}(x - x_1)$ for $x_1 \leq x \leq x_2$ lies below $\log x$. If we introduce $s = \frac{x - x_1}{x_1 - x_2}$, this relationship can be written as $\log x_1 + s(\log x_2 - \log x_1) \leq \log[x_1 + s(x_2 - x_1)]$ or, after rearranging we obtain $(1 - s) \log x_1 + s \log x_2 \leq \log[(1 - s)x_1 + sx_2]$.

If we introduce $r_i = \Sigma_j |\langle u_i | v_j \rangle|^2 q_j$ then the last inequality immediately gives $\Sigma_j |\langle u_i | v_j \rangle|^2 \cdot \log q_j \leq \log r_i$. Using this, in turn, in Eq. (3.44) gives

$$S(\rho \parallel \sigma) \geq \sum_i p_i \log \frac{p_i}{r_i}. \qquad (3.45)$$

Since $\log x \ln 2 = \ln x \leq x - 1$, the right-hand-side of this equation satisfies

$$\Sigma_i p_i \log \frac{p_i}{r_i} = -\Sigma_i p_i \log \frac{r_i}{p_i} \geq \Sigma_i p_i (1 - \frac{r_i}{p_i}) \cdot \frac{1}{\ln 2} = \Sigma_i (p_i - r_i) \cdot \frac{1}{\ln 2} = 0. \qquad (3.46)$$

In light of Eq. (3.45), this just completes the proof of the Klein inequality, Eq. (3.43). □

3.7.3 The von Neumann Entropy: A Second Set of Properties

The Klein inequality is a very powerful tool in studying the properties of the von Neumann entropy further. So, after this little mathematical detour we return to the properties of $S(\rho)$.

- **For a d−dimensional system**

$$0 \leq S(\rho) \leq \log d. \qquad (3.47)$$

Proof The lower bound is obvious from the definition of the von Neumann entropy. We can obtain the upper bound by setting $\sigma = \frac{1}{d} I$ in the relative entropy. Then the Klein inequality implies $S(\rho \parallel \sigma) = -S(\rho) - \Sigma_i \langle u_i | \rho \log(\frac{1}{d} I) | u_I \rangle = -S(\rho) + \log d$ and due to the Klein ineqaulity $-S(\rho) + \log d \geq 0$ which just gives us the upper bound for the von Neumann entropy. □

- **Suppose p_i are probabilities, $|i\rangle$ orthonormal states for system A and $\{\rho_i\}$ a set of density matrices for system B. Then**

$$S(\Sigma_i p_i |i\rangle \langle i| \otimes \rho_i) = H(\{p_i\}) + \Sigma_i p_i S(\rho_i), \qquad (3.48)$$

where $H(\{p_i\}) = -\Sigma_i p_i \log p_i$ is the Shannon entropy associated with the probability distribution $\{p_i\}$.

Proof For fixed i, let $\{u_{ij}\}$ be the eigenstates of ρ_i with eigenvalues $\{\lambda_{ij}\}$",
i.e.,"}. Using the basis $|i\rangle \otimes |u_{ij}\rangle$ to take the trace in the definition of the entropy
gives

$$S(\Sigma_i p_i |i\rangle\langle i| \otimes \rho_i) = -\Sigma_i \Sigma_j p_i \lambda_{ij} \log(p_i \lambda_{ij})$$

$$= -\Sigma_i \Sigma_j p_i \lambda_{ij} (\log p_i + \log \lambda_{ij})$$

$$= -\Sigma_i p_i \log p_i + \Sigma_i p_i S(\rho_i), \qquad (3.49)$$

which just proves Eq. (3.48). $\square$

- **Subadditivity of the entropy.**
  Here we set out to prove the subadditivity property of the entropy, which states
  that if $\rho_A = Tr_B \rho_{AB}$ and $\rho_B = Tr_A \rho_{AB}$, then

$$S(\rho_{AB}) \le S(\rho_A) + S(\rho_B). \qquad (3.50)$$

Proof In order to prove this property we again apply Klein's inequality, this time
with $\rho = \rho_{AB}$ and $\sigma = \rho_A \otimes \rho_B$. We then have

$$S(\rho_{AB} \| \rho_A \otimes \rho_B) = Tr(\rho_{AB} \log \rho_{AB}) - Tr[\rho_{AB} \log(\rho_A \otimes \rho_B)] \ge 0, \qquad (3.51)$$

and also

$$Tr(\rho_{AB} \log \rho_{AB}) = Tr[\rho_{AB}(\log \rho_A \otimes I_B + I_A \otimes \log \rho_B)]$$

$$= Tr(\rho_A \log \rho_A) + Tr(\rho_B \log \rho_B). \qquad (3.52)$$

Putting now Eqs. (3.51) and (3.52) together gives

$$- S(\rho_{AB}) + S(\rho_A) + S(\rho_B) \ge 0, \qquad (3.53)$$

which proves the result. $\square$

   A note is in order here. In a different context, the subadditivity inequality,
Eq. (3.50), is known as the triangle inequality. This is one of the most important
properties of quantities that are considered to be proper measures.
- Finally, we can put all of the previous properties together to prove a result
  that will actually say something about the effect of local measurement on
  entanglement.
  If $p_i \ge 0$ and $\Sigma_i p_i = 1$ and ρ_i are density operators, then

$$S(\Sigma_i p_i \rho_i) \ge \Sigma_i p_i S(\rho_i). \qquad (3.54)$$

Proof We assume that ρ_i are density matrices of system A. Let us introduce an
auxiliary system B, with an orthonormal basis $\{|i\rangle\}$ and define $\rho_{AB} = \Sigma_i p_i \rho_i \otimes$
$|i\rangle\langle i|$. From this definition it follows that $\rho_A = \Sigma_i p_i \rho_i$ and $\rho_B = \Sigma_i p_i |i\rangle\langle i|$, and
also $S(\rho_A) = S(\Sigma_i p_i \rho_i)$ and $S(\rho_B) = H(\{p_i\})$.

Applying the property given in Eq. (3.48) to this case we have that $S(\rho_{AB}) = H(\{p_i\}) + \Sigma_i p_i S(\rho_i)$. Applying the triangle inequality, Eq. (3.50) we finally have

$$H(\{p_i\}) + \Sigma_i p_i S(\rho_i) \leq S(\Sigma_i p_i \rho_i) + H(\{p_i\}), \tag{3.55}$$

which proves the theorem.   □

We have seen so far that two of the four permissible local operations, namely appending an additional system and applying local unitary transformations, have no effect on entanglement. Now, we are in the position to look at the effect of local measurements.

3.7.4   The Effect of Local Measurements on Entanglement

For the following considerations we again assume that Alice and Bob initially share the pure state $|\psi\rangle_{AB}$ and Alice performs a measurement on her particle. The possible outcomes of the measurement are labeled by k and the corresponding orthogonal projectors by P_k^A. In other words, P_k^A are the spectral projectors of the observable measured. She gets the result k with probability $p_k = {}_{AB}\langle\psi|P_k^A|\psi\rangle_{AB}$ and after this outcome was detected the state collapses to the unnormalized state $P_k^A|\psi\rangle_{AB\,AB}\langle\psi|P_k^A$.

If Alice does not communicate the result of her measurement to Bob, then Bob's density matrix cannot change, because otherwise superluminal communication would be possible. So, in this case, after the measurement, Bob's density matrix is

$$\rho_B = Tr_A\{\Sigma_k p_k P_k^A|\psi\rangle_{AB\,AB}\langle\psi|P_k^A \cdot \frac{1}{{}_{AB}\langle\psi|P_k^A|\psi\rangle_{AB}}\}$$

$$= Tr_A\{\Sigma_k P_k^A|\psi\rangle_{AB\,AB}\langle\psi|P_k^A\}. \tag{3.56}$$

Clearly, $Tr_A\{\Sigma_k P_k^A|\psi\rangle_{AB\,AB}\langle\psi|P_k^A\} = Tr_A(|\psi\rangle_{AB\,AB}\langle\psi|) = \rho_B$, so we also see from the mathematics that Bob's density matrix is unchanged by the measurement.

If Alice does communicate her result to Bob, then Bob's density matrix can change, and so can the entanglement. In some cases it may even increase. The average entanglement will, however, always decrease. We define the average entanglement as $E = \Sigma_k p_k E(|\psi^{(k)}\rangle_{AB})$ where $|\psi^{(k)}\rangle_{AB} = ({}_{AB}\langle\psi|P_k^A|\psi\rangle_{AB})^{-1/2} P_k|\psi\rangle_{AB}$, since Alice and Bob share the state $|\psi^{(k)}\rangle_{AB}$ with probability p_k. Let $\rho_B^{(k)} = Tr_A(|\psi^{(k)}\rangle_{AB\,AB}\langle\psi^{(k)}|)$, then $\Sigma_k p_k E(|\psi^{(k)}\rangle_{AB}) = \Sigma_k p_k S(\rho_B^{(k)})$. Now, using Eq. (3.56), we have that

$$\Sigma_k p_k \rho_B^{(k)} = Tr_A\{\Sigma_k p_k|\psi^{(k)}\rangle_{AB\,AB}\langle\psi^{(k)}|\} = \rho_B, \tag{3.57}$$

and also $E(|\psi\rangle_{AB}) = S(\rho_B)$. Finally, using the subadditivity property, $S(\rho_B) \geq \Sigma_k p_k S(\rho_B^{(k)})$, we obtain

$$E(|\psi\rangle_{AB}) \geq \Sigma_k p_k E(|\psi^{(k)}\rangle). \tag{3.58}$$

Therefore, local measurements can only decrease the average entanglement shared by Alice and Bob.

3.7.5 Towards the Entanglement of Mixed States

Before looking at the last operation, throwing away part of the system, we have to define the entanglement of mixed states. This is because if we start with a pure state and throw away—that is to say trace out—part of it, we generally end up with a mixed state.

The basic idea of approaching this problem is the following. Let us start with a bipartite mixed state ρ_{AB} and decompose it into pure states, $\rho_{AB} = \Sigma_k p_k |\psi^{(k)}\rangle_{AB AB}\langle\psi^{(k)}|$. Alice and Bob want to create n copies of ρ_{AB}, how many singlet pairs will they need to do it? They can do it by creating np_k copies of $|\psi^{(k)}\rangle_{AB}$, for each k, and combining all the particles will erase the information of which singlet pair went with which value of k.

In order to create np_k copies of $|\psi^{(k)}\rangle_{AB}$, they will need $np_k E(|\psi^{(k)}\rangle_{AB})$ singlet pairs, so to create n copies of ρ_{AB}, they will need $\Sigma_k np_k E(|\psi^{(k)}\rangle_{AB})$ singlet pairs. We can then define the entanglement of ρ_{AB} as

$$E(\rho_{AB}) = \Sigma_k E(|\psi^{(k)}\rangle_{AB}). \tag{3.59}$$

However, there is a problem. The pure state decomposition of ρ_{AB} is not unique. We are interested in the minimum number of singlets to form ρ_{AB}, so we define the entanglement of formation of ρ_{AB} as

$$E(\rho_{AB}) = \inf \Sigma_k p_k E(\rho_{AB}), \tag{3.60}$$

where the infenum (largest lower bound) is taken over all possible pure state decompositions of ρ_{AB}. Finding this, except in some special cases, is hard.

3.7.6 The Effect of Throwing Away Part of the System Locally on Entanglement

Equipped with the concept of the entanglement of formation, which can be considered to be one possible entanglement measure for mixed states, we can finally turn our attention to addressing the problem: How does throwing away part of a system locally affects entanglement?

In order to proceed with answering this last remaining question, let us assume that Alice and Bob initially share the pure state $|\psi\rangle_{AA'B}$. Alice now throws away system A', so that they are now sharing $\rho_{AB} = Tr_{A'}(|\psi\rangle_{AA'B} \cdot {}_{AA'B}\langle\psi|)$. Then the following theorem holds.

Theorem *The entanglement of the composite system cannot be increased by throwing away part of the system locally,*

$$E(\rho_{AB}) \leq E(|\psi\rangle_{AA'B}). \tag{3.61}$$

Proof We have $\rho_B = Tr_A(\rho_{AB}) = Tr_{AA'}(|\psi\rangle_{AA'B} \cdot {}_{AA'B}\langle\psi|)$ and also $E(|\psi\rangle_{AA'B}) = S(\rho_B)$. To calculate $E(\rho_{AB})$ we decompose it into pure states, $\rho_{AB} = \Sigma_k p_k |\psi_k\rangle\langle\psi_k|$. Then $E(\rho_{AB}) \leq \Sigma_k p_k E(|\psi_k\rangle\langle\psi_k|) = \Sigma_k p_k S(\rho_{Bk}))$, where $\rho_{Bk} = Tr_A(|\psi_k\rangle\langle\psi_k|)$ and $\Sigma_k p_k \rho_{Bk} = \rho_B$. Finally, we have that

$$E(|\psi\rangle_{AA'B}) = S(\Sigma_k p_k \rho_{Bk}) \geq \Sigma_k p_k S(\rho_{Bk}) \geq E(\rho_{AB}), \tag{3.62}$$

which, when reading backwards, proves the theorem. $\square$

What we have shown by all of the manipulations in this section is that if Alice and Bob start by sharing a pure state, they cannot increase their shared entanglement by LOCC. This result can be extended in a straightforward way to the case in which they are initially sharing a mixed state.

3.7.7 Bound Entanglement

It should be pointed out that not all entangled states can be distilled. That is, there are some entangled states from which a singlet state cannot be obtained by LOCC for any number of copies of the original state. Such states are called bound entangled. It can be shown that if a bipartite entangled state has a positive partial transpose, the state is bound entangled.

Let us give an example of a bound entangled state. In order to show that it is entangled, we will need an entanglement condition known as the range criterion. It states that if a density matrix ρ on $\mathcal{H}_A \otimes \mathcal{H}_B$ is separable, then there exists a family of product vectors, $|\psi_{Ak}\rangle \otimes |\psi_{Bk}\rangle$ such that it spans the range of ρ and the vectors $|\psi_{Ak}\rangle \otimes |\psi_{Bk}^*\rangle$ span the range of ρ^{T_B}, where $|\psi_{Bk}^*\rangle$ is the complex conjugate of $|\psi_{Bk}\rangle$, and the complex conjugation is performed in the same basis as the partial transpose.

In order to construct the bound entangled state we will need something called an unextendable product basis. This is a set of orthogonal product vectors in $\mathcal{H}_A \otimes \mathcal{H}_B$ that has fewer elements than the dimension of the space and is such that there is no product vector in the space that is orthogonal to all of the vectors in the set. An example for the case of two qutrits is

$$|v_0\rangle = \frac{1}{\sqrt{2}}|0\rangle(|0\rangle - |1\rangle) \qquad |v_2\rangle = \frac{1}{\sqrt{2}}|2\rangle(|1\rangle - |2\rangle)$$

$$|v_1\rangle = \frac{1}{\sqrt{2}}(|0\rangle - |1\rangle)|2\rangle \qquad |v_3\rangle = \frac{1}{\sqrt{2}}(|1\rangle - |2\rangle)|0\rangle$$

$$|v_4\rangle = \frac{1}{3}(|0\rangle + |1\rangle + |2\rangle)(|0\rangle + |1\rangle + |2\rangle). \tag{3.63}$$

Now define the projection $P = \sum_{j=0}^{4} |v_j\rangle\langle v_j|$. The claim is that the density matrix

$$\rho = \frac{1}{4}(I - P), \tag{3.64}$$

is a bound entangled state. First, it is entangled due to the range criterion. If there were a product vector in the range of ρ that would mean that the unextendable product basis could be extended, which it cannot. Hence, by the range criterion ρ cannot be separable, so it must be entangled. The next step is to show that ρ^{T_B} is positive. In this case, $\rho = \rho^{T_B}$, so that ρ^{T_B} is clearly positive. Therefore, by the result mentioned in the first paragraph of this section, ρ is bound entangled.

3.8   Concurrence and Negativity

In this section we will look at two entanglement measures, concurrence and negativity. Concurrence is closely related to the entanglement of formation, and while it is easy to calculate for two qubits, it becomes quite difficult to find for larger systems. It does work for all entangled states; a nonzero concurrence implies the state is entangled, and if the concurrence is zero the state is separable. Negativity is easier to compute, but only works for states whose partial transpose is not positive; the negativity of bound entangled states is zero. Consequently, if the negativity of a state is not zero, the state is entangled, but if the negativity is zero, we can conclude nothing about the entanglement of the state.

Let us begin with concurrence. In the case of two qubits it is possible to find the entanglement of formation of a general state explicitly. In order to do this, we have to introduce a quantity called concurrence. However, before we can define the concurrence, we must first define the so-called tilde states for a bipartite qubit pure state $|\psi\rangle_{AB}$, as

$$|\tilde{\psi}\rangle_{AB} = (\sigma_y \otimes \sigma_y)|\psi^*\rangle. \tag{3.65}$$

In this expression $|\psi^*\rangle$ is the complex conjugate of $|\psi\rangle$ in the standard basis, i.e. if $|\psi\rangle = \Sigma_{j,k=0}^{1} c_{jk}|j\rangle|k\rangle$, then $|\psi^*\rangle = \Sigma_{j,k=0}^{1} c_{jk}^*|j\rangle|k\rangle$.

The concurrence $C(|\psi\rangle_{AB})$ of $|\psi\rangle_{AB}$ is then defined as

$$C(|\psi\rangle) = |\langle\psi|\tilde{\psi}\rangle|. \tag{3.66}$$

Why does this quantity have anything to do with entanglement? Let us look at the single-qubit state, $|\psi\rangle = \alpha|0\rangle + \beta|1\rangle$, and its conjugate, $|\psi^*\rangle = \alpha^*|0\rangle + \beta^*|1\rangle$. Using $\sigma_y|0\rangle = i|1\rangle$ and $\sigma_y|1\rangle = -i|0\rangle$, we get $|\tilde{\psi}\rangle = \sigma_y|\psi^*\rangle = i(\alpha^*|1\rangle - \beta^*|0\rangle)$ from where

$$\langle\psi|\tilde{\psi}\rangle = 0 \tag{3.67}$$

follows directly for single qubit states. Let us now consider the Schmidt decompositions of $|\psi\rangle_{AB}$ and $|\tilde{\psi}\rangle_{AB}$, $|\psi\rangle_{AB} = \Sigma_{j=1}^2 \sqrt{\lambda_j}|u_j\rangle_A|v_j\rangle_B$ and $|\tilde{\psi}\rangle_{AB} = \Sigma_{j=1}^2 \sqrt{\lambda_j}|\tilde{u}_j\rangle_A|\tilde{v}_j\rangle_B$. In view of the orthogonality property of single qubit states, Eq. (3.67), we have that $|\tilde{u}_1\rangle \propto |u_2\rangle$ and $|\tilde{u}_2\rangle \propto |u_1\rangle$, so $\langle\psi|\tilde{\psi}\rangle = \sqrt{\lambda_1\lambda_2}(\langle u_1|\tilde{u}_2\rangle\langle v_1|\tilde{v}_2\rangle + \langle u_2|\tilde{u}_1\rangle\langle v_2|\tilde{v}_1\rangle)$. Setting $|u_1\rangle = \alpha|0\rangle + \beta|1\rangle$ and $|u_2\rangle = e^{i\phi_A}(\beta^*|0\rangle - \alpha^*|1\rangle)$ gives $\langle u_1|\tilde{u}_2\rangle = ie^{-i\phi_A}$ and $\langle u_2|\tilde{u}_1\rangle = -ie^{-i\phi_A}$ and, similarly, $\langle v_1|\tilde{v}_2\rangle = ie^{-i\phi_B}$ and $\langle v_2|\tilde{v}_1\rangle = -ie^{-i\phi_B}$, so all of the relevant inner products are simple phase factors. This leads to $\langle\psi|\tilde{\psi}\rangle = \sqrt{\lambda_1\lambda_2}(-2e^{-i(\phi_A+\phi_B)})$, which finally yields

$$C(|\psi\rangle) = 2\sqrt{\lambda_1\lambda_2} \tag{3.68}$$

for the concurrence, with $\lambda_1 + \lambda_2 = 1$. The state is maximally entangled when $\lambda_1 = \lambda_2 = \frac{1}{2}$, giving the maximum value of $C = 1$. For a product state, on the other hand, $\lambda_1 = 0$ or $\lambda_2 = 0$, giving $C = 0$, so C is a monotonically increasing function of entanglement.

The entanglement of $|\psi\rangle_{AB}$ can be expressed as a function of C,

$$E(|\psi\rangle_{AB}) = \mathcal{E}(C(|\psi\rangle_{AB})), \tag{3.69}$$

where

$$\mathcal{E}(C) = h\left(\frac{1 + \sqrt{1 - C^2}}{2}\right), \tag{3.70}$$

and

$$h(x) = -x\log x - (1-x)\log(1-x) \tag{3.71}$$

is the binary entropy associated with the probability distribution $\{x, 1-x\}$.

Finally, let us look at mixed states. Suppose $\rho = \Sigma_k p_k|\psi_k\rangle\langle\psi_k|$, then

$$C(\rho) = \inf \Sigma_k p_k C(|\psi_k\rangle), \tag{3.72}$$

where the infenum is taken over all possible pure state decompositions of ρ. The function $\mathcal{E}(C)$ is monotonically increasing, so

$$\mathcal{E}(C(\rho)) = \inf \mathcal{E}(\Sigma_k p_k C(|\psi_k\rangle)), \tag{3.73}$$

and it is also convex, therefore

$$\inf \mathcal{E}(\Sigma_k p_k C(|\psi_k\rangle)) \le \inf \Sigma_k p_k \mathcal{E}(C(|\psi_k\rangle)) = E(\rho). \tag{3.74}$$

From here it follows that

$$\mathcal{E}(C(\rho)) \le E(\rho). \tag{3.75}$$

To close this long chapter we list two more properties of the concurrence without presenting their proofs [W. Wooters, PRL 80, 2245 (1998)].

1. This last inequality is actually an equality.
2. There is an explicit formula for $C(\rho)$. Let us first define ρ^* as the complex conjugate of ρ in the standard basis and then $\tilde{\rho} = (\sigma_y \otimes \sigma_y)\rho^*(\sigma_y \otimes \sigma_y)$. Let λ_i for $i = 1, \ldots, 4$ denote the square roots of the eigenvalues of $\rho\tilde{\rho}$, arranged in decreasing order. Then

$$C(\rho) = \max\{0, \lambda_1 - \lambda_2 - \lambda_3 - \lambda_4\}. \tag{3.76}$$

We now move on to negativity. For a density matrix, ρ, defined on $\mathcal{H}_A \otimes \mathcal{H}_B$, its negativity is defined as

$$N(\rho) = \frac{1}{2}(\|\rho^{T_A}\|_1 - 1). \tag{3.77}$$

The norm in the above equation is the trace norm, which for an Hermitian operator, is the sum of the absolute values of its eigenvalues. We shall show shortly that the negativity is just the sum of the absolute values of the negative eigenvalues of ρ^{T_A}. The intuition behind using it as an entanglement measure is that the greater the sum of the absolute values of the negative eigenvalues, the more ρ^{T_A} deviates from being a positive operator and, therefore, the more ρ deviates from being separable.

Now let us suppose that ρ^{T_A} has positive eigenvalues $\{\lambda_j^+\}$ and negative eigenvalues $\{\lambda_j^-\}$. The trace is unaffected by the partial transpose, so $\text{Tr}(\rho^{T_A}) = 1$, and

$$\sum_j \lambda_j^+ + \sum_j \lambda_j^- = 1. \tag{3.78}$$

We also have that

$$\|\rho^{T_A}\|_1 = \sum_j \lambda_j^+ + \sum_j |\lambda_j^-|$$

$$= 1 + 2\sum_j |\lambda_j^-|. \tag{3.79}$$

Substituting this into the formula for negativity, we find, $N(\rho) = \sum_j |\lambda_j^-|$, as previously stated.

The entanglement of formation, $E(\rho)$, satisfies the condition that if ρ' is obtained from ρ by LOCC, then $E(\rho') \leq E(\rho)$. Entanglement measures with this property are called entanglement monotones. We will state, but not prove, that negativity is an entanglement monotone.

3.9 Multipartite Entanglement

When going beyond two subsystems, one finds that things get complicated quickly. In fact, there is much about multipartite entanglement that is not understood, and this field is a currently active area of research. Consequently, in this section we will confine ourselves to three-qubit states. That is enough to get an idea of some of the issues involved.

The first issue to be addressed is what is meant by a separable state. A pure three-qubit state is called fully separable if it is a product of three single-qubit states

$$|\Psi\rangle_{ABC} = |\psi_1\rangle_A \otimes |\psi_2\rangle_B \otimes |\psi_3\rangle_C. \tag{3.80}$$

A mixed state is fully separable if it can be expressed as a convex combination of pure states each of which is fully separable. A pure three-qubit state is biseparable if it is a product of a one qubit state and a two qubit state, e.g. $|\Psi\rangle_{ABC} = |\psi_1\rangle_{AB}|\psi_2\rangle_C$, where here $|\psi_1\rangle_{AB}$ is a possibly entangled two-qubit state. There are three classes of biseparable states, those separable across the partition $AB|C$, those separable across the partition $A|BC$, and those separable across the partition $AC|B$. A biseparable mixed state is one that can be expressed as a convex combination of biseparable pure states, and the states in the sum need not be separable across the same partition. Finally, a state is genuinely tripartite entangled if it is neither fully separable nor biseparable. Two important examples of pure states that are genuinely tripartite entangled are the GHZ (Greenberger, Horne, Zeilinger) state

$$|\Psi_{GHZ}\rangle = \frac{1}{\sqrt{2}}(|000\rangle + |111\rangle), \tag{3.81}$$

and the W state

$$|\Psi_W\rangle = \frac{1}{\sqrt{3}}(|100\rangle + |010\rangle + |001\rangle). \tag{3.82}$$

A fundamental difference between these two states is that a two-qubit reduced density matrix of the GHZ state is separable, while that of the W state is not.

With the increase in complexity added by additional subsystems, it would be useful if we could find a way of classifying multipartite states. In general this also becomes complicated, but in the case of three qubit states, a simple classification exists. The classification is based on an extension of LOCC called stochastic LOCC (or SLOCC). We say that a state $|\Psi\rangle$ can be transformed into a state $|\Phi\rangle$ by SLOCC if there is a nonzero probability that $|\Psi\rangle$ can be transformed into a state $|\Phi\rangle$ by LOCC. That is, we demand only that the LOCC operations that we use to make the transformation succeed with some nonzero probability, not necessarily with certainty. It turns out that $|\Psi\rangle$ can be transformed into $|\Phi\rangle$ by SLOCC if and only if

$$|\Phi\rangle = A \otimes B \otimes C|\Psi\rangle, \tag{3.83}$$

where A, B, and C are invertible single-qubit operators. It also turns out that there are two kinds of pure three-qubit states, those that can be transformed into $|\Psi_{GHZ}\rangle$ by SLOCC, which we shall call GHZ-class states, and those that can be transformed into $|\Psi_W\rangle$ by SLOCC, which we shall call W-class states. Going to mixes states, a state is of the W class if it can be expressed as a convex combination of W-class pure states. Otherwise it is of the GHZ class.

Another property of three-qubit entanglement is monogamy. What this means is that if two of the qubits are highly entangled, they can only be weakly entangled with the third. This is made precise by the Coffman–Kundu–Wootters (CKW) inequality, which states that

$$C_{A:B}^2 + C_{A:C}^2 \le C_{A:BC}^2, \tag{3.84}$$

where $C_{A:B}$ is the concurrence between qubits A and B, $C_{A:C}$ is the concurrence between qubits A and C, and $C_{A:BC}$ is the concurrence between qubits A and system consisting of qubits B and C. Since we can only compute the concurrence between two qubits, this last quantity, $C_{A:BC}$, seems to present problems. However, only two of the four dimensions of the system BC, matter, so it can be treated as an effective qubit. We shall see how this works shortly. This monogamy result can be extended to n qubits, but does not hold for systems of higher dimension.

Let us apply the CKW inequality to a W state. Tracing out qubit C, we find that

$$\rho_{AB} = \frac{1}{3}[(|01\rangle + |10\rangle)(\langle 01| + \langle 10|) + |00\rangle\langle 00|]. \tag{3.85}$$

From this we find that $C_{A:B} = 2/3$. Similarly, from ρ_{AC}, we find that $C_{A:C} = 2/3$. To find $C_{A:BC}$, we express $|\Psi_W\rangle$ as follows

$$|\Psi_W\rangle = \sqrt{\frac{2}{3}}|0\rangle_A|v_0\rangle_{BC} + \frac{1}{\sqrt{3}}|1\rangle_A|v_1\rangle_{BC}, \qquad (3.86)$$

where $|v_0\rangle = (1/\sqrt{2})(|01\rangle + |10\rangle)$ and $|v_0\rangle = |00\rangle$. Note that the state is written in Schmidt form for the split $A|BC$, and because A is a qubit, there are only two Schmidt vectors for the system BC. Therefore, we can treat BC as a qubit and use Eq. (3.68) to calculate the concurrence. We find that $C_{A:BC} = 2\sqrt{2}/3$. Substituting these values into the CKW inequality, we see that both sides are equal to 8/9, so that for the W state, the inequality becomes an equality.

3.10 Problems

1. Wigner's inequality is another inequality that will be satisfied by a local hidden-variable theory but can be violated by quantum mechanics. In order to derive it consider the following situation. A source sends one particle to Alice and one to Bob. Alice measures one of three observables, a_j, $j = 1, 2, 3$ and Bob also measures one of three, b_j, $j = 1, 2, 3$. Each of these observables gives the value 1 or -1. The source has the property that whenever Alice and Bob measure corresponding observables, that is the value of j is the same, then the results are anticorrelated. For example, if Alice measures a_1 and Bob measures b_1, then they will never get the same result, but if Alice measures a_1 and Bob measures b_2, then they can get any result. Let $p(a_j = m, b_k = n)$ denote the probability that when Alice measures a_j and Bob measures b_k, that Alice gets the value m and Bob gets n. Wigner's inequality states that if this source is described by a local hidden variable theory, then

$$p(a_1 = 1, b_2 = 1) + p(a_2 = 1, b_3 = 1) \geq p(a_1 = 1, b_3 = 1).$$

We want to prove this statement.

(a) If the source is described by a local hidden variable theory, then it can be described by a joint probability distribution $P(a_1, a_2, a_3; b_1, b_2, b_3)$. Using the constraint on the source, that measurements of corresponding observables must be anticorrelated, show that we must have that

$$P(a_1, a_2, a_3; b_1, b_2, b_3) = 0,$$

unless $a_j = -b_j$, $j = 1, 2, 3$.

(b) Use your result in part (a) to prove Wigner's inequality.

(c) We now want to show that quantum mechanics can satisfy the constraint yet violate the inequality. We choose

$$a_1 = \frac{1}{2}\sigma_z + \frac{\sqrt{3}}{2}\sigma_x \; a_2 = \sigma_z \; a_3 = \frac{1}{2}\sigma_z - \frac{\sqrt{3}}{2}\sigma_x,$$

and similarly for b_j. Here, $\sigma_x|0\rangle = |1\rangle$ and $\sigma_x|1\rangle = |0\rangle$. The source produces particles in the singlet state

$$|\Phi_-\rangle = \frac{1}{\sqrt{2}}(|0\rangle|1\rangle - |1\rangle|0\rangle),$$

and one of these particles goes to Alice and the other to Bob. This selection of observables and source satisfy the constraint. Show that they do not satisfy Wigner's inequality.

2. Another test of local hidden-variable theories is one devised by Lucien Hardy. The setup is the same as for the Bell test, a source emits two particles, one to Alice, who measures either a_1 or a_2, and one to Bob, who measures either b_1 or b_2. All measurements produce the result 1 or -1. In this case, the source has the property that $P(a_1 = -1, b_1 = -1) = 0$, $P(a_1 = 1, b_2 = -1) = 0$, and $P(a_2 = -1, b_1 = 1) = 0$.

(a) Show that if the source is described by a local hidden-variable theory that $P(a_2 = -1, b_2 = -1) = 0$.

(b) Suppose that the source emits two spin-1/2 particles, and that Alice and Bob measure the spins of their particles. a_1 and b_1 correspond to measuring σ_z and a_2 and b_2 correspond to measuring σ_x. Let $|u\rangle$ denote a spin up state and $|d\rangle$ denote a spin down state. Show that the two-particle state

$$|\Psi\rangle_{ab} = \frac{1}{\sqrt{3}}(|u, u\rangle_{ab} + |d, u\rangle_{ab} + |u, d\rangle_{ab})$$

satisfies the conditions of Hardy's test, that is, that the three requisite probabilities are zero, but that $P(a_2 = -1, b_2 = -1) \neq 0$. Therefore, quantum mechanics violates the condition derived from local realism.

3. Teleportation does not just work for qubits, it works for a quantum system of any dimension. Suppose we want to teleport an N-dimensional quantum state. Let $\{|j\rangle | j = 0, \ldots N - 1\}$ be an orthonormal basis for our N-dimensional space, and define the states

$$|\chi_{n,m}\rangle = \frac{1}{\sqrt{N}} \sum_{j=0}^{N-1} e^{2\pi i jn/N}|j\rangle|j + m \; (mod \; N)\rangle,$$

for two N-dimensional quantum systems. These will take the place of the Bell states for our more general teleportation procedure.

(a) Show that $\langle \chi_{n,m} | \chi_{n',m'} \rangle = \delta_{n,n'} \delta_{m,m'}$.

(b) We start with the state $|\phi\rangle_{A'} |\chi_{0,0}\rangle_{AB}$, where

$$|\phi\rangle = \sum_{j=0}^{N-1} \alpha_j |j\rangle,$$

is the state we want to teleport. Show that by measuring the $A'A$ particles in the $|\chi_{n,m}\rangle$ basis and using the result of this measurement to perform the correct unitary transformation on the B particle, we can transfer the state $|\phi\rangle$ onto the B particle.

4. We want to consider one step of an entanglement concentration procedure due to A. Sanpera and C. Macchiavello. Start with two pairs of particles, where each pair is in the state consisting of the mixture of qubit Bell states

$$\rho = p |\Psi_+\rangle \langle \Psi_+| + (1-p) |\Psi_-\rangle \langle \Psi_-|,$$

where $p > 1/2$ and $|\Psi_{\pm}\rangle = (|00\rangle \pm |11\rangle)/\sqrt{2}$. Let us call the pairs AB and $A'B'$, where Alice has particles A and A' and Bob has particles B and B'. Alice applies $\exp(i\pi\sigma_x/4)$ to each of her particles, and Bob applies $\exp(-i\pi\sigma_x/4)$ to both of his. Alice then sends both of her particles into a C-NOT gate, where A is the control qubit and A' is the target, and Bob sends his into a C-NOT gate with B as the control qubit and B' as the target. Alice and Bob now measure particles A' and B' in the $\{|0\rangle, |1\rangle\}$ basis, and keep the pair AB if their results agree. Show that if their results agree, then the proportion of the Bell state $|\Psi_+\rangle$ in the pair AB has increased, so that the pair has become more entangled.

5. Consider the following two-qubit density matrix

$$\rho_{AB} = p |\Phi_-\rangle_{AB} {}_{AB}\langle \Phi_-| + \frac{1-p}{4} I,$$

where $|\Phi_-\rangle_{AB} = (|0\rangle_A |1\rangle_B - |1\rangle_A |0\rangle_B)/\sqrt{2}$.

(a) Use the positive-partial transpose condition to find out for what values of p this density matrix is entangled.

(b) Find the concurrence of this density matrix as a function of p.

6. A general bipartite qubit state can be written in the standard basis as $|\psi\rangle_{AB} = a_{00}|00\rangle + a_{01}|01\rangle + a_{10}|10\rangle + a_{11}|11\rangle$. The coefficients can be arranged to form a matrix in a natural way

$$A = \begin{pmatrix} a_{00} & a_{01} \\ a_{10} & a_{11} \end{pmatrix}.$$

Show that $C(|\psi\rangle_{AB}) = 2|\det A|$, where $\det A$ is the determinant of the matrix A and $|\ldots|$ is the absolute value of the quantity inside.

7. Consider the two qubit state

$$\rho = p|\psi\rangle\langle\psi| + \frac{(1-p)}{4}I,$$

where $|\psi\rangle = a|01\rangle + b|10\rangle$. Using the partial transpose condition, find the values of p for which the partial transpose of ρ will have a negative eigenvalue. For p in that range, use the eigenvector corresponding to the negative eigenvalue to construct an entanglement witness for ρ. Express the entanglement witness as a 4×4 matrix in the computational basis.

8. Define the two mode state $(1/\sqrt{2})(a_1^\dagger + a_2^\dagger)|0\rangle$. Consider the mixed state

$$\rho = p|\psi_{01}\rangle\langle\psi_{01}| + \frac{1-p}{4}P_{01},$$

where $0 \le p \le 1$ and P_{01} is the projection operator onto the space spanned by the vectors $\{|0 > 1|0 > 2, |0 > 1|1 > 2, |1 > 1|0 > 2, |1 > 1|1 > 2\}$. Using the entanglement condition

$$|\langle a_1 a_2^\dagger\rangle|^2 > \langle a_1^\dagger a_1 a_2^\dagger a_2\rangle,$$

find a range of p for which ρ is definitely entangled.

9. Decoherence, a process that introduces noise into quantum systems, is a result of the quantum system of interest becoming entangled with the environment. In order to see what effects this can have, let's go back to Deutsch's algorithm. Suppose that after the first Hadamard gate, the control qubit becomes entangled with the environment, so that the whole state becomes

$$\frac{1}{2}(|0\rangle_1|\eta_0\rangle_e + |1\rangle_1|\eta_1\rangle_e)(|0\rangle_2 - |1\rangle_2),$$

where 1 denotes the control qubit, 2 denotes the target qubit, and e denotes the environment. Let us assume, for convenience, that the overlap between the environment states, $s = {}_e\langle\eta_0|\eta_1\rangle_e$ is real and positive, $0 \le s \le 1$. We want to find the probability of error as a function of s. If there is no decoherence, if the function is constant we will find the control qubit in the state $|0\rangle_1$, and if the function is balanced, we will find it in the state $|1\rangle_1$. With decoherence this is no longer true. Find the probability that if the function is constant the control qubit is in the state $|1\rangle_1$, and the probability that if the function is balanced the control qubit is in the state $|0\rangle_1$.

Further Reading

1. C.H. Bennett, S.J. Wiesner, Communication via one- and two-particle operators on Einstein-Podolsky-Rosen states. Phys. Rev. Lett. **69**, 2881 (1992)
2. C.H. Bennett, G. Brassard, C. Crepeau, R. Jozsa, A. Peres, W. Wootters, Teleporting an unknown quantum state via dual classical and EPR channels. Phys. Rev. Lett. **70**, 1895 (1993)
3. C.H. Bennett, D.P. DiVincenzo, J.A. Smolin, W.K. Wootters, Mixed-state entanglement and quantum error correction. Phys. Rev. A **54**, 3824 (1996)
4. C.H. Bennett, H.J. Bernstein, S. Popescu, B. Schumacher, Concentrating partial entanglement by local operations. Phys. Rev. A **53**, 2046 (1996)
5. L.-M. Duan, G. Giedke, J.I. Cirac, P. Zoller, Inseparability criterion for continuous variable systems. Phys. Rev. Lett. **84**, 2722 (2000)
6. O. Gühne, G. Toth, Entanglement detection. Phys. Rep. **474**, 1 (2009)
7. M. Hillery, B. Yurke, Bell's theorem and beyond. Quantum Semiclassical Opt. **7**, 215 (1995)
8. M. Hillery, M.S. Zubairy, Entanglement conditions for two-mode states. Phys. Rev. Lett. **96** (2006)
9. R. Horodecki, P. Horodecki, M. Horodecki, K. Horodecki, Quantum entanglement. Rev. Mod. Phys. **81**, 865 (2009)
10. R. Simon, Peres-Horodecki separability criterion for continuous variable states. Phys. Rev. Lett. **84**, 2726 (2000)
11. G. Vidal, R. Werner, Phys. Rev. A **65**, 032314 (2002)
12. W.K. Wootters, Entanglement of formation of an arbitrary state of two qubits. Phys. Rev. Lett. **80**, 2245 (1998)

Chapter 4
Generalized Quantum Dynamics

Time evolution in textbook quantum mechanics is represented by unitary maps $|\psi\rangle \rightarrow U|\psi\rangle$ and $\rho \rightarrow U\rho U^\dagger$, where $U = e^{-itH}$. This is not the most general evolution possible. We can couple our system to another one, evolve both with a unitary operator that will, in general, create entanglement between the two systems, and then trace out the second system. The resulting evolution for the original system alone will be non-unitary, in general, and can be described by a non-unitary quantum map. Here we would like to see how to describe such maps, and look at some of their properties. In addition, we will point out a map, the cloning map, which is impossible to realize.

4.1 Quantum Maps or Superoperators

4.1.1 Quantum Maps and Their Kraus Representation

Let us introduce the unitary map $|\psi\rangle_A \otimes |\psi\rangle_B \rightarrow U_{AB}(|\psi\rangle_A \otimes |\psi\rangle_B)$ where $|\psi\rangle_A \otimes |\psi\rangle_B \in \mathcal{H}_A \otimes \mathcal{H}_B$. We apply the identity operator $I_A \otimes I_B$ to the unitary map, where $I_B = \Sigma_m |m\rangle_{BB}\langle m|$ and $\{|m\rangle_B\}$ is an orthonormal basis for $\mathcal{H}_B$, yielding

$$U_{AB}(|\psi\rangle_A \otimes |\psi\rangle_B) = I_A \otimes \Sigma_m |m\rangle_{BB}\langle m|(U_{AB}|\psi\rangle_A \otimes |\psi\rangle_B). \tag{4.1}$$

We can introduce a shorthand for the expression appearing in the right-hand-side,

$$_B\langle m|U_{AB}(|\psi\rangle_A \otimes |\psi\rangle_B) \in \mathcal{H}_A \equiv A_m |\psi\rangle_A, \tag{4.2}$$

which defines the operator A_m. In terms of A_m, Eq. (4.1) can be written as

$$U_{AB}(|\psi\rangle_A \otimes |\psi\rangle_B) = \Sigma_m A_m |\psi\rangle_A \otimes |m\rangle_B, \tag{4.3}$$

© Springer Nature Switzerland AG 2021

J. A. Bergou et al., *Quantum Information Processing*, Graduate Texts in Physics, https://doi.org/10.1007/978-3-030-75436-5_4

where $\Sigma_m A_m^\dagger A_m = I_A$ as can be easily seen from its definition. We have just obtained the following mapping

$$
\begin{aligned}
\rho_A &= |\psi\rangle_{AA}\langle\psi| \\
&\rightarrow Tr_B\{\Sigma_m \Sigma_{m'} A_m |\psi\rangle_A \otimes |m\rangle_{BB}\langle m'| \otimes {}_A\langle\psi|A_{m'}^\dagger\} \\
&= \Sigma_m A_m \rho A_m^\dagger.
\end{aligned}
\tag{4.4}
$$

This gives us what is called the operator sum, or Kraus representation of the quantum map T (or superoperator T),

$$
T(\rho) = \Sigma_m A_m \rho A_m^\dagger.
\tag{4.5}
$$

In the next section we will present a systematic study of the most important properties of quantum maps.

4.1.2   Properties of Quantum Maps

Note that T has a number of important and useful properties.

1. T maps hermitian operators to hermitian operators.
2. T is trace preserving.
3. T maps positive operators to positive operators.

These properties follow directly from the definition of T.

We defined the quantum map and the corresponding Kraus representation as the remainder of a unitary map, defined on a larger Hilbert space, after tracing out part of the system. The converse is also true. Given a Kraus representation, it is possible to find a larger Hilbert space, $\mathcal{H}_A \otimes \mathcal{H}_B$, a vector $|\phi\rangle_B \in \mathcal{H}_B$ and a unitary operator U_{AB} such that

$$
A_m |\psi\rangle_A = {}_B\langle m|U_{AB}(|\psi\rangle_A \otimes |\phi\rangle_B).
\tag{4.6}
$$

We now prove this statement constructively. Let us choose $\mathcal{H}_A$ to have dimension N and $\mathcal{H}_B$ to have dimension M. Further, let $\{|m_B\rangle\}$ be an orthonormal basis for $\mathcal{H}_B$, and choose $|\phi_B\rangle$ to be an arbitrary state of $\mathcal{H}_B$. Let us then define a transformation U_{AB} via

$$
U_{AB}(|\psi\rangle_A \otimes |\phi\rangle_B) = \sum_m A_m |\psi\rangle_A \otimes |m\rangle_B \,,
\tag{4.7}
$$

which implies the Eq. (4.6). U_{AB} is inner product preserving,

$$\left(\sum_{m'} {}_A\langle \psi' | \otimes {}_B\langle m' | A_{m'}^\dagger \right) \left(\sum_m A_m |\psi\rangle_A \otimes |m\rangle_B \right) = \sum_m {}_A\langle \psi' | A_m^\dagger A_m | \psi \rangle_A$$

$$= {}_A\langle \psi' | \psi \rangle_A , \qquad (4.8)$$

so it is unitary on the one-dimensional subspace spanned by $|\phi_B\rangle$ and it can be extended to a full unitary operator on $\mathcal{H}_A \otimes \mathcal{H}_B$ because, e.g., on the subspace that is orthogonal to $|\phi_B\rangle$ it can be the identity.

The Kraus representation of a superoperator is not unique. Let $\{|m'_B\rangle\}$ be a different orthonormal basis for $\mathcal{H}_B$, and

$$B_{m'} = \langle m'_B | U_{AB} (|\psi_A\rangle \otimes |\phi_B\rangle), \qquad (4.9)$$

and we have

$$T(\rho_A) = \Sigma_{m'} B_{m'} \rho_A B_{m'}^\dagger . \qquad (4.10)$$

If $|m'_B\rangle = \Sigma_m U_{m'm} |m_B\rangle$ then $\langle m'_B| = \Sigma_m U_{mm'}^\dagger \langle m_B|$ and we have that $B_{m'} = \Sigma_m U_{mm'}^\dagger A_m$. We shall eventually show that any two Kraus representations for the same superoperator are related in this way.

First we want to show that a superoperator satisfying conditions 1–3 has a Kraus representation. Actually, we have to replace 3 by a stronger definition:

$3'$. T is completely positive.

This means the following. We know that T maps bounded operators to bounded operators, T: bounded operators on $\mathcal{H}_A$ → bounded operators on $\mathcal{H}_A$. Let us append an ancilla space $\mathcal{H}_B$ to $\mathcal{H}_A$ and extend T to a bounded operator on this larger space, so it is in the set of bounded operators $\mathcal{B}(\mathcal{H}_A \otimes \mathcal{H}_B)$, by $T \to T \otimes I_B$. If, for any $\mathcal{H}_B$, $T \otimes I_B$ is positive, then we say that T is completely positive.

Physically what this means is the following. T describes the evolution of the system A, and system B does not evolve. T is completely positive if $\rho_A \otimes \rho_B \to T(\rho_A) \otimes \rho_B$ is a density matrix for any ρ_A and ρ_B. An example of a map that is positive, but not completely positive is the transpose. It preserves eigenvalues, so it is positive. On the other hand $(\rho)_A^T \otimes I_B$ is just a partial transpose and we have seen that the partial transpose is not positive.

We now want to prove that a superoperator satisfying 1–3$'$ has a Kraus representation. Before proceeding, we note the following method that we will use in the proof. Let A be an operator on $\mathcal{H}_A$, where $dim\mathcal{H}_A = N$. Suppose $dim\mathcal{H}_B \geq N$ and let $\{|j_A\rangle\}$ and $\{|j_B\rangle\}$ orthonormal bases of $\mathcal{H}_A$ and $\mathcal{H}_B$, respectively. Consider the state

$$|\psi_{AB}\rangle = \Sigma_{j=1}^N |j_A\rangle \otimes |j_B\rangle . \qquad (4.11)$$

If $|\phi_A\rangle \in \mathcal{H}_A$, we can express it in terms of $|\psi_{AB}\rangle$ as a "partial inner product" with a vector $|\phi_B^*\rangle \in \mathcal{H}_B$, where $|\phi_A\rangle = \Sigma_{j=1}^N c_j |j_A\rangle$ and $|\phi_B^*\rangle = \Sigma_{j=1}^N c_j^* |j_B\rangle$. Then

$$\langle\phi_B^*|\psi_{AB}\rangle = (\Sigma_{j=1}^N c_j \langle j_B|)\Sigma_{j'=1}^N \frac{1}{\sqrt{N}}|j_A'\rangle \otimes |j_B'\rangle = \frac{1}{\sqrt{N}}|\phi_A\rangle. \tag{4.12}$$

The mapping $|\phi_A\rangle \to |\phi_B^*\rangle$ is antilinear and norm preserving. We can calculate the effect of $A \otimes I_B$ on $|\psi_{AB}\rangle$ similarly,

$$\langle\phi_B^*|(A \otimes I_B)|\psi_{AB}\rangle = (\Sigma_{j=1}^N c_j \langle j_B|)\Sigma_{j'=1}^N \frac{1}{\sqrt{N}}A|j_A'\rangle \otimes |j_B'\rangle$$

$$= \frac{1}{\sqrt{N}}A(\Sigma_{j=1}^N c_j |j_A\rangle) = \frac{1}{\sqrt{N}}A|\phi_A\rangle. \tag{4.13}$$

Equipped with this method we can now proceed with the proof. Suppose T is a superoperator satisfying 1, 2, and 3'. T acts on $\mathcal{B}(\mathcal{H}_A)$. $T \otimes I_B$ acting on $\mathcal{B}(\mathcal{H}_A) \otimes \mathcal{B}(\mathcal{H}_B)$ is positive. This implies that if $\rho_{AB} = |\psi_{AB}\rangle\langle\psi_{AB}|$ and

$$\rho_{AB}' = (T \otimes I_B)(\rho_{AB}), \tag{4.14}$$

then ρ_{AB}' is also a density matrix. It can be expanded as an ensemble of pure states $\rho_{AB}' = \Sigma_\mu q_\mu |\phi_{AB,\mu}\rangle\langle\phi_{AB,\mu}|$. By a derivation entirely similar to those in Eqs. (4.12) and (4.13), we obtain

$$T(|\phi_A\rangle\langle\phi_A|) = N\langle\phi_B^*|(T \otimes I_B)(\rho_{AB})|\phi_B^*\rangle$$

$$= N\Sigma_\mu q_\mu \langle\phi_B^*|\phi_{AB,\mu}\rangle\langle\phi_{AB,\mu}|\phi_B^*\rangle. \tag{4.15}$$

Now define $A_\mu : |\phi_A\rangle \to \sqrt{Nq_\mu}\langle\phi_B^*|\phi_{AB,\mu}\rangle$. A_μ is a linear operator on $\mathcal{H}_A$, and we have

$$T(|\phi_A\rangle\langle\phi_A|) = \Sigma_\mu A_\mu |\phi_A\rangle\langle\phi_A| A_\mu^\dagger, \tag{4.16}$$

which we can extend for any density matrix ρ_A to

$$T(\rho_A) = \Sigma_\mu A_\mu \rho_A A_\mu^\dagger. \tag{4.17}$$

Because T is trace preserving for any ρ_A, we have that $\Sigma_\mu Tr(\rho_A A_\mu^\dagger A_\mu) = 1$ from where $\Sigma_\mu A_\mu^\dagger A_\mu = I$ follows.

We actually need to show that

$$T(\mathcal{M}_A) = \Sigma_\mu A_\mu \mathcal{M}_A A_\mu^\dagger \tag{4.18}$$

for any $\mathcal{M}_A \in \mathcal{B}(\mathcal{H}_A)$. We can do this by showing it is true for a basis of $\mathcal{B}(\mathcal{H}_A)$. Such an operator basis is given by $\{(|j_A\rangle\langle k_A|) \mid j, k = 1, \ldots, N\}$. We know the above equation is true for any operator of the form $\mathcal{M}_A = \Sigma_n c_n |\phi_{A,n}\rangle\langle\phi_{A,n}|$. Defining

$$
\begin{aligned}
|\phi_{A,1}\rangle = \tfrac{1}{\sqrt{2}}(|j_A\rangle + |k_A\rangle) & \quad |\phi_{A,3}\rangle = \tfrac{1}{\sqrt{2}}(|j_A\rangle + i|k_A\rangle) \\
|\phi_{A,2}\rangle = \tfrac{1}{\sqrt{2}}(|j_A\rangle - |k_A\rangle) & \quad |\phi_{A,4}\rangle = \tfrac{1}{\sqrt{2}}(|j_A\rangle - i|k_A\rangle)
\end{aligned}
\tag{4.19}
$$

we find

$$
|j_A\rangle\langle k_A| = \frac{1}{2}(|\phi_{A,1}\rangle\langle\phi_{A,1}| - |\phi_{A,2}\rangle\langle\phi_{A,2}|) + \frac{i}{2}(|\phi_{A,3}\rangle\langle\phi_{A,3}| - |\phi_{A,4}\rangle\langle\phi_{A,4}|).
\tag{4.20}
$$

From here it follows immediately

$$
T(|\phi_A\rangle\langle\phi_A|) = \Sigma_\mu A_\mu |j_A\rangle\langle k_A| A_\mu^\dagger.
\tag{4.21}
$$

This completes the proof.

We now want to use the construction of the operators A_μ to make some statements about their properties.

4.1.3 Properties of the Kraus Operators

The first question we address is: How many Kraus operators do we need? The mapping in Eq. (4.14) maps $\mathcal{H}_A \otimes \mathcal{H}_B$, where $\mathcal{H}_B = span\{|j_B\rangle\}$, into itself, and this space has dimension N^2. Diagonalizing ρ'_{AB} we will find at most N^2 vectors in the expansion of ρ'_{AB}. Therefore, there is a Kraus representation with at most N^2 operators.

The next question we address is the uniqueness of the Kraus representation. It is clear that the Kraus representation is not unique, because the decomposition of ρ'_{AB} is not unique. We want to see how different Kraus representations of the same superoperator are related. The idea here is to show that each Kraus representation is related to a decomposition of ρ'_{AB}, and then use the theorem about different decompositions of density matrices.

If we have that for any $\mathcal{M}_A \in \mathcal{B}(\mathcal{H}_A)$

$$
T(\mathcal{M}_A) = \Sigma_\mu A_\mu \mathcal{M}_A A_\mu^\dagger,
\tag{4.22}
$$

then

$$(T \otimes I_B)(|\psi_{AB}\rangle\langle\psi_{AB}|) = (T \otimes I_B)\left(\frac{1}{N}\Sigma_{j,j'=1}^{N}|j_A\rangle \otimes |j_B\rangle\langle j_A'| \otimes \langle j_B'|\right)$$

$$= \frac{1}{N}\Sigma_\mu\Sigma_{j,j'}A_\mu|j_A\rangle \otimes |j_B\rangle\langle j_A'|A_\mu^\dagger \otimes \langle j_B'|. \quad (4.23)$$

Define $\sqrt{q_\mu}|\phi_{AB,\mu}\rangle = \frac{1}{\sqrt{N}}\Sigma_j A_\mu|j_A\rangle \otimes |j_B\rangle$ where $\|\phi_\mu\| = 1$, so then

$$\rho_{AB}' = \Sigma_\mu q_\mu|\phi_{AB,\mu}\rangle\langle\phi_{AB,\mu}|. \quad (4.24)$$

Now suppose we have two different Kraus representations for T, $T(\mathcal{M}_A) = \Sigma_\mu A_\mu \mathcal{M}_A A_\mu^\dagger$ and $T(\mathcal{M}_A) = \Sigma_\mu D_\mu \mathcal{M}_A D_\mu^\dagger$. These each give us a decomposition of ρ_{AB}', $\rho_{AB}' = \Sigma_\mu q_\mu|\phi_{AB,\mu}\rangle\langle\phi_{AB,\mu}| = \Sigma_\nu q_\nu'|\phi_{AB,\nu}'\rangle\langle\phi_{AB,\nu}'|$, where $\sqrt{q_\mu}|\phi_{AB,\mu}\rangle = \frac{1}{\sqrt{N}}\Sigma_j A_\mu|j_A\rangle \otimes |j_B\rangle$ and $\sqrt{q_\nu'}|\phi_{AB,\nu}'\rangle = \frac{1}{\sqrt{N}}\Sigma_j D_\mu|j_A\rangle \otimes |j_B\rangle$.

We know that there exists a unitary matrix $U_{\nu\mu}$ such that

$$\sqrt{q_\nu'}|\phi_{AB,\nu}'\rangle = \Sigma_\mu U_{\nu\mu}\sqrt{q_\mu}|\phi_{AB,\mu}\rangle, \quad (4.25)$$

or

$$\Sigma_j D_\nu|j_A\rangle \otimes |j_B\rangle = \Sigma_\mu\Sigma_j U_{\nu\mu}A_\mu|j_A\rangle \otimes |j_B\rangle. \quad (4.26)$$

We can read out from here that

$$D_\nu|j_A\rangle = \Sigma_\mu U_{\nu\mu}A_\mu|j_A\rangle, \quad (4.27)$$

but $\{|j_A\rangle\}$ is a basis, so, finally,

$$D_\nu = \Sigma_\mu U_{\nu\mu}A_\mu. \quad (4.28)$$

From here, we can conclude that any two Kraus representations of the same superoperator are related by a unitary matrix.

4.2   An Example: The Depolarizing Channel

We shall now look at an example of a superoperator on qubits, the depolarizing channel. A quantum channel is, in general, a quantum map that maps density matrixes to density matrixes. The idea of the depolarizing channel is that qubit has probability of $(1 - p)$ of nothing happening, $\frac{p}{3}$ of σ_x acting on it (bit flip), $\frac{p}{3}$ of σ_z acting on it (phase flip), and $\frac{p}{3}$ of σ_y acting on it (both). One way to do this is to tensor our qubit Hilbert space with a four-dimensional "environment" Hilbert space, $\mathcal{H}_A \otimes \mathcal{H}_E$. We have a unitary operator acting on this tensor product Hilbert space

$$U_{AE}|\psi_A\rangle \otimes |0_E\rangle = \sqrt{1-p}|\psi_A\rangle \otimes |0_E\rangle + \sqrt{\frac{p}{3}}\left(\sigma_x|\psi_A\rangle \otimes |1_E\rangle\right.$$
$$\left. + \sigma_y|\psi_A\rangle \otimes |2_E\rangle + \sigma_z|\psi_A\rangle \otimes |3_E\rangle\right), \tag{4.29}$$

and, after tracing out the environment we get

$$T(|\psi_A\rangle\langle\psi_A|) = Tr_E(U_{AE}|\psi_A\rangle \otimes |0_E\rangle\langle\psi_A| \otimes \langle 0_E|U_{AE}^{-1})$$
$$= (1-p)|\psi_A\rangle\langle\psi_A| + \frac{p}{3}\sigma_x|\psi_A\rangle\langle\psi_A|\sigma_x$$
$$+ \frac{p}{3}\sigma_y|\psi_A\rangle\langle\psi_A|\sigma_y + \frac{p}{3}\sigma_z|\psi_A\rangle\langle\psi_A|\sigma_z. \tag{4.30}$$

From here we can read out the Kraus operators: $A_0 = \sqrt{1-p}I$, $A_1 = \sqrt{\frac{p}{3}}\sigma_x$, $A_2 = \sqrt{\frac{p}{3}}\sigma_y$, and $A_3 = \sqrt{\frac{p}{3}}\sigma_z$. It is easy to check that $\Sigma_{\mu=0}^3 A_\mu^\dagger A_\mu = I$, which is a direct consequence of the unitarity of U_{AE}.

It is interesting to see what happens to the Bloch sphere under this mapping. Let

$$\rho = \frac{1}{2}(I + \mathbf{n} \cdot \sigma), \tag{4.31}$$

and use

$$\sigma_j \sigma_k \sigma_j = \begin{cases} -\sigma_k & k \neq j \\ \sigma_j & j = k \end{cases}. \tag{4.32}$$

Then

$$T(\sigma_x) = (1-p)\sigma_x + \frac{p}{3}\sigma_x - \frac{2p}{3}\sigma_x = (1 - \frac{4p}{3})\sigma_x, \tag{4.33}$$

and, similarly $T(\sigma_y) = (1 - \frac{4p}{3})\sigma_y$ and $T(\sigma_z) = (1 - \frac{4p}{3})\sigma_z$. So, finally

$$T(\rho) = \frac{1}{2}(I + \mathbf{n}' \cdot \sigma), \tag{4.34}$$

where $\mathbf{n}' = (1 - \frac{4p}{3})\mathbf{n}$.

This tells us that the map representing the depolarizing channel just causes the entire Bloch sphere to contract by a factor of $|1 - (4p/3)|$.

4.3   An Impossible Map

4.3.1   The Cloning Map and the No-Cloning Theorem

It is also useful to know that certain maps are impossible. One of them is the "cloning" map that would duplicate quantum states. Suppose we have a device that does copy qubit quantum states. A general input for such a device is of the form $|\psi_a\rangle \otimes |0_b\rangle \otimes |Q_c\rangle$ where $|\psi_a\rangle$ is the state of qubit a to be copied, $|0_b\rangle$ is a blank initial state of qubit b that becomes the copy, and $|Q_c\rangle$ is an ancillary state which can be interpreted as the initial state of the copier. What we want is the unitary map

$$U(|\psi_a\rangle \otimes |0_b\rangle \otimes |Q_c\rangle) = |\psi_a\rangle \otimes |\psi_b\rangle \otimes |Q_{\psi,c}\rangle \qquad (4.35)$$

Since the copier must work for arbitrary states, so, in particular, it must copy basis states

$$U(|0_a\rangle \otimes |0_b\rangle \otimes |Q_c\rangle) = |0_a\rangle \otimes |0_b\rangle \otimes |Q_{0,c}\rangle,$$
$$U(|1_a\rangle \otimes |0_b\rangle \otimes |Q_c\rangle) = |1_a\rangle \otimes |1_b\rangle \otimes |Q_{1,c}\rangle. \qquad (4.36)$$

These relations determine how a general state is copied. If $|\psi\rangle = \alpha|0\rangle + \beta|0\rangle$, then multiplying the first equation by α, the second by β, and adding them together gives

$$U(|\psi_a\rangle \otimes |0_b\rangle \otimes |Q_c\rangle) = \alpha|0_a\rangle \otimes |0_b\rangle \otimes |Q_{0,c}\rangle + \beta|1_a\rangle \otimes |1_b\rangle \otimes |Q_{1,c}\rangle, \qquad (4.37)$$

and this is not the same as Eq. (4.35). This is known as the no cloning theorem. It is a direct consequence of the linearity of quantum mechanics, and implies that quantum information is very different from classical information, which can be copied. The fact that quantum information cannot be copied can also be used to our advantage. In particular, it is one of the reasons why quantum cryptography works.

The no-cloning theorem only forbids the existence of a machine that makes perfect copies. We will see in a later chapter that it is possible to construct a cloner that makes approximate copies.

4.3.2   Faster Than Light Communication II

If cloning were possible, superluminal communication would be, too. This was, in fact, the way the no-cloning theorem was discovered. A faster-than-light communication scheme was proposed by Nick Herbert that relied on cloning. Herbert didn't talk about a cloner, but he assumed that a laser amplifier acts as a cloner, which, in fact, it does not. However, if one did have a perfect cloner, then Herbert's scheme would work, and clearly something had to be wrong with it. Wootters and Zurek identified cloning as the problem, and the proof of the no-cloning theorem we have given above is due to them. The story behind Nick Herbert and the no-cloning theorem is an interesting one, and we have provided a reference that gives an account of it.

To demonstrate how a perfect cloner would enable superluminal communication, suppose that Alice and Bob intially share an ebit, $|\psi_{AB}\rangle = \frac{1}{\sqrt{2}}(|0_A\rangle|1_B\rangle - |1_A\rangle|0_B\rangle)$. In the $|\pm\rangle = \frac{1}{\sqrt{2}}(|0\rangle \pm |1\rangle)$ basis, the same state can be written as $|\psi_{AB}\rangle = -\frac{1}{\sqrt{2}}(|+_A\rangle|-_B\rangle - |-_A\rangle|+_B\rangle)$. Alice now measures her particle either in the $\{|0\rangle, |1\rangle\}$ basis or in the $\{|+\rangle, |-\rangle\}$ basis, while Bob clones his particle, making $2N$ copies. He then measures N in the $\{|0\rangle, |1\rangle\}$ basis and N in the $\{|+\rangle, |-\rangle\}$ basis. The basis in which all measurements produce the same results tells Bob which basis Alice measured in.

4.4 Problems

1. We want to show that a superoperator is invertible if and only if it is unitary, i.e. $M(B) = UBU^\dagger$, for any $B \in \mathcal{B}(\mathcal{H})$. It is clear that if M is unitary, then it is invertible. We need to now show the opposite, i.e. that if M is invertible, then it is unitary. Let

$$M(|\psi\rangle\langle\psi|) = \sum_\mu M_\mu|\psi\rangle\langle\psi|M_\mu^\dagger.$$

The superoperator N is the inverse of M if $N \circ M = I$, or

$$\sum_{\mu,\nu} N_\nu M_\mu|\psi\rangle\langle\psi|M_\mu^\dagger N_\nu^\dagger = |\psi\rangle\langle\psi|,$$

for all $|\psi\rangle$.

(a) Use the fact that

$$\sum_{\mu,\nu} |\langle\psi|N_\nu M_\mu|\psi\rangle|^2 = 1,$$

which is implied by the above equation, and the normalization conditions on the operators M_μ and N_ν to show that $N_\nu M_\mu = \lambda_{\nu\mu} I$.

(b) Use the result in part (a) to show that $M_{\mu'}^\dagger M_\mu$ is a multiple of the identity for any μ and μ'.

(c) Use the result in part (b) to show that M is unitary.

2. (a) Let $|\psi_1\rangle$ and $|\psi_2\rangle$ be two one qubit states. Find a value of ϕ for which the map

$$|\psi_1\rangle|\psi_2\rangle \rightarrow \frac{1}{\sqrt{2}}(|\psi_2\rangle|\psi_1\rangle + e^{i\phi}|\psi_1\rangle|\psi_2\rangle),$$

   can be realized as a unitary operation.

   (b) The above transformation can be used to spread the information in a single qubit over two qubits. Suppose $|\psi_1\rangle = |0\rangle$ and $|\psi_2\rangle = |\psi\rangle$. Now suppose we lose one of the qubits. Find the reduced density matrix of the remaining qubit and its fidelity to the state $|\psi\rangle$. By spreading the information contained in one qubit over two, we retain some information about the original qubit even though one of the qubits is lost.

   (c) For a general one qubit state $|\psi\rangle$, find a value of ϕ so that the transformation

$$|0\rangle|0\rangle|\psi\rangle \rightarrow \frac{1}{\sqrt{3}}[|0\rangle|0\rangle|\psi\rangle + e^{i\phi}(|0\rangle|\psi\rangle|0\rangle + |\psi\rangle|0\rangle|0\rangle)].$$

   can be realized by a unitary operator.

3. A CNOT gate is a rather versatile device that can be used to realize a number of maps.

   (a) Suppose the input state is $(\sqrt{p_0}|0\rangle + \sqrt{p_1}|1\rangle) \otimes |\psi\rangle$, where $p_0 + p_1 = 1$, $|\psi\rangle$ is a general one-qubit state, and the first qubit is the control qubit, and the second is the target qubit. We send this state through the CNOT gate and trace out the control qubit. Find the Kraus operators for the resulting map on the target qubit.

   (b) Now consider the input state $|\psi\rangle \otimes (1/\sqrt{2})(e^{i\theta}|0\rangle + e^{-i\theta}|1\rangle)$. Send this state through the CNOT and trace out the target qubit. Find the Kraus operators for the resulting map on the control qubit.

4. The SWAP operator on two qubits acts as $S|\psi\rangle_a \otimes |\phi\rangle_b = |\phi\rangle_a \otimes |\psi\rangle_b$. A partial SWAP operator is given by $P(\theta) = \cos\theta I_{ab} + i\sin\theta S$, where I_{ab} is the identity operator on two qubits. It can be thought of as an operator that partially exchanges the information between two qubits.

   (a) Show that if we consider the two-qubit density matrix given by $\rho_a \otimes \xi_b$, where ρ_a and ξ_b are one-qubit density matrices, that

$$\rho_a' = \text{Tr}_b[P(\theta)\rho_a \otimes \xi_b P^\dagger(\theta)],$$

   is given by

$$\rho_a' = \cos^2\theta\rho_a + \sin^2\theta\xi_a + i\cos\theta\sin\theta[\xi_a, \rho_a].$$

(b) Expressing ρ_a and ρ_a' in Bloch form

$$\rho_a = \frac{1}{2}I_a + \mathbf{r} \cdot \boldsymbol{\sigma}, \qquad \rho_a' = \frac{1}{2}I_a + \mathbf{r}' \cdot \boldsymbol{\sigma},$$

find $\mathbf{r}'$ in terms of $\mathbf{r}$.

5. Time-dependent quantum maps can often be described by a master equation. This is a differential equation for the density matrix of a system. Let us consider a master equation for a qubit

$$\frac{d\rho}{dt} = \frac{\gamma}{2}(2\sigma^{(-)}\rho\sigma^{(+)} - \sigma^{(+)}\sigma^{(-)}\rho - \rho\sigma^{(+)}\sigma^{(-)}),$$

where $\sigma^{(+)}|0\rangle = |1\rangle$, $\sigma^{(+)}|1\rangle = 0$, $\sigma^{(-)} = (\sigma^{(+)})^\dagger$, and γ is a constant that governs how fast the process takes place. If our qubit consists of the ground and excited states of an atom, the above master equation describes the state of the qubit as the excited state decays. The Kraus operators for the map generated by this master equation are of the form

$$A_1 = \begin{pmatrix} 1 & 0 \\ 0 & f_1(t) \end{pmatrix} \qquad A_2 = \begin{pmatrix} 0 & f_2(t) \\ 0 & 0 \end{pmatrix}.$$

Find $f_1(t)$ and $f_2(t)$. The resulting map is an example of an amplitude-damping channel.

Further Reading

1. John Preskill, Lecture notes for Physics, vol. 219. http://www.theory.caltech.edu/people/preskill/ph229/
2. W.K. Wootters, W.H. Zurek, A single quantum cannot be cloned. Nature **299**, 802 (1982)
3. D. Kaiser, *How the Hippies Saved Physics* (W. W. Norton and Company, New York, 2011)

Chapter 5
Quantum Measurement Theory

5.1 Outline

Measurements are an integral part of quantum information processing. Reading out the quantum information at the end of the processing pipeline is equivalent to learning what final state the system is in at the output since information is encoded in the state. In fact, information is the state itself. Since finding out the state of a system can be done only by performing measurements on it, we need a thorough understanding of the quantum theory (and practice) of measurements. To this end we will begin by a simplified model of a quantum measurement, due essentially to von Neumann, and from this model we'll read out the postulates of standard quantum measurement theory. Then, by analyzing the underlying assumptions we'll show that some of the postulates can be replaced by more relaxed ones and this will lead us to the concept of generalized measurements (POVMs) which are particularly useful in measurement optimization problems. Next, by invoking Neumark's theorem we will show how to actually implement POVMs experimentally. As illustrations of these general concepts we will study two state discrimination strategies in some detail, namely the unambiguous discrimination and the minimum error discrimination of two quantum states. To close this chapter we will analyze the B92 protocol for quantum key distribution (QKD). QKD is the crucial ingredient of most quantum cryptographic protocols and in the B92 proposal all of the concepts of this chapter come together in a particularly clean and instructive way.

© Springer Nature Switzerland AG 2021

J. A. Bergou et al., *Quantum Information Processing*, Graduate Texts in Physics,
https://doi.org/10.1007/978-3-030-75436-5_5

5.2   Standard Quantum Measurements

We begin with a brief review of a simplified model of a quantum measurement. Let us assume that we want to measure a physical quantity to which, in quantum mechanics, there corresponds a Hermitian operator X. The measurement then consists of the following process. We couple this observable to a so-called pointer variable, the states of which we assume to be macroscopically distinguishable. This is equivalent to assuming that the states of the pointer variable are essentially classical and it is our basic assumption that classical states can be readily measured. For example, our pointer can be a freely propagating heavy particle and the pointer variable that we observe is simply its position. The initial state of the pointer is a narrow but not too narrow wave packet. What we mean by this is the following. On the one hand, the wave packet must be narrow enough so that the possible pointer positions are clearly distinguishable; there is no overlap among them. On the other hand, the wave packet should not be narrower than necessary because if it is it will spread too fast during the time of the measurement. Let us make this a little more quantitative. From the uncertainty principle, $\Delta x \Delta p \simeq \hbar$, we obtain $\Delta p \simeq \hbar/\Delta x$ which leads to an uncertainty $\Delta v = \hbar/m\Delta x$ in the speed of the pointer particle. Thus, during the time of the measurement the spread of the initial wave packet increases as

$$\Delta x(t) = \Delta x + \frac{\hbar t}{m\Delta x} . \tag{5.1}$$

This expression is a minimum for a given measurement time t if the spread of the initial wave packet is $\Delta x_{opt} = \sqrt{\frac{\hbar t}{m}}$, yielding

$$\Delta x_{min}(t) \equiv \Delta x_{SQL} = 2\sqrt{\frac{\hbar t}{m}} , \tag{5.2}$$

where SQL stands for Standard Quantum Limit. The initial wave packet should not be prepared narrower than Δx_{opt} which in most cases is not a serious restriction since m is large.

Next, we introduce a coupling between the system and the pointer. The full Hamiltonian is given by

$$H = H_0 + \frac{P^2}{2m} + \hbar g X P , \tag{5.3}$$

where, on the right hand side, H_0 is the Hamiltonian of the system, the next term is the kinetic energy of the pointer and the last term is the coupling between the system and the pointer, g being the coupling constant. Since we want to observe the position of the pointer, we choose the coupling between the complementary quantity, the canonical momentum P of the pointer, and the observable X of the system that we want to measure. For simplicity, we assume that the observable X commutes with the unperturbed Hamiltonian H_0 of the system. The above Hamiltonian leads to the time evolution described by the unitary operator

$$U(t) = e^{-igtXP} .$$ (5.4)

The observable X is Hermitian, so it does have a spectral representation, $X = \sum_j \lambda_j P_j = \sum_j \lambda_j |j\rangle\langle j|$ where λ_j are the (real) eigenvalues, $|j\rangle$ the corresponding eigenstates and $P_j = |j\rangle\langle j|$ the projector on the subspace spanned by $|j\rangle$. The eigenstates form a complete set in the Hilbert space of the system which is equivalent to saying that the projectors span the identity, $\sum_j P_j = 1$. Using this last relation we can write the time evolution operator as

$$U(t) = \sum_j e^{-ix_j P} |j\rangle\langle j| ,$$ (5.5)

where we introduced the notation

$$x_j = gt\lambda_j .$$ (5.6)

Now, we assume that the joint system-pointer system was initially prepared in the state $\sum_j c_j|j\rangle \otimes |\psi(x)\rangle$ where $|\psi^S\rangle = \sum_j c_j|j\rangle$ is an arbitrary initial state of the system and $|\psi(x)\rangle$ is the initial state of the pointer which we assume to be a well-localized wave packet around $x = 0$, as discussed above. If we apply the time evolution operator, Eq. (5.5), to this initial state we obtain the state of the joint system-pointer system after the measurement time t,

$$|\psi^{SP}\rangle = \sum_j c_j|j\rangle|\psi(x - x_j)\rangle .$$ (5.7)

What we see from here is that there is a very strong correlation between the state of the pointer and the state of the system. We assume that the pointer is essentially classical so we will always find it in one of the new positions at $x = x_j$. When it is found at $x = x_j$ the state of the system is $c_j|j\rangle$. Since $x_j = gt\lambda_j$ is uniquely related to the eigenvalue λ_j, we can say that by observing the position of the pointer after the measurement we have measured the observable X and found one of its eigenvaules λ_j as the measurement result. Furthermore the (non-normalized) state of the system, if this particular value was found, is just $c_j|j\rangle$. Taking the inner product of $|\psi^S\rangle = \sum_j c_j|j\rangle$ with $|j\rangle$ gives $c_j = \langle j|\psi^S\rangle$, which tells us that the non-normalized postmeasurement state is $|j\rangle\langle j|\psi^S\rangle = P_j|\psi^S\rangle$. The normalized state after the measurement, if the particular outcome λ_j was found, is $|\phi_j\rangle = \frac{P_j|\psi^S\rangle}{|c_j|}$.

These findings are summarized by the postulates of quantum measurement theory in a more formal way. However, before we list these postulates we want to have an expression for the resolution of the above measurement. Obviously, we can resolve the different pointer positions if their distance is larger than the Standard Quantum Limit, $\Delta x_j = x_{j+1} - x_j = g t \Delta \lambda_j \geq x_{SQL}$. When we use the relation between the pointer position and the eigenvalues, Eq. (5.6), we will find the resolution limit, as

$$\Delta \lambda_j \geq \frac{2}{g} \sqrt{\frac{\hbar}{mt}}, \qquad (5.8)$$

which is the minimum separation of the eigenvalues that can be resolved in a quantum measurement. As expected, with increasing measurement time the resolution improves.

We are now in the position to read out the postulates of the quantum measurement theory from the preceding discussion. Let us assume that we are measuring the observable X which has the spectral representation $X = \sum_j \lambda_j |j\rangle\langle j|$. From the hermiticity of X it follows that the eigenvalues λ_j are real. For simplicity we assume that the eigenvalues are nondegenerate and the corresponding eigenvectors, $\{|j\rangle\}$, form a complete orthonormal basis set. Then

1. The projectors $P_j = |j\rangle\langle j|$ span the entire Hilbert space, $\sum_j P_j = 1$.
2. From the orthogonality of the states we have $P_i P_j = P_i \delta_{ij}$. In particular, $P_i^2 = P_i$ from where it follows that the eigenvalues of any projector are 0 and 1.
3. A measurement of X yields one of the eigenvalues λ_j.
4. The state of the system after the measurement is $|\phi_j\rangle = \dfrac{P_j|\psi\rangle}{\sqrt{\langle \psi | P_j | \psi \rangle}}$ if the outcome is λ_j.
5. The probability that this particular outcome is found as the measurement result is $p_j = ||P_j\psi\rangle||^2 = \langle \psi | P_j^2 | \psi \rangle = \langle \psi | P_j | \psi \rangle$ where we used the property 2.
6. If we perform the measurement but we do not record the results, the postmeasurement state can be described by the density operator $\rho = \sum_j p_j |\phi_j\rangle\langle\phi_j| = \sum_j P_j |\psi\rangle\langle\psi| P_j$.

These six postulates adequately describe what happens to the system during the measurement if it was initially in a pure state. If the system is initially in the mixed state ρ the last three postulates are to be replaced by their immediate generalizations:

4a. The state of the system after the measurement is $\rho_j = \dfrac{P_j \rho P_j}{Tr(P_j \rho P_j)} = \dfrac{P_j \rho P_j}{Tr(P_j \rho)}$ if the outcome is λ_j.
5a. The probability that this particular outcome is found as the measurement result is $p_j = Tr(P_j \rho P_j) = Tr(P_j^2 \rho) = Tr(P_j \rho)$ where, again, we used the property 2.
6a. If we perform the measurement but we do not record the results, the post-measurement state can be described by the density operator $\tilde{\rho} = \sum_j p_j \rho_j = \sum_j P_j \rho P_j$.

Of course, 4a–6a reduce to 4–6 for the pure state density matrix $\rho = |\psi\rangle\langle\psi|$. Therefore, in what follows we use the density matrix to describe a general (pure or mixed) quantum state unless we want to emphasize that the state is pure.

Let us summarize the message of these postulates. They essentially tell us that the measurement process is random, we cannot predict its outcome. What we can predict is the spectrum of the possible outcomes and the probability that a particular outcome is found in an actual measurement. This leads us to the ensemble interpretation of quantum mechanics. The state $|\psi\rangle$ (or ρ for mixed states) describes not a single system but an ensemble of identically prepared systems. If we perform the same measurement on each member of the ensemble we can predict the possible measurement results and the probabilities with which they occur but we cannot predict the outcome of an individual measurement, except, of course, when the probability of a certain outcome is 0 or 1. With the help of these postulates we can then calculate the moments of the probability distribution, $\{p_j\}$, generated by the measurement. The first moment is the average of a large number of identical measurements performed on the initial ensemble. It is called the expectation value of X and is denoted as $\langle X \rangle$,

$$\langle X \rangle = \sum_j \lambda_j p_j = \sum_j \lambda_j Tr(P_j \rho) = Tr(X\rho)\,, \tag{5.9}$$

where we used the spectral representation of X. The second moment, $\langle X^2 \rangle = Tr(X^2 \rho)$, is related to the variance σ,

$$\sigma^2 = \langle (X - \langle X \rangle)^2 \rangle = \langle X^2 \rangle - \langle X \rangle^2\,. \tag{5.10}$$

Higher moments can also be calculated in a straightforward manner but typically the first and second moments are the most important ones to consider.

5.3   Positive Operator Valued Measures: POVM's

Now we are in the position to put the postulates of standard measurement theory under closer scrutiny. What the last three postulates provide us with is, in fact, an algorithm to generate probabilities. The generated probabilities are non-negative, $0 \leq p_j \leq 1$, and the probability distribution is normalized to unity, $\sum_j p_j = 1$ which is a consequence of the first two postulates. Furthermore, the number of possible outcomes is bounded by the number of terms in the orthogonal decomposition of the identity operator of the Hilbert space. Obviously, one cannot have more orthogonal projections than the dimensionality, N_A, of the Hilbert space of the system, so $j \leq N_A$. It would, however, be often desirable to have more outcomes than the dimensionality while keeping the positivity and normalization of the probabilities. We will first show that this is formally possible: if we relax the above rather restrictive postulates and replace them with more flexible ones we can still obtain a meaningful probability generating algorithm. Then we will show that there are physical processes that fit these more general postulates.

Let us begin with the formal considerations and take a closer look at Postulate 5a (or 5) which is the one that gives us the prescription for the generation of probabilities. We notice that in order to get a positive probability by this prescription it is sufficient if P_j^2 is a positive operator, we do not need to require the positivity of an underlying P_j operator. So let us try the following. We introduce a positive operator, $\Pi_j \geq 0$, which is the generalization of P_j^2, and prescribe $p_j = Tr(\Pi_j \rho)$. Of course, we want to ensure that the probability distribution generated by this new prescription is still normalized. Inspecting the postulates we can easily figure out that normalization is a consequence of Postulate 1 and, therefore, require that $\sum_j \Pi_j = I$, that is the positive operators still represent a decomposition of the identity. We will call a decomposition of the identity in terms of positive operators, $\sum_j \Pi_j = I$, a POVM (Positive Operator Valued Measure) and $\Pi_j \geq 0$ the elements of the POVM. These generalizations will form the core of our new postulates $1'$ and $5'$.

As observed in the previous paragraph, for a POVM to exist we do not have to require orthogonality and positivity of the underlying P_j operators. Therefore, the underlying operators that, via Postulates 4 (or 4a) and 6 (or 6a), determine the postmeasurement state can be just about any operators, even non-Hermitian ones. For projective measurements orthogonality was essentially a consequence of Postulate 2, which was our most constraining postulate because it restricted the number of terms in the decomposition of the identity to at most the dimensionality of the system. Let us now see how far we can get by abandoning it.

If we abandon Postulate 2 then the operators that generate the probability distribution are no longer the same as the ones that generate the postmeasurement states and we have a considerable amount of freedom in choosing them. Let us denote the operators that generate the postmeasurement state by A_j, they are the generalizations of the orthogonal projectors, P_j. In other words, we define the non-normalized postmeasurement state by $A_j|\psi\rangle$ and the corresponding normalized state after the measurement by $|\phi\rangle = A_j|\psi\rangle / \sqrt{\langle \psi | A_j^\dagger A_j | \psi \rangle}$. This expression will form the essence of our new Postulate $4'$. It immediately tells us that Π_j has the structure $\Pi_j = A_j^\dagger A_j$ which by construction is a positive operator. Let us now use our freedom in designing the postmeasurement state. First note that, since the POVM elements are positive operators, $\Pi_j^{1/2}$ exists. Obviously, this is a possible choice for A_j. So is

$$A_j = U_j \Pi_j^{1/2} , \tag{5.11}$$

where U_j is an arbitrary unitary operator. This is the most general form of the detection operators, satisfying $A_j^\dagger A_j = \Pi_j$ and the above expression corresponds to their polar decomposition. What we see is that the POVM elements Π_j determine the absolute value operators through $|A_j| = \Pi_j^{1/2}$ but leave their unitary part open. The A_j operators represent a generalization of the projectors P_j whereas Π_j is a generalization of P_j^2. The set $\{A_j\}$ is called the set of detection operators and these operators figure prominently in our new postulates $2'$, $4'$ and $6'$ replacing the corresponding ones of the standard measurements.

With this we completed our goal that we set out to do at the beginning of this section, namely the generalization of all of the postulates of the standard measurement theory to more flexible ones while keeping the spirit of the old ones. It is now time to list our new postulates.

$1'$. We consider the decomposition of the identity, $\sum_j \Pi_j = 1$, in terms of positive operators, $\Pi_j \geq 0$. Such a decomposition is called a POVM (Positive Operator Valued Measure) and the Π_j the elements of the POVM.

$2'$. The elements of the POVM, Π_j, can be expressed in terms of the detection operators A_j as $\Pi_j = A_j^\dagger A_j$ where, in general, the detection operators are non-Hermitian ones, restricted only by the requirement $\sum_j A_j^\dagger A_j = I$. Then, by construction, the POVM elements are positive operators. Conversely, for a given POVM the detection operators can be expressed in terms of the POVM elements as $A_j = U_j \Pi_j^{1/2}$ where U_j is an arbitrary unitary operator.

$3'$. A detection yields one of the alternatives corresponding to an element of the POVM.

$4'$. The state of the system after the measurement is $|\phi_j\rangle = \dfrac{A_j|\psi\rangle}{\sqrt{\langle\psi|A_j^\dagger A_j|\psi\rangle}}$ if it was initially in the pure state $|\psi\rangle$, and $\rho_j = \dfrac{A_j\rho A_j^\dagger}{Tr(A_j\rho A_j^\dagger)} = \dfrac{A_j\rho A_j^\dagger}{Tr(A_j^\dagger A_j\rho)}$ if it was initially in the mixed state ρ. The inclusion of the arbitrary unitary operator U_j into the detection opertor gives us a great deal of flexibility in designing the postmeasurement state.

$5'$. The probability that this particular alternative is found as the measurement result is $p_j = Tr(A_j\rho A_j^\dagger) = Tr(A_j^\dagger A_j\rho) = Tr(\Pi_j\rho)$ where we used the cyclic property of the trace operation.

$6'$. If we perform the measurement but we do not record the results, the post-measurement state is described by the density operator $\tilde{\rho} = \sum_j p_j\rho_j = \sum_j A_j\rho A_j^\dagger$.

Very often we are not concerned with the state of the system after such operation is performed but only with the resulting probability distribution. For this, it is sufficient to consider Postulates $1'$ and $5'$ defining the probability of finding alternative j as the detection result. Note, that at no step did we require the orthogonality of the Π_j's. Since orthogonality is no longer a requirement, the number of terms in this decomposition of the identity is not bounded by N_A. In fact, the number of terms can be arbitrary. Obviously, what we arrived at is a generalization of the von Neumann projective measurement. It is a surprising generalization as it tells us that just about any operation that satisfies Postulates $1'$ and $2'$ is a legitimate operation that generates a valid probability distribution. It is also a rather natural generalization of the standard quantum measurement since it provides us with a well-defined algorithm that generates a well-behaved probability distribution. So this procedure can be regarded as a *generalized measurement* and, indeed, for most purposes it is a sufficient generalization of the standard quantum measurement.

A further note is in place here. Very often a projector projects on a one dimensional subspace, which is spanned by the vector $|\omega_j\rangle$, in which case it can be written as $P_j = |\omega_j\rangle\langle\omega_j|$. The corresponding generalization to a non-Hermitian detection operator can be written as $A_j = c_j|\tilde{\omega}_j\rangle\langle\omega_j|$ where $\langle\omega_j|\omega_j\rangle = \langle\tilde{\omega}_j|\tilde{\omega}_j\rangle = 1$, and c_j is a complex number inside the unit circle $|c_j|^2 \leq 1$, and $\langle\omega_j|\tilde{\omega}_j\rangle$ is arbitrary. Then

$$\Pi_j = |c_j|^2|\omega_j\rangle\langle\omega_j|\,, \tag{5.12}$$

but $\langle\omega_j|\omega_k\rangle \not\propto \delta_{ij}$ and, hence, it is explicit that the POVM is not an orthogonal decomposition of the identity. Since $A_j|\psi\rangle = c_j\langle\omega_j|\psi\rangle|\tilde{\omega}_j\rangle$, we see that $|\tilde{\omega}_j\rangle$ is proportional to the postmeasurement state $|\phi_j\rangle$, and $p_j = \langle\psi|\Pi_j|\psi\rangle = \langle\psi|A_j^\dagger A_j|\psi\rangle = |c_j|^2|\langle\omega_j|\psi\rangle|^2$. So, we have that

$$|c_j|^2 = \frac{p_j}{|\langle\omega_j|\psi\rangle|^2}\,, \tag{5.13}$$

and

$$\Pi_j = \frac{p_j}{|\langle\omega_j|\psi\rangle|^2}|\omega_j\rangle\langle\omega_j|\,. \tag{5.14}$$

Of course, up to this point all this is just a formal mathematical generalization of the standard quantum measurement. The important question is, how can we implement such a thing physically? In the next section we set out to answer this question and then we will study examples of POVMs.

5.4 Neumark's Theorem and the Implementation of a POVM via Generalized Measurements

First, let us take a look at what happens if we couple our system to another system called ancilla, let them evolve, and then measure the ancilla. The Hilbert space of this larger system is $\mathcal{H}_A \otimes \mathcal{H}_B$, using the tensor product extension, where the Hilbert space of the original system is $\mathcal{H}_A$ and the Hilbert space of the ancilla is $\mathcal{H}_B$. We want to gain information about the state of the system that we now denote as $|\psi_A\rangle$. We assume that the system and the ancilla are initially independent; their joint initial state is $|\psi_A\rangle \otimes |\psi_B\rangle$. Let $\{|m_B\rangle\}$ be an orthonormal basis for $\mathcal{H}_B$, and U_{AB} a unitary operator acting on $\mathcal{H}_A \otimes \mathcal{H}_B$. The probability p_m of measuring $|m_B\rangle$ is then given by

$$p_m = \|(I_A \otimes |m_B\rangle\langle m_B|)U_{AB}(|\psi_A\rangle \otimes |\psi_B\rangle)\|^2 . \tag{5.15}$$

Define

$$A_m|\psi_A\rangle \equiv \langle m_B|U_{AB}(|\psi_A\rangle \otimes |\psi_B\rangle) . \tag{5.16}$$

Then A_m is a linear operator on $\mathcal{H}_A$ that depends on $|m_B\rangle$, $|\psi_B\rangle$ and U_{AB}. With the help of this definition we can write the measurement probability as

$$p_m = \|A_m|\psi_A\rangle \otimes |m_B\rangle\|^2 = \langle\psi_A|A_m^\dagger A_m|\psi_A\rangle . \tag{5.17}$$

Note that

$$\sum_m \langle\psi_A|A_m^\dagger A_m|\psi_A\rangle = \sum_m ((\langle\psi_A| \otimes \langle\psi_B|)U_{AB}^\dagger|m_B\rangle\langle m_B|U_{AB}(|\psi_A\rangle \otimes |\psi_B\rangle))$$

$$= 1 . \tag{5.18}$$

Since this is true for any $|\psi_A\rangle$, we must have that

$$\sum_m A_m^\dagger A_m = I_A , \tag{5.19}$$

where I_A is the identity in $\mathcal{H}_A$.

The non-normalized state of the total 'system plus ancilla' after the measurement is $A_m|\psi_A\rangle \otimes |m_B\rangle$ so the (normalized) postmeasurement state of the system alone is

$$|\phi_A\rangle = \frac{1}{\sqrt{\langle\psi_A|A_m^\dagger A_m|\psi_A\rangle}} A_m|\psi_A\rangle . \tag{5.20}$$

Clearly, the (normalized) state of the ancilla after the measurement is $|\phi_B\rangle = |m_B\rangle$, up to an arbitrary phase factor. After the measurement is done and outcome $|m_B\rangle$ is found, the ancilla is no longer of interest and can be discarded.

The set $\{A_m^\dagger A_m\}$ thus gives a decomposition of the identity in terms of positive operators. Therefore, we can identify the set with a POVM where $\{A_m^\dagger A_m\}$ are its elements. In fact, what we see here is the first half of Neumark's theorem: If we couple our system to an ancilla, let them evolve so that they become entangled, and perform a measurement on the ancilla, which collapses the ancilla to one of the basis vectors of the ancilla space, then this procedure will also transform the system because the ancilla degrees of freedom are now entangled to the system. The transformation of the state of the system is, however, neither unitary nor a projection. It can adequately be described as a POVM so the above procedure corresponds to a POVM in the system Hilbert space. Thus, we have just found a procedure that, when we look at the system only, looks like a POVM. We now know that there are physical processes that can adequately be described as POVMs.

Next we address the question, given the set of operators $\{A_m\}$ acting on $\mathcal{H}_A$ such that $\sum_m A_m^\dagger A_m = I$, can this be interpreted as resulting from a measurement on a larger space? That is, can we find $\mathcal{H} = \mathcal{H}_A \otimes \mathcal{H}_B$, $|\psi_B\rangle$, $\{|m_B\rangle\} \in \mathcal{H}_B$ and U_{AB} acting on $\mathcal{H}$ such that

$$A_m|\psi_A\rangle = \langle m_B|U_{AB}(|\psi_A\rangle \otimes |\psi_B\rangle)) \tag{5.21}$$

holds?

The answer to this question is yes as we will now prove it constructively. Let us choose $\mathcal{H}_B$ to have dimension M and let $\{|m_B\rangle\}$ be an orthonormal basis for $\mathcal{H}_B$, and choose $|\psi_B\rangle$ to be an arbitrary but fixed initial state in $\mathcal{H}_B$. Let us further define a transformation U_{AB} via

$$U_{AB}(|\psi_A\rangle \otimes |\psi_B\rangle) = \sum_m A_m|\psi_A\rangle \otimes |m_B\rangle , \tag{5.22}$$

which implies the Eq. (5.21). U_{AB} is inner product preserving,

$$\left(\sum_{m'}\langle\psi_A'|A_{m'}^\dagger \otimes \langle m_B'|\right)\left(\sum_m A_m|\psi_A\rangle \otimes |m_B\rangle\right) = \sum_m \langle\psi_A'|A_m^\dagger A_m|\psi_A\rangle$$

$$= \langle\psi_A'|\psi_A\rangle , \tag{5.23}$$

so it is unitary on the one-dimensional subspace spanned by $|\psi_B\rangle$ and it can be extended to a full unitary operator on $\mathcal{H}_A \otimes \mathcal{H}_B$ because, e.g., on the subspace that is orthogonal to $|\psi_B\rangle$ it can be the identity.

This completes the proof of Neumark's theorem which asserts that there is a one-to-one correspondence between a POVM and the above procedure which sometimes is itself called a generalized measurement. Hence, a generalized measurement can be regarded as the physical implementation of a given POVM.

To close this section we will now illustrate these general considerations on an example. The example is an application of the minimum-error state discrimination strategy that will be discussed in the next section. Suppose one is given a qubit which is prepared equally likely in either of the following three states

$$|\psi_0\rangle = -\frac{1}{2}\left(|0\rangle + \sqrt{3}\,|1\rangle\right),$$

$$|\psi_1\rangle = -\frac{1}{2}\left(|0\rangle - \sqrt{3}\,|1\rangle\right),$$

$$|\psi_2\rangle = |0\rangle , \tag{5.24}$$

that is the probability that $|\psi_j\rangle$ (for $j = 0, 1, 2$) is prepared is $1/3$. These three states form an overcomplete set of symmetric states that is known as the trine ensemble. For the minimum error discrimination consider the operators,

$$\Pi_j = A_j^\dagger A_j = \frac{2}{3} |\psi_j\rangle\langle\psi_j| . \qquad (5.25)$$

Since $\Pi_j \geq 0$ and together they span the Hilbert space of the qubit, $\sum_{j=0}^{2} \Pi_j = 1$, we have a legitimate POVM. If we use this POVM and if we get result j, we guess that we were given $|\psi_j\rangle$. The probability of being correct is $p_j = \langle\psi_j|A_j^\dagger A_j|\psi_j\rangle = 2/3$ and the probability of making an error is $q_j = \langle\psi_{j'}|A_j^\dagger A_j|\psi_{j'}\rangle = 1/6$ (for $j \neq j'$). In fact, the above POVM is the optimal one for minimum-error discrimination, p_j takes its maximum possible value and q_j is the minimum allowed by the laws of quantum mechanics.

For a physical implementation of this optimal POVM along the lines of Neumark's theorem let us define the (non-normalized) qutrit vectors

$$|v_0\rangle = \sqrt{\frac{2}{3}}|0\rangle + \frac{1}{2\sqrt{6}}(|1\rangle + |2\rangle) ,$$

$$|v_1\rangle = \frac{1}{2\sqrt{2}}(|1\rangle - |2\rangle) ,$$

$$|u_0\rangle = \frac{1}{2\sqrt{2}}(|1\rangle - |2\rangle) ,$$

$$|u_1\rangle = \frac{1}{2}\sqrt{\frac{3}{2}}(|1\rangle + |2\rangle) , \qquad (5.26)$$

where $\{|m_B\rangle\}$ (for $m_B = 0, 1, 2$) is an orthonormal basis in the qutrit Hilbert space which is our $\mathcal{H}_B$. Note that $\langle v_0|v_1\rangle = \langle u_0|u_1\rangle = 0$ and $\|v_0\|^2 + \|v_1\|^2 = \|u_0\|^2 + \|u_1\|^2 = 1$. Let us further introduce the transformation U with the definition,

$$U|0_A\rangle|0_B\rangle = |0_A\rangle|v_{0,B}\rangle + |1_A\rangle|v_{1,B}\rangle ,$$

$$U|1_A\rangle|0_B\rangle = |0_A\rangle|u_{0,B}\rangle + |1_A\rangle|u_{1,B}\rangle , \qquad (5.27)$$

where A refers to the system and B to the ancilla. As discussed before, it is sufficient to define this transformation on a single initial state $|\psi_B\rangle$ of the ancilla which we conveniently choose as $|\psi_B\rangle = |0_B\rangle$. Then U can be extended to a full unitary transformation on the ancilla space by choosing it to be the identity on the subspace orthogonal to $|0_B\rangle$.

Obviously, any system state can be represented as $|\psi_A\rangle = \alpha|0_A\rangle + \beta|1_A\rangle$ with $|\alpha|^2 + |\beta|^2 = 1$. Multiplying the first of the equations in (5.27) by α and the second by β, adding them together and taking the $|m_B\rangle$ component of the resulting expression, yields

$$\langle m_B | U(|\psi_A\rangle|0_B\rangle) = |0_A\rangle(\alpha\langle m|v_0\rangle + \beta\langle m|u_0\rangle) + |1_A\rangle(\alpha\langle m|v_1\rangle + \beta\langle m|u_1\rangle) \,.$$
$$(5.28)$$

Using $\alpha = \langle 0_A|\psi_A\rangle$ and $\beta = \langle 1_A|\psi_A\rangle$, this expression defines A_m as the operator acting on $|\psi_A\rangle$. Then A_m is explicitly given by

$$A_m = |0_A\rangle(\langle 0_A|\langle m|v_0\rangle + \langle 1_A|\langle m|u_0\rangle) + |1_A\rangle(\langle 0_A|\langle m|v_1\rangle + \langle 1_A|\langle m|u_1\rangle) \,. \quad (5.29)$$

Finally, a comparison to Eq. (5.24) reveals that A_j can be written in the compact form,

$$A_j = \sqrt{\frac{2}{3}}|\psi_j\rangle\langle\psi_j| \,. \tag{5.30}$$

A direct substitution shows that Eq. (5.25) is satisfied by this A_j. Thus, we have just found a physical implementation of the optimal POVM for the minimum-error discrimination of the trine states.

What we have done here can be summarized in the following way. We needed three outcomes of our generalized measurement so we introduced a three-dimensional ancilla space, called a qutrit. Then we unitarily entangled our system with the ancilla and, after this interaction, a projective measurement on the ancilla degrees of freedom was performed. The POVM then emerges as the residual effect on the original system due to entanglement of the system to the ancilla when a von Neumann measurement is performed on the ancilla only.

This method corresponds to the tensor product extension of the Hilbert space. The Hilbert space of the combined system is the tensor product of the Hilbert spaces of the two subsystems. There are two conceptually different ways of extending a Hilbert space, the tensor product extension being one of them. The other method is the direct sum extension. The extended Hilbert space is then the direct sum of the Hilbert space spanned by the states of the original system and of the Hilbert space spanned by the auxiliary states, being also called ancilla states. For the trine ensemble, it is possible to associate the three two-dimensional non-normalized detection states $\sqrt{2/3}|\psi_j\rangle$ with three orthonormal states in three dimensions, given by

$$|\tilde\psi_j\rangle = \sqrt{\frac{2}{3}}\,|\psi_j\rangle + \sqrt{\frac{1}{3}}\,|2\rangle, \tag{5.31}$$

where the ancillary basis state $|2\rangle$ is orthogonal to the two basis states, $|0\rangle$ and $|1\rangle$, of the system. By performing the von Neumann measurement that consists of the three projections $|\tilde\psi_j\rangle\langle\tilde\psi_j|$ ($j = 0, 1, 2$) in the enlarged, i.e. three-dimensional Hilbert space, the required generalized measurement is realized in the original two-dimensional Hilbert space of the qubit. In effect, the direct sum extension of the

Hilbert space relies on the assumption that the original qubit secretly consists of two components of a qutrit.

5.5 Examples: Strategies for State Discrimination

As examples of a measurement optimization task we will consider two schemes for the optimal discrimination of quantum states. The first is unambiguous discrimination and the second is discrimination with minimum error. We will see that the optimum measurement for the first strategy is a POVM while the optimum measurement for the second is a standard von Neumann measurement. The two main discrimination strategies evolved rather differently from the very beginning. Unambiguous discrimination started with pure states and only very recently was it extended to discriminating among mixed quantum states. Minimum-error discrimination addressed the problem of discriminating between two mixed quantum states from the very beginning and the result for two pure states follows as a special case. The two strategies are, in a sense, complementary to each other. Unambiguous discrimination is relatively straightforward to generalize for more than two states, at least in principle, but it is difficult to treat mixed states. The error-minimizing approach, initially developed for two mixed states, is hard to generalize for more than two states.

5.5.1 Unambiguous Discrimination of Two Pure States

Unambiguous discrimination is concerned with the following problem. An ensemble of quantum systems is prepared so that each individual system is prepared in one of two known states, $|\psi_1\rangle$ or $|\psi_2\rangle$ with probability η_1 or η_2 (such that $\eta_1 + \eta_2 = 1$), respectively. The preparation probabilities are called a priori probabilities or, simply, priors. The states are, in general, not orthogonal, $\langle\psi_1|\psi_2\rangle \neq 0$ but linearly independent. The preparer, Alice, then draws a system at random from this ensemble and hands it over to an observer, called Bob, whose task is to determine which one of the two states he is given. The observer also knows how the ensemble was prepared, i.e. has full knowledge of the two possible states and their priors but does not know the actual state that was drawn. All he can do is to perform a single measurement or perhaps a POVM on the individual system he receives.

In the unambiguous discrimination strategy the observer is not allowed to make an error, i.e. he is not permitted to conclude that he was given one state when actually he was given the other. First we show that this can not be done with 100% probability of success. To this end, let us assume the contrary and assume we have two detection operators, Π_1 and Π_2, that together span the Hilbert space of the two states,

$$\Pi_1 + \Pi_2 = I \ . \tag{5.32}$$

For unambiguous detection we also require that

$$\Pi_1|\psi_2\rangle = 0\,,$$

$$\Pi_2|\psi_1\rangle = 0\,, \tag{5.33}$$

so that the first detector never clicks for the second state and vice versa, and we can identify the detector clicks with one of the states unambiguously. The probability of successfully identifying the first state if it is given is $p_1 = \langle\psi_1|\Pi_1|\psi_1\rangle$ and the probability of successfully identifying the second state if it is given is $p_2 = \langle\psi_2|\Pi_2|\psi_2\rangle$. Multiplying (5.32) with $\langle\psi_1|$ from the left and $|\psi_1\rangle$ from the right and taking into account (5.33), gives $p_1 = 1$ and, similarly, we obtain $p_2 = 1$, and it appears as though we could have perfect unambiguous discrimination. However, multiplying (5.32) with $\langle\psi_1|$ from the left and $|\psi_2\rangle$ from the right and taking into account (5.33) again, gives $0 = \langle\psi_1|\psi_2\rangle$ which can be satisfied for orthogonal states only. In fact, we have just proved that perfect discrimination of nonorthogonal quantum states is not possible.

Equation (5.32) allows two alternatives only, it assumes that we can have two operators that unambiguously identify the two states all the time. Since this is impossible, we are forced to modify this equation and have to allow for one other alternative. We introduce a third POVM element, Π_0, such that Eq. (5.33) is still satisfied but (5.32) is modified to

$$\Pi_1 + \Pi_2 + \Pi_0 = I\,. \tag{5.34}$$

The first and second POVM elements will continue to unambiguously identify the first and second state, respectively. However, Π_0 can click for both states and, thus, this POVM element corresponds to an inconclusive detection result. It should be emphasized that this outcome is not an error, we will never identify the first state with the second and vice versa, we simply will not make any conclusion in this case. We can now introduce success and failure probabilities in such a way that $\langle\psi_1|\Pi_1|\psi_1\rangle = p_1$ is the probability of successfully identifying $|\psi_1\rangle$, and $\langle\psi_1|\Pi_0|\psi_1\rangle = q_1$ is the probability of failing to identify $|\psi_1\rangle$, (and similarly for $|\psi_2\rangle$). For unambiguous discrimination we have $\langle\psi_2|\Pi_1|\psi_2\rangle = \langle\psi_1|\Pi_2|\psi_1\rangle = 0$ from (5.33). Using this, we obtain from Eq. (5.34) $p_1 + q_1 = p_2 + q_2 = 1$. This means that if we allow inconclusive detection results to occur with a certain probability then in the remaining cases the observer can conclusively determine the state of the individual system.

It is rather easy to see that a simple von Neumann measurement can accomplish this task. Let us denote the Hilbert space of the two given states by $\mathcal{H}$ and introduce the projector P_1 for $|\psi_1\rangle$ and $\bar{P}_1$ for the orthogonal subspace, such that $P_1 + \bar{P}_1 = I$, the identity in $\mathcal{H}$. Then we know for sure that $|\psi_2\rangle$ was prepared if in the measurement of $\{P_1, \bar{P}_1\}$ a click in the $\bar{P}_1$ detector occurs. A similar conclusion for $|\psi_1\rangle$ can be reached with the roles of $|\psi_1\rangle$ and $|\psi_2\rangle$ reversed. Of course, when a click along P_1 (or P_2) occurs then we learn nothing about which state was prepared, this outcome thus corresponding to the inconclusive result. In the von Neumann set-ups one of the alternatives is missing. We either identify one state or we get an inconclusive result but we miss the other state completely. This scenario is actually allowed by (5.34).

We now turn our attention to the determination of the optimum measurement strategy for unambiguous discrimination. It is the strategy, or measurement set-up, for which the average failure probability is minimum (or, equivalently, the average success probability is maximum). We want to determine the operators in (5.34) explicitly. If we introduce $|\psi_j^\perp\rangle$ as the vector orthogonal to $|\psi_j\rangle$ $(j = 1, 2)$ then the condition of unambiguous detection, Eq. (5.33), mandates the choices

$$\Pi_1 = c_1 |\psi_2^\perp\rangle\langle\psi_2^\perp| , \tag{5.35}$$

and

$$\Pi_2 = c_2 |\psi_1^\perp\rangle\langle\psi_1^\perp| . \tag{5.36}$$

Here c_1 and c_2 are positive coefficients to be determined from the condition of optimum.

Inserting these expressions in the definition of p_1 and p_2 gives $c_1 = p_1/|\langle\psi_1|\psi_2^\perp\rangle|^2$ and a similar expression for c_2. Finally, introducing $\cos\Theta = |\langle\psi_1|\psi_2\rangle|$ and $\sin\Theta = |\langle\psi_1|\psi_2^\perp\rangle|$, we can write the detection operators as

$$\Pi_1 = \frac{p_1}{\sin^2\Theta} |\psi_2^\perp\rangle\langle\psi_2^\perp| ,$$
$$\Pi_2 = \frac{p_2}{\sin^2\Theta} |\psi_1^\perp\rangle\langle\psi_1^\perp| . \tag{5.37}$$

Now, Π_1 and Π_2 are positive semi-definite operators by construction. However, there is one additional condition for the existence of the POVM which is the positivity of the inconclusive detection operator,

$$\Pi_0 = I - \Pi_1 - \Pi_2 . \tag{5.38}$$

This is a simple 2 by 2 matrix in $\mathcal{H}$ and the corresponding eigenvalue problem can be solved analytically. Non-negativity of the eigenvalues leads, after some tedious but straightforward algebra, to the condition

$$q_1 q_2 \geq |\langle\psi_1|\psi_2\rangle|^2 , \tag{5.39}$$

where $q_1 = 1 - p_1$ and $q_2 = 1 - p_2$ are the failure probabilities for the corresponding input states.

Equation (5.39) represents the constraint imposed by the positivity requirement on the optimum detection operators. The task we set out to solve can now be formulated as follows. Let

$$Q = \eta_1 q_1 + \eta_2 q_2 \qquad (5.40)$$

denote the average failure probability for unambiguous discrimination. We want to minimize this failure probability subject to the constraint, Eq. (5.39). Due to the relation, $P = \eta_1 p_1 + \eta_2 p_2 = 1 - Q$, the minimum of Q also gives us the maximum probability of success. Clearly, for optimum the product $q_1 q_2$ should be at its minimum allowed by (5.39), and we can then express q_2 with the help of q_1 as $q_2 = \cos^2 \Theta / q_1$. Inserting this expression in (5.40) yields

$$Q = \eta_1 q_1 + \eta_2 \frac{\cos^2 \Theta}{q_1} \, , \qquad (5.41)$$

where q_1 can now be regarded as the independent parameter of the problem. Optimization of Q with respect to q_1 gives $q_1^{POVM} = \sqrt{\eta_2/\eta_1} \cos \Theta$ and $q_2^{POVM} = \sqrt{\eta_1/\eta_2} \cos \Theta$. Finally, substituting these optimal values into Eq. (5.40) gives the optimum failure probability,

$$Q^{POVM} = 2\sqrt{\eta_1 \eta_2} \cos \Theta \, . \qquad (5.42)$$

Let us next see how this result compares to the average failure probabilities of the two possible unambiguously discriminating von Neumann measurements that were described at the beginning of this section. The average failure probability for the first von Neumann measurement, with its failure direction along $|\psi_1\rangle$, can be written by simple inspection as

$$Q_1 = \eta_1 + \eta_2 |\langle \psi_1 | \psi_2 \rangle|^2 \, , \qquad (5.43)$$

since $|\psi_1\rangle$ gives a click with probability 1 in this direction but it is only prepared with probability η_1 and $|\psi_2\rangle$ gives a click with probability $|\langle \psi_1 | \psi_2 \rangle|^2$ but it is only prepared with probability η_2.

By entirely similar reasoning, the average failure probability for the second von Neumann measurement, with its failure direction along $|\psi_2\rangle$, is given by

$$Q_2 = \eta_1 |\langle \psi_1 | \psi_2 \rangle|^2 + \eta_2 \, . \qquad (5.44)$$

What we can observe is that Q_1 and Q_2 are given as the arithmetic mean of two terms and Q^{POVM} is the geometric mean of the same two terms for either case. So, one would be tempted to say that the POVM performs better always. This, however, is not quite the case, it does so only when it exists. The obvious condition for the POVM solution to exist is that both $q_1^{POVM} \leq 1$ and $q_2^{POVM} \leq 1$. Using $\eta_2 = 1 - \eta_1$, a little algebra tells us that the POVM exists in the range $\cos^2 \Theta / (1 +$

$\cos^2 \Theta) \leq \eta_1 \leq 1/(1 + \cos^2 \Theta)$. If η_1 is smaller than the lower boundary, the POVM goes over to the first von Neumann measurement and if η_1 exceeds the upper boundary the POVM goes over to the second von Neumann measurement. This can be easily seen from Eqs. (5.37) and (5.38) since $p_1 = 1 - q_1 = 0$ for $q_1 = 1$ and Π_0 becomes a projection along $|\psi_1\rangle$ (and correspondingly for $p_2 = 0$).

These findings can be summarized as follows. The optimal failure probability, Q^{opt}, is given as

$$Q^{opt} = \begin{cases} Q^{POVM} & \text{if } \frac{\cos^2 \Theta}{1+\cos^2 \Theta} \leq \eta_1 \leq \frac{1}{1+\cos^2 \Theta}, \\ Q_1 & \text{if } \eta_1 < \frac{\cos^2 \Theta}{1+\cos^2 \Theta}, \\ Q_2 & \text{if } \frac{1}{1+\cos^2 \Theta} < \eta_1. \end{cases} \tag{5.45}$$

The optimum POVM operators are given by

$$\Pi_1 = \frac{1 - q_1^{opt}}{\sin^2 \Theta} |\psi_2^\perp\rangle\langle\psi_2^\perp|,$$

$$\Pi_2 = \frac{1 - q_2^{opt}}{\sin^2 \Theta} |\psi_1^\perp\rangle\langle\psi_1^\perp|. \tag{5.46}$$

These expressions show explicitly that $\Pi_1 = 0$ and Π_2 is the projector $|\psi_1^\perp\rangle\langle\psi_1^\perp|$ when $q_1^{opt} = 1$ and $q_2^{opt} = \cos^2 \Theta$, i.e. the POVM goes over smoothly into a projective measurement at the lower boundary and, similarly, into the other von Neumann projective measurement at the upper boundary.

Figure 5.1 displays the failure probabilities, Q_1, Q_2, and Q^{POVM} vs. η_1 for a fixed value of the overlap, $\cos^2 \Theta$.

Fig. 5.1 Failure probability, Q, vs. the prior probability, η_1. Dashed line: Q_1, dotted line: Q_2, solid line: Q^{POVM}. For the figure we used the following representative value: $|\langle\psi_1|\psi_2\rangle|^2 = 0.1$. For this the optimal failure probability, Q_{opt} is given by Q_1 for $0 < \eta_1 < 0.09$, by Q^{POVM} for $0.09 \leq \eta_1 \leq 0.9$ and by Q_2 for $0.9 < \eta_1$

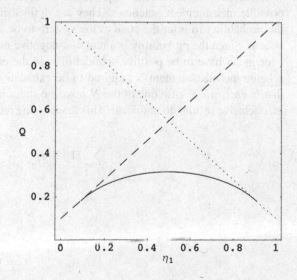

The above result is very satisfying from a physical point of view. The POVM delivers a lower failure probability in its entire range of existence than either of the two von Neumann measurements. At the boundaries of this range it merges smoothly with the one von Neumann measurement that has a lower failure probability at that point. Outside this range the state preparation is dominated by one of the states and the optimal measurement becomes a von Neumann projective measurement, using the state that is prepared less frequently as its failure direction.

5.5.2 *Minimum Error Discrimination of Two Quantum States*

In the previous section we have required that, whenever a definite answer is returned after a measurement on the system, the result should be unambiguous, at the expense of allowing inconclusive outcomes to occur. For many applications in quantum communication, however, one wants to have conclusive results only. This means that errors are unavoidable when the states are non-orthogonal. Based on the outcome of the measurement, in each single case then a guess has to be made as to what the state of the quantum system was. In the optimal strategy we want to minimize the probability of making a wrong guess, hence this procedure is known as *minimum error* discrimination. The problem is to find the optimum measurement that minimizes the probability of errors.

Let us state the optimization problem a little more precisely. In the most general case, we want to distinguish, with minimum probability of error, among N given states of a quantum system (where $N \geq 2$). The states are given by the density operators ρ_j ($j = 1, 2, \ldots, N$) and the jth state occurs with the given a priori probability η_j, such that $\sum_{j=1}^{N} \eta_j = 1$. The measurement can be formally described with the help of a POVM, where the POVM elements, Π_j, correspond to the possible measurement outcomes. They are defined in such a way that $\mathrm{Tr}(\rho \Pi_j)$ is the probability to infer the state of the system to be ρ_j if it has been prepared in a state ρ. Since the probability is a real non-negative number, the detection operators once again have to be positive-semidefinite. In the error-minimizing measurement scheme the measurement is required to be exhaustive and conclusive in the sense that in each single case one of the N possible states is identified with certainty and inconclusive results do not occur. This leads to the requirement

$$\sum_{j=1}^{N} \Pi_j = I_{D_S},$$ (5.47)

where I_{D_S} denotes the identity operator in the D_S-dimensional physical state space of the quantum system. The overall probability P_{err} to make an erroneous guess for any of the incoming states is then given by

$$P_{\text{err}} = 1 - P_{\text{corr}} = 1 - \sum_{j=1}^{N} \eta_j \text{Tr}(\rho_j \Pi_j) \qquad (5.48)$$

with $\sum_j \eta_j = 1$. Here we introduced the probability P_{corr} that the guess is correct. In order to find the minimum-error measurement strategy, one has to determine the POVM that minimizes the value of P_{err} under the constraint given by Eq. (5.47). By inserting these optimum detection operators into Eq. (5.48), the minimum error probability $P_{\text{err}}^{\min} \equiv P_E$ is determined. The explicit solution to the error-minimizing problem is not trivial and analytical expressions have been derived only for a few special cases.

For the case that only two states are given, either pure or mixed, the minimum error probability, P_E, was derived in the mid 70s by Helstrom in the framework of quantum detection and estimation theory. We find it more instructive to start by analyzing the two-state minimum-error measurement with the help of an alternative method that allows us to gain immediate insight into the structure of the optimum detection operators, without applying variational techniques. Starting from Eq. (5.48) and making use of the relations $\eta_1 + \eta_2 = 1$ and $\Pi_1 + \Pi_2 = I_{D_S}$ that have to be fulfilled by the a priori probabilities and the detection operators, respectively, we see that the total probability to get an erroneous result in the measurement is given by

$$P_{\text{err}} = 1 - \sum_{j=1}^{2} \eta_j \text{Tr}(\rho_j \Pi_j) = \eta_1 \text{Tr}(\rho_1 \Pi_2) + \eta_2 \text{Tr}(\rho_2 \Pi_1). \qquad (5.49)$$

This can be alternatively expressed as

$$P_{\text{err}} = \eta_1 + \text{Tr}(\Lambda \Pi_1) = \eta_2 - \text{Tr}(\Lambda \Pi_2), \qquad (5.50)$$

where we introduced the Hermitian operator

$$\Lambda = \eta_2 \rho_2 - \eta_1 \rho_1 = \sum_{k=1}^{D_S} \lambda_k |\phi_k\rangle\langle\phi_k|. \qquad (5.51)$$

Here the states $|\phi_k\rangle$ denote the orthonormal eigenstates belonging to the eigenvalues λ_k of the operator Λ. The eigenvalues are real, and without loss of generality we can number them in such a way that

$$
\begin{aligned}
\lambda_k &< 0 \qquad \text{for} \qquad 1 \le k < k_0, \\
\lambda_k &> 0 \qquad \text{for} \qquad k_0 \le k \le D, \\
\lambda_k &= 0 \qquad \text{for} \qquad D < k \le D_S.
\end{aligned}
\tag{5.52}
$$

By using the spectral decomposition of Λ, we get the representations

$$
P_{\text{err}} = \eta_1 + \sum_{k=1}^{D_S} \lambda_k \langle \phi_k | \Pi_1 | \phi_k \rangle = \eta_2 - \sum_{k=1}^{D_S} \lambda_k \langle \phi_k | \Pi_2 | \phi_k \rangle.
\tag{5.53}
$$

Our optimization task now consists in determining the specific operators Π_1, or Π_2, respectively, that minimize the right-hand side of Eq. (5.53) under the constraint that

$$
0 \le \langle \phi_k | \Pi_j | \phi_k \rangle \le 1 \qquad (j = 1, 2)
\tag{5.54}
$$

for all eigenstates $|\phi_k\rangle$. The latter requirement is due to the fact that $\text{Tr}(\rho \Pi_j)$ denotes a probability for any ρ. From this constraint and from Eq. (5.53) it immediately follows that the smallest possible error probability, $P_{\text{err}}^{\min} \equiv P_E$, is achieved when the detection operators are chosen in such a way that the equations $\langle \phi_k | \Pi_1 | \phi_k \rangle = 1$ and $\langle \phi_k | \Pi_2 | \phi_k \rangle = 0$ are fulfilled for eigenstates belonging to negative eigenvalues, while eigenstates corresponding to positive eigenvalues obey the equations $\langle \phi_k | \Pi_1 | \phi_k \rangle = 0$ and $\langle \phi_k | \Pi_2 | \phi_k \rangle = 1$. Hence the optimum POVM operators can be written as

$$
\Pi_1 = \sum_{k=1}^{k_0-1} |\phi_k\rangle\langle\phi_k|, \qquad \Pi_2 = \sum_{k=k_0}^{D_S} |\phi_k\rangle\langle\phi_k|,
\tag{5.55}
$$

where the expression for Π_2 has been supplemented by projection operators onto eigenstates belonging to the eigenvalue $\lambda_k = 0$, in such a way that $\Pi_1 + \Pi_2 = I_{D_S}$. Obviously, provided that there are positive as well as negative eigenvalues in the spectral decomposition of Λ, the minimum-error measurement for discriminating two quantum states is a von Neumann measurement that consists in performing projections onto the two orthogonal subspaces spanned by the set of states $\{|\phi_1\rangle, \ldots, |\phi_{k_0-1}\rangle)\}$, on the one hand, and $\{|\phi_{k_0}\rangle, \ldots, |\phi_{D_S}\rangle)\}$, on the other hand. An interesting special case arises when negative eigenvalues do not exist. In this case it follows that $\Pi_1 = 0$ and $\Pi_2 = I_{D_S}$ which means that the minimum error probability can be achieved by always guessing the quantum system to be in the state ρ_2, without performing any measurement at all. Similar considerations hold true in the absence of positive eigenvalues so a measurement does not always aid minimum-error discrimination. By inserting the optimum detection operators into Eq. (5.50) the minimum error probability is found to be

$$P_E = \eta_1 - \sum_{k=1}^{k_0-1} |\lambda_k| = \eta_2 - \sum_{k=k_0}^{D} |\lambda_k|. \tag{5.56}$$

Taking the sum of these two alternative representations and using $\eta_1 + \eta_2 = 1$, we arrive at

$$P_E = \frac{1}{2}\left(1 - \sum_k |\lambda_k|\right) = \frac{1}{2}\left(1 - \text{Tr}|\Lambda|\right), \tag{5.57}$$

where $|\Lambda| = \sqrt{\Lambda^\dagger \Lambda}$. Together with Eq. (5.48) this immediately yields the well-known Helstrom formula for the minimum error probability in discriminating ρ_1 and ρ_2,

$$P_E = \frac{1}{2}\left(1 - \text{Tr}|\eta_2\rho_2 - \eta_1\rho_1|\right) = \frac{1}{2}\left(1 - \|\eta_2\rho_2 - \eta_1\rho_1\|_1\right). \tag{5.58}$$

In the special case that the states to be distinguished are the pure states $|\psi_1\rangle$ and $|\psi_2\rangle$, this expression reduces to

$$P_E = \frac{1}{2}\left(1 - \sqrt{1 - 4\eta_1\eta_2|\langle\psi_1|\psi_2\rangle|^2}\right). \tag{5.59}$$

This expression, which is the one found in textbooks, can be cast to the equivalent form,

$$P_E = \eta_{min}\left(1 - \frac{2\eta_{max}(1 - |\langle\psi_1|\psi_2\rangle|^2)}{\eta_{max} - \eta_{min} + \sqrt{1 - 4\eta_{min}\eta_{max}|\langle\psi_1|\psi_2\rangle|^2}}\right), \tag{5.60}$$

where η_{min} (η_{max}) is the smaller (greater) of the prior probabilities, η_1 and η_2. This form lends itself to a transparent interpretation. The first factor on the right-hand-side is what we would get if we always guessed the state that is prepared more often, without any measurement at all. Thus, the factor multiplying η_{min} is the result of the optimized measurement.

The set-up of the detectors that achieve the optimum error probabilities is particularly simple for the case of equal a priori probabilities. Two orthogonal detectors, placed symmetrically around the two pure states, will do the task. The simplicity is particularly striking when one compares this set-up to the corresponding POVM set-up for optimal unambiguous discrimination.

Finally, we present an interesting relation, without proof, that is always satisfied between the minimum-error probability of the minimum-error detection and the optimal failure probability of unambiguous detection. It reads as

$$P_E \leq \frac{1}{2} Q^{opt}. \tag{5.61}$$

This means that for two arbitrary states (mixed or pure), prepared with arbitrary a priori probabilities, the smallest possible failure probability in unambiguous discrimination is at least twice as large as the smallest probability of errors in minimum-error discrimination of the same states.

5.6   Sequential Measurements

So far, we have only discussed situations in which a system is measured, and that is the end of the story. The formalism, however, provides us with the state of the system after the measurement, which is, in general, different from the pre-measurement state. This allows us to analyze a situation in which the same system is measured several times. One possible use of measuring the system multiple times is to try to gain more information about it. This leads to the question, is it possible to gain further information about the initial state of a system by measuring it more than once? We will provide an example to show that the answer to this question can be "yes". Another reason for making multiple measurements on the same system is that we might want different parties, each of whom makes a measurement, to gain information about the initial state of the system. We will show how unambiguous state discrimination can be adapted to achieve this.

5.6.1   Simple Example

Suppose Alice is given a qubit in the state $|\psi_{in}\rangle$. She then measures it in the basis

$$|\psi(\theta, \phi)\rangle = \cos(\theta/2)|0\rangle + e^{i\phi}\sin(\theta/2)|1\rangle,$$
$$|\psi_\perp(\theta, \phi)\rangle = -\sin(\theta/2)|0\rangle + e^{i\phi}\cos(\theta/2)|1\rangle, \tag{5.62}$$

for some $0 \leq \theta \leq \pi$ and $0 \leq \phi \leq 2\pi$. Clearly, since she is measuring the qubit in the state $|\psi_{in}\rangle$ directly, she will gain information about the initial state. She then passes the qubit on to Bob without telling him anything. In particular, Bob does not know the value of either θ or ϕ or the result of Alice's measurement. Can Bob gain any information about $|\psi_{in}\rangle$, even though he does not know what measurement Alice made or her result?

In order to find out, let's find the density matrix of the qubit after Alice has measured it. With probability $|\langle\psi_{in}|\psi(\theta, \phi)\rangle|^2$ the state will be $|\psi(\theta, \phi)\rangle$ and with probability $|\langle\psi_{in}|\psi_\perp(\theta,\phi)\rangle|^2$ it will be $|\psi_\perp(\theta,\phi)\rangle$. Bob does not know the basis Alice used to measure, so we must average over the angles. Therefore, Bob's density matrix is

$$\rho_b = \frac{1}{4\pi}\int_0^{2\pi} d\phi \int_0^\pi d\theta \sin\theta [|\langle\psi_{in}|\psi(\theta,\phi)\rangle|^2 |\psi(\theta,\phi)\rangle\langle\psi(\theta,\phi)|$$

$$+ |\langle\psi_{in}|\psi_\perp(\theta,\phi)\rangle|^2 |\psi_\perp(\theta,\phi)\rangle\langle\psi_\perp(\theta,\phi)|],$$

$$= \frac{1}{3}|\psi_{in}\rangle\langle\psi_{in}| + \frac{1}{3}I. \tag{5.63}$$

This density matrix describes a noisy version of the initial state of the qubit, so, despite Alice's unknown measurement, it does contain information about $|\psi_{in}\rangle$. Consequently, by measuring the qubit, Bob can gain information about its initial state.

5.6.2 Sequential Unambiguous Discrimination

Now that we have seen that information about the initial state can survive a measurement, we will see how that can work in unambiguous state discrimination. We would like more than one party to be able to determine which of two states was sent by having them measure the same qubit. In particular, Alice prepares a qubit in either $|\psi_1\rangle$ or $|\psi_2\rangle$ and sends it to Bob. Bob performs an unambiguous discrimination measurement on the qubit, and sends it on to Charlie, who also performs an unambiguous discrimination measurement on the qubit. We want both Bob and Charlie to have a nonzero chance of identifying the state, and the probability of both of them succeeding to be a maximum. We will assume that no classical communication can take place between Bob and Charlie after Bob performs his measurement. That means that Charlie never knows whether Bob's measurement succeeded or failed. The key to making this procedure work is twofold. First, the state discrimination Bob performs cannot be optimal, otherwise he would have extracted all of the quantum information carried by the qubit, and there would be none left for Charlie to measure. Second, the states Bob sends to Charlie must be independent of whether Bob's measurement succeeded or failed.

To begin we assume that Alice prepares qubits in $|\psi_1\rangle$ or $|\psi_2\rangle$ with equal probability. Without loss of generality, the overlap of the two possible states, $s = \langle\psi_1|\psi_2\rangle$ is taken to be real $(0 \le s \le 1)$ and we choose the phase of $|\psi_1^\perp\rangle$, the vector orthogonal to $|\psi_1\rangle$, so that

$$|\psi_2\rangle = s|\psi_1\rangle + \sqrt{1-s^2}|\psi_1^\perp\rangle$$

$$|\psi_2^\perp\rangle = \sqrt{1-s^2}|\psi_1\rangle - s|\psi_1^\perp\rangle. \tag{5.64}$$

Both Bob's and Charlie's measurements are described by POVM's. Each POVM has three elements, one, Π_1, corresponding to the detection of $|\psi_1\rangle$, the second, Π_2, corresponding to the detection of $|\psi_2\rangle$, and the third, Π_0, corresponding to the failure of the measurement. Each element is a positive operator on the two-dimensional qubit Hilbert space, and their sum is the identity operator.

The requirement that errors are not allowed mandates that the POVM elements describing Bob's measurement are of the form $\Pi_1^B = c_1|\psi_2^\perp\rangle\langle\psi_2^\perp|$ and $\Pi_2^B = c_2|\psi_1^\perp\rangle\langle\psi_1^\perp|$ for the conclusive outcomes and

$$\Pi_0^B = I - \Pi_1^B - \Pi_2^B \tag{5.65}$$

for the inconclusive one, since the three elements add to the identity. Here c_1 and c_2 are positive constants yet to be determined, subject to the constraint $\Pi_0 \geq 0$. Π_1 and Π_2 are positive by construction.

The probability that Bob unambiguously detects $|\psi_i\rangle$ if it is sent is given by $p_i = \langle\psi_i|\Pi_i^B|\psi_i\rangle$, for $i = 1, 2$ and the probability that the measurement fails if $|\psi_i\rangle$ is sent is given by $q_i = \langle\psi_i|\Pi_0^B|\psi_i\rangle$. Note that the probability that $|\psi_j\rangle$ is detected if $|\psi_i\rangle$ is sent is zero for $i \neq j$, so $p_i + q_i = 1$. These relations allow us to express c_i in terms of the more physical success and failure probabilities,

$$c_i = \frac{p_i}{1 - s^2} = \frac{1 - q_i}{1 - s^2}. \tag{5.66}$$

We will have to know the states after Bob's measurement, since they will be the input states for Charlie's measurement. They can be expressed in terms of the detection operators A_j that are related to the corresponding POVM elements by $\Pi_j^B = A_j^\dagger A_j$ for $j = 0, 1, 2$. If $|\psi_i\rangle$ is the state before the measurement, then if we obtain the result i for the measurement ($i = 1, 2$ success), the post-measurement state (success state) $|\phi_i\rangle$ is given by

$$|\phi_i\rangle = \frac{A_i|\psi_i\rangle}{\|A_i\psi_i\|}, \tag{5.67}$$

and if we obtain the result 0 for the measurement, the post-measurement state (failure state) $|\chi_i\rangle$ is given by

$$|\chi_i\rangle = \frac{A_0|\psi_i\rangle}{\|A_0\psi_i\|}, \tag{5.68}$$

The operators A_j can be chosen in the form $A_j = U_j(\Pi_j^B)^{1/2}$, where U_j can be any unitary operator. Thus, we have quite a bit of freedom in choosing these operators and, consequently, Bob's post-measurement states. In our case they can be expressed as $A_1 = \sqrt{c_1}|\phi_1\rangle\langle\psi_2^\perp|$ and $A_2 = \sqrt{c_2}|\phi_2\rangle\langle\psi_1^\perp|$.

We can now see what happens after Bob's measurement. If Alice sent $|\psi_i\rangle$, then Bob will send Charlie the state $|\phi_i\rangle$ with probability p_i or the state $|\chi_i\rangle$ with probability q_i. However, it is the case that for unambiguous discrimination to be possible, the states to be discriminated must be linearly independent, and since we are in a two-dimensional space, Charlie can only discriminate between two possible pure states. This implies, since Charlie will not know whether Bob's measurement succeeded or failed, that we must have $|\phi_i\rangle = |\chi_i\rangle$ which, in turn, implies

$$A_0 = \sqrt{a_1}|\phi_1\rangle\langle\psi_2^\perp| + \sqrt{a_2}|\phi_2\rangle\langle\psi_1^\perp| \,, \tag{5.69}$$

where a_1 and a_2 are constants to be determined. Therefore, if Alice sent $|\psi_1\rangle$, Charlie will receive $|\phi_1\rangle$, whether Bob's measurement succeeded or not, and if Alice sent $|\psi_2\rangle$, Charlie will receive $|\phi_2\rangle$, again whether Bob's measurement succeeded or not. Charlie's task, then, is to optimally discriminate between $|\phi_1\rangle$ and $|\phi_2\rangle$. Further, since $\langle\psi_i|A_0^\dagger A_0|\psi_i\rangle = q_i$, we have that

$$a_i = q_i/(1 - s^2) \,. \tag{5.70}$$

We now have two different expressions for Π_0, Eq. (5.65) and $A_0^\dagger A_0$ from Eq. (5.69), so we still have to check their compatibility. In the $\{|\psi_1\rangle, |\psi_1^\perp\rangle\}$ basis the operator Π_0^B, Eq. (5.65), takes the form

$$\Pi_0^B = \begin{pmatrix} 1 - c_1 + c_1 s^2 & c_1 s\sqrt{1 - s^2} \\ c_1 s\sqrt{1 - s^2} & 1 - c_1 s^2 - c_2 \end{pmatrix} \,. \tag{5.71}$$

It is easy to obtain the eigenvalues and corresponding eigenvectors explicitly. For our purposes, however, the conditions of non-negativity of Π_0, $\text{Tr}(\Pi_0) = 2 - c_1 - c_2 \geq 0$ and $\det \Pi_0 = 1 - c_1 - c_2 + c_1 c_2 (1 - s^2) \geq 0$, are more useful. The second is the stronger of the two conditions. When it is satisfied the first one is always met. Using Eq. (5.66), the condition on the failure probabilities takes the form,

$$1 \geq q_1 q_2 \geq s^2 \,. \tag{5.72}$$

If we now calculate $\Pi_0 = A_0^\dagger A_0$ from Eq. (5.69) with a_i from (5.70), we find that the two expressions agree if

$$q_1 q_2 = \frac{s^2}{t^2} \tag{5.73}$$

where we introduced $\langle\phi_1|\phi_2\rangle \equiv t$, which we can assume is real and positive. The condition Eq. (5.73) is clearly compatible with Eq. (5.72) provided $t = \langle\phi_1|\phi_2\rangle \geq s = \langle\psi_1|\psi_2\rangle$.

The emerging picture is now the following. Bob extracts some information about the two possible inputs, $|\psi_1\rangle$ and $|\psi_2\rangle$. By doing so he produces states with a greater overlap, $t > s$, that is to say the resulting states are harder to distinguish. Charlie's task, then, is to optimally discriminate between $|\phi_1\rangle$ and $|\phi_2\rangle$. Since an optimized measurement extracts all of the remaining information, Charlie's post-measurement states can carry no further information about the initial preparation so for all inputs and outcomes they are collapsed to the same common state. The failure probabilities for Bob's measurement must satisfy the constraint given by Eq. (5.73). Charlie's failure probabilities must satisfy an entirely similar constraint that we can most easily obtain by replacing s with t and t with 1 in (5.73), since we notice that for his measurement t is the overlap of the input states and the overlap of the post-measurement states is 1. The two constraints are given together as [upper index B (C): Bob (Charlie)]

$$q_1^B q_2^B = \frac{s^2}{t^2}, \qquad q_1^C q_2^C = t^2. \tag{5.74}$$

Let us now examine the probability of both measurements succeeding. If $|\psi_1\rangle$ is sent, the joint probability of success is $P_1 = p_1^B p_1^C = (1 - q_1^B)(1 - q_1^C)$ and if $|\psi_2\rangle$ is sent, the joint success probability is $P_2 = p_2^B p_2^C = (1 - q_2^B)(1 - q_2^C)$, so the average joint success probability is

$$P_S = \frac{1}{2}[(1 - q_1^B)(1 - q_1^C) + (1 - q_2^B)(1 - q_2^C)]. \tag{5.75}$$

since each state has a prior probability of $1/2$.

This is the quantity we want to optimize under the two constraints given in (5.74). We will also impose the condition that the failure probabilities for both states be the same, i.e. $q_1^B = q_2^B$ and $q_1^C = q_2^C$. The optimization is now straightforward and can be done by, e.g., using the method of Lagrange multipliers, with the result $q_1^B = q_2^B = q_1^C = q_2^C = \sqrt{s}$ and $t = \sqrt{s}$. Using the optimal values in (5.75), we finally obtain

$$P_S^{(opt)} = (1 - \sqrt{s})^2. \tag{5.76}$$

From this equation we see that there is a nonzero probability that both Bob and Charlie can find out, without error, which state Alice sent. Thus a sequential measurement can be used to distribute quantum information among several parties. The cost is that the success probability is reduced. If there is only Bob, he can make an optimal unambiguous state discrimination measurement, and his probability of success is $1 - s$. With both Bob and Charlie present, the success probability is reduced to $(1 - \sqrt{s})^2$.

5.7 Problems

1. Find the eigenvalues of the POVM element, given in Eq. (5.38), corresponding to inconclusive outcomes and show that the condition of their positivity can be cast to the form given in Eq. (5.39).
2. For optimum unambiguous discrimination between two pure quantum states, the POVM elements are given in Eq. (5.46) and by $\Pi_0 = I - \Pi_1 - \Pi_2$. Find an implementation in terms of a generalized measurement via Neumark's theorem, introducing the ancilla by using the tensor product extension of the Hilbert space.
3. The derivation of the general formula for the minimum error probability is given in the text.

   (a) Show that for the special case of two pure states, $\rho_1 = |\psi_1\rangle\langle\psi_1|$ and $\rho_2 = |\psi_2\rangle\langle\psi_2|$, Eq. (5.58) reduces to Eq. (5.59).

   (b) The general expression for the optimal detection operators is given in Eq. (5.55). Find their explicit expression for the pure state case of part (a).

4. (a) Show that Q^{POVM} in (5.42) and P_E in (5.59) satisfy the inequality (5.61).

   (b) If you are very ambitious, prove the inequality.

5. (a) Let us consider the so-called trine states

$$|\psi_1\rangle = |0\rangle \quad |\psi_2\rangle = -\tfrac{1}{2}(|0\rangle + \sqrt{3}|1\rangle)$$

$$|\psi_3\rangle = -\tfrac{1}{2}(|0\rangle - \sqrt{3}|1\rangle).$$

These are states of a single qubit. We are given a qubit that is guaranteed to be in one of these three states, and we want to find a POVM that does the following. If we obtain result 1 (corresponding to operators A_1 and $A_1^\dagger$), then we know that the qubit we were given was not in state $|\psi_1\rangle$. If we get result 2, it was not in state $|\psi_2\rangle$, and if we get result 3, then it was not in state $|\psi_3\rangle$. Find a POVM that does this.

   (b) Now let us look at the four states (the tetrad states)

$$|\psi_1\rangle = \frac{1}{\sqrt{3}}(-|0\rangle + \sqrt{2}e^{-2\pi i/3}|1\rangle) \quad |\psi_2\rangle = \frac{1}{\sqrt{3}}(-|0\rangle + \sqrt{2}e^{2\pi i/3}|1\rangle)$$

$$|\psi_3\rangle = \frac{1}{\sqrt{3}}(-|0\rangle + \sqrt{2}|1\rangle) \qquad |\psi_4\rangle = |0\rangle.$$

We want to consider the minimum-error detection scenario for these states. That is, we are given a qubit in one of these four states, and we want to find which one, with the requirement that our probability of making a mistake is the smallest possible. The POVM that accomplishes this is given by the operators $A_j = (1/\sqrt{2})|\psi_j\rangle\langle\psi_j|$, where $j = 1, \ldots 4$. Verify that

$$\sum_{j=1}^{4} A_j^\dagger A_j = I,$$

and find the probability that a state will be correctly identified. Also find the probability that an error will be made, that is that we are given $|\psi_j\rangle$ but we identify it as $|\psi_{j'}\rangle$, where $j \neq j'$.

6. When we derived POVM's we used a Hilbert space that was a tensor product between the space for the system we wanted to measure and the space for an ancilla. It is also possible to derive a POVM by considering the direct sum of two Hilbert spaces. In particular, if we are measuring states that are confined to a subspace of a larger space, then we can describe projective measurements on the entire space as POVM's on the subspace. Let us see how this works by means of an example.

(a) Consider again the trine states, but now let us suppose that they are states of a qutrit rather than of a qubit. The entire Hilbert space, $\mathcal{H}_3$, has the orthonormal basis, $\{|0\rangle, |1\rangle, |2\rangle\}$, and the subspace, S, in which the trine states lie, is spanned by the basis elements $|0\rangle$ and $|1\rangle$. The POVM operators for the minimum error scenario are given by $A_j = \sqrt{2/3}\,|\psi_j\rangle\langle\psi_j|$ (the states $|\psi_j\rangle$, for $j = 1, 2, 3$ are given in part (a) of problem 1). Find one-dimensional projections P_j, acting on $\mathcal{H}_3$ that satisfy

$$\langle\psi|P_j|\psi\rangle = \langle\psi|A_j^\dagger A_j|\psi\rangle,$$

for any state $|\psi\rangle \in S$.

(b) Suppose that we want to measure the projections P_j, and that we can easily measure the projections corresponding to the basis states $\{|0\rangle, |1\rangle, |2\rangle\}$. We can then measure the projections P_j by measuring the projections $|j\rangle\langle j|$ if we can find a unitary transformation, U, such that $|j\rangle\langle j| = U P_j U^{-1}$. This implies that

$$|\langle j|U|\psi\rangle|^2 = \langle\psi|P_j|\psi\rangle,$$

so that the probability of measuring $|j\rangle$ in the transformed state is the same as that of measuring P_j in the original state. Find such a unitary operator in this case.

Further Reading

1. S.M. Barnett, S. Croke, Quantum state discrimination. Adv. Opt. Phot. **1**, 238 (2009)
2. J.A. Bergou, Tutorial review: discrimination of quantum states. J. Mod. Opt. **57**, 160 (2010)
3. J. Bergou, E. Feldman, M. Hillery, Extracting information from a qubit by multiple observers: toward a theory of sequential state discrimination. Phys. Rev. Lett. **111**, 100501 (2013)
4. J. Preskill, Lecture notes for Physics 219. http://www.theory.caltech.edu/people/preskill/ph229/
5. P. Rapčan, J. Calsamiglia, R. Muñoz-Tapia, E. Bagan, V. Bužek, Scavenging quantum information: multiple observations of quantum systems. Phys. Rev. A **84**, 032326 (2012)

Chapter 6
Quantum Cryptography

6.1 Outline

Quantum communication is the most advanced area of quantum information processing and quantum computing. This is where the most fundamental features of quantum mechanics are only a short step away from spectacular practical applications. We have already seen two such applications: Dense coding and teleportation. In this chapter we shall deal with what is arguably the most successful area of all of quantum information and quantum computing: Quantum cryptography.

Cryptography is the art of secret communication. It has been around since ancient times. What distinguishes quantum cryptography from classical one is that classical information can be copied at will. Therefore, no classical cryptographic protocol is entirely secure, although there are classical cryptographic protocols that are very hard to break in practice. Quantum information, on the other hand, i.e., unknown quantum states, cannot be cloned (cf. the no-cloning theorem). Quantum cryptography enables two parties, traditionally called Alice and Bob, to exchange information in a provably secure way. The security stems from the fact that if an eavesdropper, traditionally called Eve, tries to intercept the messages, her presence can be detected by the unavoidable disturbance she causes by trying to access the information.

In the next section of this chapter we give a very brief introduction to the ideas behind classical cryptography. At the heart of any provably secure cryptographic protocol lies the process of establishing a secret key. Quantum key distribution (QKD) solves this problem using fundamental principles of quantum mechanics. Alice and Bob can then use the secret key to encode and decode their messages. In the following two sections we will briefly describe a number of QKD protocols that clearly show what fundamental features of quantum mechanics are being used as resources.

© Springer Nature Switzerland AG 2021

J. A. Bergou et al., *Quantum Information Processing*, Graduate Texts in Physics,
https://doi.org/10.1007/978-3-030-75436-5_6

Quantum cryptography is now a highly developed subject, and what we will present here just scratches the surface. Our intent is to provide an introduction to some of the basic ideas on which the subject is based.

6.2   The One Time Pad

The first documented cases of secret communication date back about thirty centuries. Since then its history can be described as the ongoing struggle between code makers and code breakers. Sometimes code makers outsmart the code breakers, sometimes the code breakers are ahead in the game. With quantum mechanics, code makers finally seem to be gaining the upper hand.

To be precise, the word code refers to the particular kind of secret communication where a word or even a full sentence is replaced with a word, a number or a symbol. Initially very popular, its use has decreased over time to give way to the cipher, which acts at the level of the smallest building blocks: the letters. In a cipher letters are being replaced by letters, numbers or symbols.

If letters of a message are simply rearranged, we are talking about transposition. The alternative to transposition is substitution when letters are being substituted. An early example for a substitution cipher is the *Caesar shift*, used by Julius Caesar for the purpose of his military correspondence. In this cipher each letter in the message is simply replaced with the letter that is three places further down the alphabet. So, for example, *Caesar* becomes *Fdhvdu*.

This code is very easy to break. We can make things a little more difficult for an eavesdropper if we use different shifts for different messages, that is we do not always shift by three. Then, in addition to the message, the shift has to be sent to the receiver. This is the simplest example of a key. The key, in this case the shift, enables the receiver to decrypt the message. Things get much harder for the eavesdropper if we use different shifts for different letters in a single message. This, of course, makes the key much longer; rather than one number (shift) for each message, we have one for each letter in a message. The advantage, however, is that if we have a random key, and only use it once, the code is unbreakable. This procedure is known as a one-time pad.

This simple example for a cipher shows two distinct ingredients of encryption: the algorithm and the key. The algorithm specifies the encryption and decryption procedures, but in order to use it, one has to have a key known by both the sender and receiver. The algorithm can be publicly known, so that the security of the system depends on restricting knowledge of the key. Thus, a major problem in cryptography is how to distribute a secure key to only the legitimate users. This is the problem of key distribution. One can resort to couriers with briefcases handcuffed to their wrists, but in the electronic age, something more sophisticated is needed. What we will show in this chapter, is that quantum mechanics can be used to distribute secure keys. The resulting prescriptions are known as quantum key distribution protocols.

6.3 The B92 Quantum Key Distribution Protocol

Both of the state discrimination strategies discussed in the previous chapter come very nicely together in the so called B92 quantum key distribution (QKD) protocol. In 1992 Charles Bennett proposed using the unambiguous discrimination of two nonorthogonal states as the basis of a form of quantum cryptography. Quantum cryptography is a method of generating a secure shared key by quantum mechanical means that is discarded after being used only once. So, it is the quantum version of the one-time pad cipher.

In cryptography the sender is often called Alice and the receiver Bob. We will use this nomenclature in what follows. As we have seen, the question is, then, how to generate, or distribute, a secret key between Alice and Bob. The B'92 protocol provides one possible solution to this problem, using quantum mechanical means. Here is how it works.

1. Alice generates a random sequence of 0's and 1's, that is a random classical bit string.
2. Alice encodes each data bit in a qubit, $|\psi_0\rangle = |0\rangle$ if the corresponding bit is 0, and $|\psi_1\rangle = \frac{1}{\sqrt{2}}(|0\rangle + |1\rangle)$ if the corresponding bit is 1. This way she generates a random string of qubits.
3. Alice then sends the resulting string of qubits to the receiver, Bob.
4. Bob applies optimum unambiguous state discrimination strategy to each qubit he receives. Using Eq. (5.42), the success probability for Bob's measurement is $P = 1 - Q^{POVM} = 1 - \frac{1}{\sqrt{2}} \approx 0.293$.
   From now on, Alice and Bob will exchange only classical information.
5. Bob tells Alice, over a public classical channel, in which instances the discrimination succeeded but not the result.
6. They keep only those bits when the discrimination was successful, and delete those when it failed. After this they share the so-called *raw key*.

The raw key is the same for Alice and Bob, since Alice knows what she sent in those instances when Bob successfully identified the state of the qubit and Bob was using unambiguous discrimination, so there is no error. This is true as long as an eavesdropper and noise are both absent.

So, why is this procedure secure if there is an eavesdropper? Suppose the eavesdropper, called Eve, has intercepted a qubit. She cannot determine whether it is in the state $|\psi_0\rangle$ or $|\psi_1\rangle$. One thing she can do is to apply the optimum unambiguous state discrimination procedure. Then she will fail with a probability of $\frac{1}{\sqrt{2}} \approx 71\%$. When she does, she has no idea what state was sent, so she must guess which one to send to Bob. Since the two states are prepared with equal probability, Eve will guess half the time right and half the time wrong. This means that the probability that Bob will receive a wrong bit is $\frac{1}{2\sqrt{2}} \approx 35.3\%$. These errors can easily be detected if Alice and Bob add one more step to their protocol.

7'. Alice and Bob publicly compare some of their bits. If there are no errors there is no eavesdropper and they keep the remaining bits. If there are errors, in the range of 35%, there is likely to be an eavesdropper. They then simply throw out all bits and try again.

But wait a second. Eve's goal, besides learning as much as possible about the key, is also to introduce as few errors as possible. There are unavoidable errors in any communication scheme, partly due to the imperfections of the communication channel and partly due to the imperfect detection. Eve's goal is to remain below this unavoidable noise level in order to avoid being detected. So, suppose she has intercepted the particle but she now chooses the minimum error strategy to determine which state was sent. Using Eq. (5.59), her error rate will now be $\frac{1}{2}(1 - \frac{1}{\sqrt{2}}) \approx 14.6\%$ which is much less than the error rate that she introduces if she uses the unambiguous discrimination strategy. In addition, she still learns the key with a fidelity of about 85%. However, even this rather low error rate can still be detected if Alice and Bob modify the last step of their protocol.

7. Over a classical communication channel, Alice and Bob publicly compare some of their bits. If there are no errors there is no eavesdropper and they keep the remaining bits. If there are errors, in the range of 14%, there is likely to be an eavesdropper. They then simply throw out all bits and try again.

This requirement is much more stringent than the one in Step 7 of the original protocol. It is still possible to detect the presence of an eavesdropper but the requirements on the channel quality and detector efficiency are much more demanding than in the case when Eve uses the same strategy as Bob. So, here we had an example where one state discrimination strategy is optimal for the intended recipient and the other for the eavesdropper and to analyze the worst case scenario for Alice and Bob we have to consider all of their possibilities. There are many other QKD protocols but this one is perhaps the clearest example of how important optimal detection strategies are for quantum communication.

6.4 The BB 84 Protocol

The first, and most famous, quantum key distribution protocol was developed by Bennett and Brassard in 1984 [2], and is known as the BB 84 protocol. It is actually possible to buy commercial quantum cryptography systems that make use of this protocol. It is a four-state protocol in which Alice and Bob make use of two sets of bases to establish a shared key.

Alice sends qubits to Bob, each state being a member of one of two sets of orthonormal bases, the z basis, $\{|0\rangle, |1\rangle\}$ or the x basis $\{|+x\rangle, |-x\rangle\}$, where $|\pm x\rangle = (|0\rangle \pm |1\rangle)/\sqrt{2}$. She decides which state to send at random, i.e. she chooses a basis at random and a state from that basis at random. The states $|0\rangle$ and $|+x\rangle$ correspond to a bit value of 0 and $|1\rangle$ and $|-x\rangle$ correspond to a bit value of 1. Upon receiving the qubit, Bob measures it in one of the two bases, choosing which basis to use at random. If he uses the same basis as the one Alice chose, he will obtain the same state that Alice sent. For example, if Alice sends $|0\rangle$ and Bob measures in the z basis, he will obtain $|0\rangle$. If, however, Bob chooses the wrong basis, his results will be random. If Alice sent $|0\rangle$ and Bob measures in the x basis, he will obtain $|+x\rangle$ with a probability of $1/2$ and $|-x\rangle$ with a probability of $1/2$. After measuring a qubit, Bob announces over a public channel which basis he used, but not the result of the measurement. Alice then tells Bob whether the basis he used was the same as the one she chose. If they agree, they keep the bit value corresponding to that qubit. If they disagree, they throw out that bit.

In the intercept-resend attack, the eavesdropper, Eve, captures the qubit that Alice sent, measures it, and then, based on her measurement result, prepares another qubit to send on to Bob. Her problem is that she does not know in which basis to measure Alice's qubit, so she has to guess. If she guesses correctly, and Alice and Bob use the same basis, she knows the value of that key bit, and she has not been detected. However, with a probability of $1/2$ she will guess incorrectly, and measure the qubit in the wrong basis, and obtain a random result, which she will then use to prepare a qubit to send to Bob. For example, suppose Alice sends $|0\rangle$ but Eve measures in the x basis. She will, with a probability of $1/2$, obtain $|+x\rangle$ and send that on to Bob, and will obtain $|-x\rangle$, also with a probability of $1/2$, and send that on to Bob. Now suppose that Bob chooses the same basis as Alice, in this case the z basis. In either case, whether Eve sent $|+x\rangle$ or $|-x\rangle$, he will obtain $|0\rangle$ with a probability of $1/2$ and $|1\rangle$ with a probability of $1/2$. If he obtains $|0\rangle$, then Eve's intervention goes undetected, and she knows the value of the bit. Howerver, if Bob obtains $|1\rangle$, when he should have obtained $|0\rangle$ had there been no eavesdropper, Eve's presence will be revealed. Consequently, Eve will introduce errors in the case when Alice and Bob use the same basis. This will happen with a probability of $1/4$; a probability of $1/2$ that Eve chooses the wrong basis times a probability of $1/2$ that if she does, Bob obtains a measurement result different from what Alice sent. Alice and Bob can detect these errors by publicly comparing a subset of the bits for which they chose the same bases. If there are no errors, there was no eavesdropping, but if there are errors, there was an eavesdropper present. In that case they throw out all of the bits and start over.

Eve can try a different kind of attack in which she entangles the incoming qubit with an ancilla. When she receives Alice's qubit, she appends an ancilla qubit to it in the state $|0\rangle$ and then applies a two-qubit unitary operation, U, which acts as

$$U|0\rangle_a|0\rangle_e = |0\rangle_a|\phi_{00}\rangle_e + |1\rangle_a|\phi_{01}\rangle_e$$
$$U|1\rangle_a|0\rangle_e = |0\rangle_a|\phi_{10}\rangle_e + |1\rangle_a|\phi_{11}\rangle_e, \tag{6.1}$$

where the subscript a designates Alice's qubit and the subscript e designates Eve's qubit. Because U is unitary, the states of Eve's qubit must satisfy

$$\|\phi_{00}\|^2 + \|\phi_{01}\|^2 = 1 \qquad \|\phi_{10}\|^2 + \|\phi_{11}\|^2 = 1$$

$$\langle \phi_{00}|\phi_{10}\rangle + \langle \phi_{01}|\phi_{11}\rangle = 0. \tag{6.2}$$

After entangling her qubit with Alice's, Eve sends Alice's qubit on to Bob.

We will not do a complete analysis of this attack, but we will show that if Eve is to introduce no errors, then she can obtain no information. If no errors are to be produced when Bob measures in the z basis, then we must have $|\phi_{01}\rangle = |\phi_{10}\rangle = 0$. Now let us see what happens in the x basis. First, we have that

$$U|\pm x\rangle_a|0\rangle_e = \frac{1}{\sqrt{2}}[|0\rangle_a(|\phi_{00}\rangle_e \pm |\phi_{10}\rangle_e)$$

$$+|1\rangle_a(|\phi_{01}\rangle_e \pm |\phi_{11}\rangle_e)]. \tag{6.3}$$

Now if there are to be no errors when Bob measures in the x basis, then when Alice's qubit is $|+x\rangle_a$ the right-hand side of the above equation must be proportional to $|+x\rangle$, which implies that

$$|\phi_{00}\rangle_e + |\phi_{10}\rangle_e = |\phi_{01}\rangle_e + |\phi_{11}\rangle_e, \tag{6.4}$$

and when Alice sends $|-x\rangle_a$, the state after applying U should be proportional to $|-x\rangle$, which implies that

$$|\phi_{00}\rangle_e - |\phi_{10}\rangle_e = -(|\phi_{01}\rangle_e - |\phi_{11}\rangle_e). \tag{6.5}$$

Combining these conditions with the one we obtained from the z basis, $|\phi_{01}\rangle = |\phi_{10}\rangle = 0$, we see that we must also have $|\phi_{00}\rangle = |\phi_{11}\rangle$. These two conditions, however, imply that Eve's qubit is not entangled with Alice's qubit at all, and, therefore, Eve transfers no information about Alice's qubit to her ancilla qubit. Thus, if Eve is to introduce no errors, she will gain no information.

6.5 The E91 Protocol and Device-Independent Key Distribution

In 1991 Artur Ekert proposed a protocol based on shared entanglement rather than on one party sending particles directly to another. Suppose a source sends one qubit to Alice and another to Bob, and suppose that these qubits are in a singlet state. Alice and Bob, independently and randomly, decide to whether to measure their qubit in the x or y bases, where $|\pm y\rangle = (|0\rangle \pm i|1\rangle)/\sqrt{2}$. Alice and Bob then announce which basis they used. If they used the same basis, their result will be anti-correlated, e.g. if Alice got $| + x\rangle$, Bob will have gotten $| - x\rangle$. Since each knows what the other got, they can use this information to establish a key.

What we have described is a simplified version of Ekert's scheme. He actually proposed that each party make measurements along three axes rather than two. They use their measurement results for the cases in which they chose different bases to test whether a Bell inequality is violated or not. If Eve had taken over the source and were sending Alice and Bob particles in definite states, for example a $| + x\rangle$ to Alice and a $| - x\rangle$ to Bob, then the Bell inequality would not be violated, and Alice and Bob would detect her. The inequality would also not be violated if Eve measured the particles and then sent Alice and Bob particles in states corresponding to those measurements.

The use of Bell inequalities to validate the results of quantum cryptographic protocols has led to the field of device-independent quantum cryptography, in which the parties do not need to trust the devices they are using to establish a key. In this scenario, Alice and Bob trust neither the source of the particles nor their own measuring devices. Alice and Bob regard the system as a black box into which they put inputs, measurement choices, and receive outputs, measurement results. The system is described by a probability $P(a, b|x, y)$, where $a, b, x, y \in \{0, 1\}$, and x is Alice's measurement choice, y is Bob's, a is Alice's measurement result, and b is Bob's. For example, if Alice and Bob share the two-qubit state $(1/\sqrt{2})(|00\rangle + |11\rangle) = (1/\sqrt{2})(| + x\rangle| + x\rangle + | - x\rangle| - x\rangle)$, and $x = y = 0$ corresponds to measuring in the z basis and $x = y = 1$ corresponds to measuring in the x basis, then we have that

$$P(a, b|x, y) = \begin{cases} \frac{1}{2} & x = y \\ \frac{1}{4} & x \neq y \end{cases} \tag{6.6}$$

Suppose that $P(a, b|x, y)$ can be written in the form,

$$P(a, b|x, y) = \sum_{\lambda} D(a|x, \lambda)D(b|y, \lambda)\mu(\lambda), \tag{6.7}$$

where $\mu(\lambda)$ is a probability distribution on some variable λ. The detector functions $D(a|x, \lambda)$ and $D(b|y, \lambda)$ are either 0 or 1, and tell us which measurement result occurred for a given measurement choice and value of λ. For example, $D(a|x, \lambda)$ will be 1 for the value of a that occurs for the measurement choice x and value of λ, and 0 for the other value of a. In this case, since we can assume Eve knows λ, she will also know a and b once the measurement choices x and y are announced, and any possible key is insecure. Now if $P(a, b|x, y)$ cannot be expressed in the above form, then the correlations it embodies will violate a Bell inequality (this is implied by a result of A. Fine [5], see the references). Therefore, Bell inequality violation, and hence entanglement, is a necessary condition for device-independent quantum cryptography.

6.6 Quantum Secret Sharing

Secret sharing is a cryptographic protocol in which a secret is split into several parts with each part being given to a different party. In order to recover the secret, all of the parties have to cooperate. It is a means of providing extra security. For example, a bank manager may split the combination of the vault into two pieces and give each piece to a different person. The reasoning is that if at least one of the persons is honest, an honest person will keep a dishonest one from doing anything wrong once the vault is open. If both people are dishonest, this will not work, but the probability of encountering two dishonest people is lower than encountering one, so an extra measure of security is gained.

Classically one can split a key very easily. Suppose Alice possesses a sequence of zeroes and ones, which she wants to use as a key. She creates a random sequence of zeroes and ones, and adds it, bitwise and modulo 2, to the key sequence, to create, what we will call, a sum sequence. She sends the sum sequence to Bob and the random sequence to Charlie. In order to find Alice's key sequence, Bob and Charlie have to cooperate. In particular, if Bob and Charlie add their sequences bitwise and modulo 2, the random sequence cancels out and they are left with Alice's original sequence.

One can combine this procedure with quantum key distribution to form a quantum secret sharing protocol that provides protection against eavesdropping. Alice uses, for example, BB84, to establish keys with Bob and Charlie. The actual key she will use to encode any messages is just the sum (bitwise and modulo 2) of these two keys. Therefore, in order to decode any message that Alice sends them, Bob and Charlie will have to cooperate, in particular they will have to combine their two keys to find the one Alice is actually using.

Another way of approaching this problem is to use entanglement. Suppose Alice prepares one of two entangled states

$$|\Psi_0\rangle = \cos\theta|00\rangle + \sin\theta|11\rangle$$
$$|\Psi_1\rangle = \cos\theta|00\rangle - \sin\theta|11\rangle, \tag{6.8}$$

where $|\Psi_0\rangle$ corresponds to a bit value of 0 and $|\Psi_1\rangle$ corresponds to a bit value of 1. She sends one qubit to Bob and the other to Charlie. As we will see, Bob and Charlie have to cooperate in order to find out which state Alice sent. Bob now measures his qubit in the x basis. Define the single qubit states

$$|\psi_\pm\rangle = \cos\theta|0\rangle \pm \sin\theta|1\rangle. \tag{6.9}$$

If Alice sent $|\Psi_0\rangle$, then if Bob gets $|+x\rangle$, Charlie will have the state $|\psi_+\rangle$, and if Bob gets $|-x\rangle$, then Charlie will have the state $|\psi_-\rangle$. Similarly, if Alice sent $|\Psi_1\rangle$, then if Bob gets $|+x\rangle$, Charlie will have the state $|\psi_-\rangle$, and if Bob gets $|-x\rangle$, then Charlie will have the state $|\psi_+\rangle$. Charlie now performs optimal unambiguous state discrimination for the states $|\psi_\pm\rangle$ on his qubit. He will succeed with a probability of

$1 - |\cos(2\theta)|$. He tells Alice and Bob when his measurement succeeds and when it fails. They throw out the instances in which it failed. In the case in which Charlie's measurement succeeds, he has either the result $|\psi_+\rangle$ or the result $|\psi_-\rangle$, and Bob has either $|+x\rangle$ or $|-x\rangle$. Neither of them alone can determine which state Alice sent, but if they combine their results, they can. Therefore, the information about which state, and thereby which key bit, Alice sent is split between Bob and Charlie.

An eavesdropper is faced with the same situation as in the B92 protocol, distinguishing between two nonorthogonal states, in this case $|\Psi_0\rangle$ and $|\Psi_1\rangle$. Eve will invariably misidentify the state she receives some of the time, and she will then send the wrong state on to Bob and Charlie. That will lead to the situation in which Bob and Charlie receive a different state from the one that Alice sent, and this will result in errors in the shared key. These errors can be detected if Alice, Bob, and Charlie compare a subset of their key bits. If there are no errors, there was no eavesdropper present.

While this scheme has protection against eavesdropping from an external eavesdropper, which is what it was designed to do, it is susceptible to cheating by Charlie (we thank Erika Andersson for pointing this out). If Charlie is able to grab Bob's qubit, he can simply measure it in the $|\pm x\rangle$ basis and then send it on to Bob. Charlie then knows the secret bit, and neither Bob nor Alice knows that he does. This attack can be defended against by modifying the protocol slightly. Before any measurements are made, Bob decides, with probability $1/2$. whether to apply the operator σ_z to his qubit. Applying σ_z has the effect of switching $|\Psi_0\rangle$ and $|\Psi_1\rangle$. However, if Charlie has measured Bob's qubit this switch does not happen, and this will produce errors. Let's look at an example to see how this works in more detail. Suppose Alice sent $|\Psi_0\rangle$. If Charlie is honest, then if Bob does not apply σ_z, Charlie will end up with $|\psi_+\rangle$, but if Bob does apply σ_z, then Charlie will end up with $|\psi_-\rangle$. If Charlie has measured Bob's qubit, then the two-qubit state is no longer entangled, and no matter what Bob does, Charlie will have $|\psi_+\rangle$. If Alice Bob and Charlie then compare the results for a subset of qubits, with Charlie announcing his result before Bob, errors will be discovered and Charlie's cheating will be revealed.

6.7 Problems

1. Suppose we are using a singlet state $|\phi_-\rangle = (|0\rangle|1\rangle - |1\rangle|0\rangle)/\sqrt{2}$ in the Ekert 91 protocol, and Alice and Bob are measuring in the x and y bases. We want to find a Bell inequality that is maximally violated under these conditions. Suppose the two observables that Alice is using in the Bell inequality are σ_x and σ_y. Find two observables for Bob of the form, $\hat{\mathbf{n}}_1 \cdot \boldsymbol{\sigma}$ and $\hat{\mathbf{n}}_2 \cdot \boldsymbol{\sigma}$, where $\hat{\mathbf{n}}_1$ and $\hat{\mathbf{n}}_2$ are unit vectors in the $x - y$ plane such that the expression appearing in the resulting Bell inequality

$$|\langle \sigma_x(\hat{\mathbf{n}}_1 \cdot \boldsymbol{\sigma})\rangle + \langle \sigma_x(\hat{\mathbf{n}}_2 \cdot \boldsymbol{\sigma})\rangle + \langle \sigma_y(\hat{\mathbf{n}}_1 \cdot \boldsymbol{\sigma})\rangle - \langle \sigma_y(\hat{\mathbf{n}}_2 \cdot \boldsymbol{\sigma})\rangle|$$

is equal to $2\sqrt{2}$ for the singlet state. It is useful to first prove that for any unit vectors $\hat{\mathbf{e}}$ and $\hat{\mathbf{n}}$, $\langle\phi_-|(\hat{\mathbf{e}}\cdot\boldsymbol{\sigma})(\hat{\mathbf{n}}\cdot\boldsymbol{\sigma})|\phi_-\rangle = -\hat{\mathbf{e}}\cdot\hat{\mathbf{n}}$.

2. Alice and Bob are using the B92 protocol with states $|\psi_0\rangle$ and $|\psi_1\rangle$. Eve captures the qubit going from Alice to Bob and entangles it with an ancilla qubit and then sends Alice's original qubit on to Bob. Eve wants to measure the ancilla qubit to gain information about the qubit Alice sent to Bob. In particular, Eve uses the unitary entangling operation U to perform

$$U|\psi_0\rangle_A|0\rangle_E = |\psi_0\rangle_A|v_{00}\rangle_E + |\psi_0^\perp\rangle_A|v_{01}\rangle_E$$

$$U|\psi_1\rangle_A|0\rangle_E = |\psi_1\rangle_A|v_{11}\rangle_E + |\psi_1^\perp\rangle_A|v_{10}\rangle_E,$$

where $\langle\psi_j|\psi_j^\perp\rangle = 0$, for $j = 0, 1$, and the vectors $|v_{jk}\rangle$ for $j, k = 0, 1$ are not necessarily normalized. Show that if Eve is to create no errors she will gain no information about Alice's qubit.

3. In the original B92 protocol, Bob did not use the unambiguous discrimination POVM but instead switched randomly between two projective measurements. As before, Alice sends either $|\psi_0\rangle$ or $|\psi_1\rangle$, where $\langle\psi_0|\psi_1\rangle = s$, and s is real and $0 \le s \le 1$. Each state is sent with a probability of $1/2$. Bob measures either in the $\{\psi_0, \psi_0^\perp\}$ basis or the $\{\psi_1, \psi_1^\perp\}$ basis. Here $\langle\psi_j|\psi_j^\perp\rangle = 0$ for $j = 0, 1$. Find the procedure Bob uses to identify the state Alice sent. This procedure will only succeed part of the time, so find the probability that Bob is able to successfully identify Alice's state.

4. Another way to do quantum secret sharing is to use the GHZ state $|\Psi\rangle_{abc} = (1/\sqrt{2})(|000\rangle_{abc} + |111\rangle_{abc})$. Define the x and y bases as $|\pm x\rangle = (1/\sqrt{2})(|0\rangle \pm |1\rangle)$ and $|\pm y\rangle = (1/\sqrt{2})(|0\rangle \pm i|1\rangle)$. Alice, Bob and Charlie each have one of the qubits is the GHZ state. Show that if Alice and Bob both measure in the same basis, and Charlie measures in the x basis, then if Bob and Charlie communicate their measurement results to each other, they can determine the result of Alice's measurement. In addition, show that if Alice and Bob measure in different bases, and Charlie measures in the y basis, then, again, if Bob and Charlie communicate their measurement results to each other, they can determine the result of Alice's measurement. Therefore, Alice can establish a joint key with Bob and Charlie, but Bob and Charlie have to cooperate to obtain it.

Further Reading

1. C.H. Bennett, Quantum cryptography using any two nonorthogonal states. Phys. Rev. Lett. **68**, 3121 (1992)
2. C.H. Bennett, G. Brassard, Quantum cryptography: public key distribution and coin tossing, in *Proceedings of IEEE International Conference on Computers, Systems, and Signal Processing*, Bangalore (IEEE, New York 1984), p. 175
3. For sharing quantum secrets see R. Cleve, D. Gottesman, H.-K. Lo, For sharing quantum secrets. Phys. Rev. Lett. **83**, 648 (1999)

4. A. Ekert, Quantum cryptography based on Bell's theorem. Phys. Rev. Lett. **67**, 661 (1991)
5. A. Fine, Hidden variables, joint probabilities and the Bell inequality. Phys. Rev. Lett. **48**, 291 (1982)
6. For an example of a device-independent QKD scheme see S. Pironio, A. Acin, N. Brunner, N. Gisin, S. Massar, V. Scarani, Device-independent quantum key distribution secure against collective attacks. New J. Phys. **11**, 045021 (2009)
7. For more details on QKD see N. Gisin, G. Ribordy, W. Tittel, H. Zbinden, Quantum cryptography. Rev. Mod. Phys. **74**, 145 (2002)
8. M. Hillery, V. Bužek, A. Berthiaume, Quantum secret sharing. Phys. Rev. A **59**, 1829 (1999)
9. J. Mimih, M. Hillery, Unambiguous discrimination of special sets of multipartite states using local measurements and classical communication. Phys. Rev. A **71**, 012329 (2005)

Chapter 7
Quantum Algorithms

In this chapter we shall look at a number of quantum algorithms. We are going to compare their performance, in terms of number of steps, to classical algorithms that accomplish the same task.

7.1 The Deutsch-Jozsa Algorithm

We shall start with a generalization of the Deutsch algorithm, known as the Deutsch-Jozsa algorithm. The problem can be stated as follows: Given a Boolean function on n-digit binary numbers, $f : \{0, 1\}^n \rightarrow \{0, 1\}$, which is promised to be either constant or balanced. Determine which.

Classically, $2^{(n-1)} + 1$ function evaluations are necessary in the worst case scenario. There is a quantum algorithm that requires only one function evaluation. The corresponding quantum circuit is shown in Fig. 7.1.

In order to understand how this circuit works we shall analyze the state of the $n + 1$ qubit system at each step, i.e. at the input, after the first set of Hadamard gates, after the f-controlled-NOT gate and after the second set of Hadamard gates, which constitutes the output state generated by the circuit.

Since Hadamard gates are placed before and after the f-CNOT gate, we begin by analyzing the action of a set of Hadamard gates on a general binary number state which, in turn, is defined as follows. Let $x = x_{n-1}x_{n-2}\ldots x_0$, where $x_j \in \{0, 1\}$, stand for an n-digit binary number. Then the binary number state $|x\rangle$ of an n qubit system in the computational basis is given by $|x\rangle = |x_{n-1}\rangle \otimes |x_{n-2}\rangle \ldots \otimes |x_0\rangle$. The action of the Hadamard gate on a single qubit in the computational basis state can

© Springer Nature Switzerland AG 2021

J. A. Bergou et al., *Quantum Information Processing*, Graduate Texts in Physics, https://doi.org/10.1007/978-3-030-75436-5_7

Fig. 7.1 Quantum circuit for the Deutsch-Jozsa problem

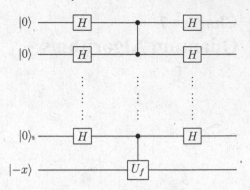

be summarized as $H|x_j\rangle \to \frac{1}{\sqrt{2}}(|0\rangle + (-1)^{x_j}|1\rangle)$, so

$$\Pi_{j=0}^{n-1}|x_j\rangle \to \left(\frac{1}{\sqrt{2}}\right)^n \prod_{j=0}^{n-1}(|0\rangle + (-1)^{x_j}|1\rangle)$$

$$= \left(\frac{1}{\sqrt{2}}\right)^n \sum_{z=0}^{2^n-1}\left(\prod_{j=\text{ such that } z_j=1}(-1)^{x_j}\right)|z\rangle$$

$$= \left(\frac{1}{\sqrt{2}}\right)^n \sum_{z=0}^{2^n-1}\left(\prod_{j=0}^{n-1}(-1)^{x_j z_j}\right)|z\rangle. \tag{7.1}$$

Notice that $\prod_{j=0}^{n-1}(-1)^{x_j z_j} = (-1)^{\sum_{j=0}^{n-1} x_j z_j} = (-1)^{\left[\sum_{j=0}^{n-1} x_j z_j \bmod 2\right]}$. Let us define the dot product as $x \cdot z \equiv \sum_{j=0}^{n-1} x_j z_j \bmod 2$, then

$$|x\rangle \to \left(\frac{1}{2}\right)^{n/2}\sum_{z=0}^{2^n-1}(-1)^{x \cdot z}|z\rangle. \tag{7.2}$$

In other words, a given n-digit binary number state is turned into an equally weighted superposition of all 2^n binary number states of the n qubits and the sign of each term is determined by the parity of the dot product between the given state and binary state in the term.

If now we apply this to the input state of the n control qubits, $|\psi_{in}\rangle = |0\rangle$, we obtain the state of the n-qubit system after the first set of Hadamards as

$$|\psi_1\rangle = \left(\frac{1}{2}\right)^{n/2}\sum_{z=0}^{2^n-1}|z\rangle. \tag{7.3}$$

Next we analyze the action of the f-CNOT gate on this state. To this end we note that the f-controlled-NOT gate acts as $|x\rangle|y\rangle \to |x\rangle|y + f(x) \, mod \, 2\rangle$ where $|x\rangle$ is the state of the n control qubits and $|y\rangle$ is the state of the single target qubit. Therefore,

$$|x\rangle \otimes \frac{1}{\sqrt{2}}(|0\rangle - |1\rangle) \to |x\rangle \otimes (|f(x)\rangle - |1 + f(x)\rangle)$$

$$= (-1)^{f(x)}|x\rangle \otimes \frac{1}{\sqrt{2}}(|0\rangle - |1\rangle). \qquad (7.4)$$

Combining this with $|\psi_1\rangle$ in Eq. (7.3), we obtain the state after the f-CNOT gate as

$$|\psi_2\rangle \otimes \frac{1}{\sqrt{2}}(|0\rangle - |1\rangle) = \left(\frac{1}{2}\right)^{(n+1)/2} \sum_{x=0}^{2^n-1} (-1)^{f(x)}|x\rangle \otimes (|0\rangle - |1\rangle). \qquad (7.5)$$

Finally, applying Eq. (7.2) to this state yields the output state after the final set of Hadamard gates as

$$|\psi_{in}\rangle \otimes \frac{1}{\sqrt{2}}(|0\rangle - |1\rangle) \to \left(\frac{1}{2}\right)^{n+1/2} \sum_{x,z=0}^{2^n-1} (-1)^{f(x)+x\cdot z}|z\rangle \otimes (|0\rangle - |1\rangle)$$

$$= |\psi_{out}\rangle \otimes \frac{1}{\sqrt{2}}(|0\rangle - |1\rangle). \qquad (7.6)$$

The amplitude of the initial state, $|\psi_{in}\rangle = |0\rangle$ in the output state is easily obtained as $\langle 0|\psi_{out}\rangle = \left(\frac{1}{2}\right)^n \sum_{x=0}^{2^n-1}(-1)^{f(x)}$, and

$$\langle 0|\psi_{out}\rangle = \begin{cases} 0 & \text{if f(x) balanced}, \\ (-1)^{f(0)} & \text{if f(x) constant} \to |\psi_{out}\rangle = (-1)^{f(0)}|0\rangle. \end{cases} \qquad (7.7)$$

Therefore measuring each of the n output qubits we have with certainty that

1. f(x) = constant if we find all qubits in their 0 state,
2. f(x) = balanced if not all of them are found in their 0 state.

Note that this is accomplished with only one function evaluation.

7.2 The Bernstein-Vazirani Algorithm

We can use the Deutsch-Jozsa circuit to solve another problem due to Bernstein and Vazirani. Suppose

$$f(x) = a \cdot x + b \, (\text{mod } 2), \qquad (7.8)$$

where $a \in \{0, 1\}^n$ and $b \in \{0, 1\}$. Our goal is to determine a (we do not know a or b). Classically, because a contains n bits of information we are going to have to evaluate $f(x)$ n times at least. One method is to evaluate it for $x = 0$, giving b, and then for $x_j = 0 \ldots 010 \ldots 0$, where the 1 is in the jth place, for $j = 1, \ldots, n$.

With this $f(x)$ our state at the output of the quantum circuit is

$$|\Psi_{out}\rangle = \left(\frac{1}{2}\right)^n \sum_{x,y=0}^{2^n-1} (-1)^b (-1)^{x \cdot (a+y)} |y\rangle . \tag{7.9}$$

where $(a + y)$ in the exponent stands for bitwise addition.

We show that $\sum_{x=0}^{2^n-1} (-1)^{x \cdot z} = 0$ unless $z \in \{0, 1\}^n = 0$. This can be seen as follows. First, we rewrite the sum as

$$\sum_{x=0}^{2^n-1} (-1)^{x \cdot z} = \sum_{x=0}^{2^n-1} \prod_{j=0}^{n-1} (-1)^{x_j z_j} = \sum_{x_{n-1}=0}^{1} \cdots \sum_{x_0=0}^{1} \prod_{j=0}^{n-1} (-1)^{x_j z_j} . \tag{7.10}$$

Suppose now $z_k = 1$, then

$$\sum_{x} (-1)^{x \cdot z} = \sum_{x_{n-1}=0}^{1} \cdots \sum_{x_{k+1}=0}^{1} \sum_{x_{k-1}=0}^{1} \cdots \sum_{x_0=0}^{1} \prod_{j=0, j \neq k}^{n-1} (-1)^{x_j z_j} (1 + (-1)) = 0 ,$$
$$\tag{7.11}$$

where the last two terms in the bracket arise from $x_k = 0$ yielding the $+1$ and $x_k = 1$ yielding the (-1). Therefore,

$$\sum_{x} (-1)^{x \cdot z} = 2^n \delta_{z,0} ,$$

and

$$|\Psi_{out}\rangle = (-1)^b |a\rangle , \tag{7.12}$$

so that measuring the n output qubits in $|\Psi_{out}\rangle$ gives us a with only one function evaluation.

7.3 Quantum Search: The Grover Algorithm

Typically, Grover's problem can be stated as the search for one marked entry in an unsorted database. Mathematically, it can be formulated as the following problem. Let $f(x) = 0$ or 1 where x is an n bit binary number. In particular

$$f(x) = \begin{cases} 1 & \text{if } x = x_0 , \\ 0 & \text{if } x \neq x_0 . \end{cases} \tag{7.13}$$

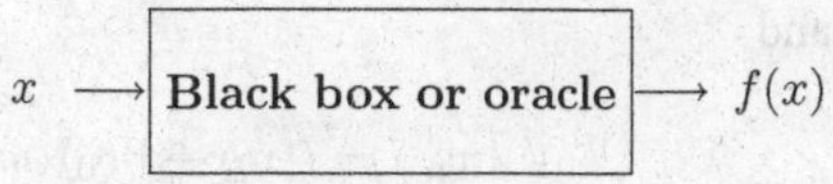

Fig. 7.2 Scheme of the
search problem

$$x \longrightarrow \boxed{\text{Black box or oracle}} \longrightarrow f(x)$$

x_0 is unknown and we would like to find it. The search is schematically depicted in Fig. 7.2.

The central question is: How many function evaluations are necessary? Classically, if $N = 2^n$, then $\mathcal{O}(N)$ evaluations are necessary. On a quantum computer it can be done with $\mathcal{O}(\sqrt{N})$ evaluations. (Our treatment is taken from R. Jozsa, quant-ph/990121.)

To this end, define the following operators:

$$U_f|x\rangle = (-1)^{f(x)}|x\rangle = (-1)^{\delta_{x,x_0}}|x\rangle \ ,$$

$$U_0|x\rangle = (-1)^{\delta_{x,0}}|x\rangle = (I - 2|0\rangle\langle 0|)|x\rangle \ ,$$

$$U_H = (H)^{\otimes n} \ . \tag{7.14}$$

An alternative form of U_f is given by $U_f = I - 2|x_0\rangle\langle x_0|$ and we already know the circuit for this operator.

Grover's algorithm consists in applying the operator $Q = -U_H U_0 U_H U_f$ to the initial state, $|w_0\rangle = U_H|0\rangle = \frac{1}{\sqrt{N}}\sum_{x=0}^{N-1}|x\rangle$, $\mathcal{O}(\sqrt{N})$ times and then measuring the state in the computational basis. The answer will be, with probability greater than $\frac{1}{2}$, x_0 (actually, with a probability close to 1).

How does this work? First define $U_{w_0} = U_H U_0 U_H = I - 2|w_0\rangle\langle w_0|$ and $S = \text{span}\{|w_0\rangle, |x_0\rangle\}$ which is a two-dimensional subspace. For any $|\psi\rangle = c_1|w_0\rangle + c_2|x_0\rangle \in S$, we have

$$Q|\psi\rangle = -U_{w_0} U_f(c_1|w_0\rangle + c_2|x_0\rangle)) = -U_{w_0}[c_1(|w_0\rangle - \frac{2}{\sqrt{N}}|x_0\rangle)) - c_2|x_0\rangle]$$

$$= c_1|w_0\rangle + \left(\frac{2}{\sqrt{N}} + c_2\right)(|x_0\rangle - \frac{2}{\sqrt{N}}|x_0\rangle)) \in S \ , \tag{7.15}$$

so that Q maps S unto itself. Therefore, all of the action in Grover's algorithm takes place in a 2D subspace. Note also that if c_1 and c_2 are real so are the coefficients of $|w_0\rangle$ and $|x_0\rangle$. As we start by applying Q to $|w_0\rangle$, we actually need only to consider $S' = \{c_1|w_0\rangle + c_2|x_0\rangle|c_1, c_2 \text{ real}\}$, i.e. S' is a real 2D subspace.

Now look at Q more closely. The operator U_f in S' is just a reflection about the line parallel to $|x_0^\perp\rangle$. Note:

$$|x_0^\perp\rangle = (|w_0\rangle + |x_0\rangle\langle x_0|w_0\rangle)/(1 - |\langle x_0|w_0\rangle|^2)^{1/2} \ ,$$

and

$$|w_0^\perp\rangle = (|x_0\rangle + |w_0\rangle\langle w_0|x_0\rangle)/(1 - |\langle x_0|w_0\rangle|^2)^{1/2} \,.$$

We also have that, in the subspace S', $|w_0\rangle\langle w_0| + |w_0^\perp\rangle\langle w_0^\perp| = I$. From here, it follows that $-U_f = -(I - 2|w_0\rangle\langle w_0|) = I - 2|w_0^\perp\rangle\langle w_0^\perp| = U_{w_0^\perp}$ and this is a reflection about the line through $|w_0\rangle$. Therefore, Q corresponds to two consecutive reflections,

$$Q = U_{w_0^\perp} U_f = (\text{reflection about } w_0)(\text{reflection about } x_0^\perp) \,. \tag{7.16}$$

The geometry of the two reflections is shown in Fig. 7.3.

Theorem 1 *Let $\mathcal{M}_1$ and $\mathcal{M}_2$ be two mirror lines in Euclidean plane $\mathbb{R}^2$ intersecting at point O and α be the angle from $\mathcal{M}_1$ to $\mathcal{M}_2$. The operation of reflection through $\mathcal{M}_1$ followed by reflection through $\mathcal{M}_2$ is a rotation by 2α about O.*

Proof The proof uses pictorial but nevertheless rigorous arguments. Let $\mathcal{M}_1$ be parallel to v_1 and $\mathcal{M}_2$ parallel to v_2. If the theorem holds, for v_1 and v_2 it holds for any superposition of them, hence for any vector. Let R_1 be the reflection through $\mathcal{M}_1$ and R_2 be the reflection through $\mathcal{M}_2$. We will now separately study what happens to v_1 and v_2, as a result of these two reflections. First, look at v_1. A reflection of v_1 through $\mathcal{M}_1 = v_1$ maps v_1 onto itself. A subsequent reflection of v_1 through $\mathcal{M}_2 = v_2$ corresponds to an effective rotation of v_1 by the angle 2α in the counterclockwise direction. The situation is shown in Fig. 7.4.

Fig. 7.3 Geometry associated with the Grover search algorithm

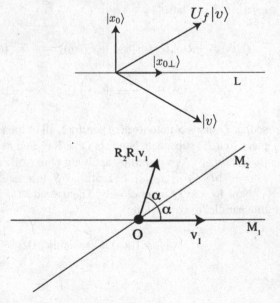

Fig. 7.4 Two subsequent reflections of v_1, $R_2 R_1$, first through $\mathcal{M}_1$ followed by another through $\mathcal{M}_2$, correspond to an effective rotation by the angle 2α

Fig. 7.5 Two subsequent
reflections of v_2, $R_2 R_1$, first
through M_1 followed by
another through M_2,
correspond to an effective
rotation by the angle 2α

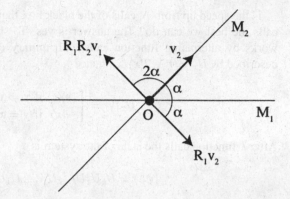

Next, look at v_2. The situation is shown in Fig. 7.5. A reflection of v_2 through
$M_1 = v_1$ rotates v_2 by the angle 2α in the clockwise direction. A subsequent
reflection of $R_1 v_2$ around $M_2 = v_2$ corresponds to an effective rotation of v_2 by
the angle 2α relative to its original orientation.

Therefore, Q is a rotation in S' by angle 2α, where α is the angle between $|w_0\rangle$
and $|x_0^\perp\rangle$. Furthermore,

$$\cos\alpha = \langle w_0|x_0^\perp\rangle = \left(1 - \frac{1}{\sqrt{N}}\right)^{\frac{1}{2}},$$

and

$$\sin\alpha = \langle w_0|x_0\rangle = \left(1 - \cos^2\alpha\right)^{\frac{1}{2}} = \frac{1}{\sqrt{N}}.$$

$\square$

Starting with the state (in the $|x_0\rangle$, $|x_{0\perp}\rangle$ basis)

$$|w_0\rangle = |x_0\rangle|\langle x_0|w_0\rangle + |x_{0\perp}\rangle\langle x_{0\perp}|w_0\rangle$$
$$= \sin\alpha|x_0\rangle + \cos\alpha|x_{0\perp}\rangle, \tag{7.17}$$

we have

$$Q^n|w_0\rangle = \sin\alpha_n|x_0\rangle + \cos\alpha_n|x_{0\perp}\rangle, \tag{7.18}$$

where $\alpha_n = (2n+1)\alpha$. We want to choose n so that α_n is close to $\pi/2$. For large N,
$\alpha \cong \frac{1}{\sqrt{N}}$ so we want $(2n+1)\frac{1}{\sqrt{N}} \cong \frac{\pi}{2}$. Therefore, n=closest integer to $\frac{\pi}{4}\sqrt{N} - \frac{1}{2}$.
Let us call this value $\bar{n}$. Then the probability of measuring $x_0 = |\langle x_0|Q^{\bar{n}}|w_0\rangle|^2 = $
$\sin^2\alpha_{\bar{n}} \cong 1$ and the probability of measuring $x \neq x_0 = |\langle x|Q^{\bar{n}}|w_0\rangle|^2 = \cos^2\alpha_{\bar{n}} = $
$\mathcal{O}(\frac{1}{N^2})$.

Is the speed up from N calls of the black box that evaluates the function, to $\sqrt{N}$ calls the best we can do? The answer is yes. To show this assume the algorithm works by alternating function calls with unitary evolution. The function call is described by $U_x = I - 2|x\rangle\langle x|$, hence

$$U_x|y\rangle = \begin{cases} |y\rangle & \text{if } y \neq x , \\ -|x\rangle & \text{if } y = x . \end{cases} \tag{7.19}$$

After k function calls the state of the system is

$$|\psi_k^x\rangle = U_k U_x U_{k-1} U_x \ldots U_1 U_x |\psi_{in}\rangle . \tag{7.20}$$

Then the strategy is as follows. Compare $|\psi_k^x\rangle$ to $|\psi_k\rangle = U_k U_{k-1} \ldots U_1 |\psi_{in}\rangle$ to show that if the probability of finding x is great, in particular, if $|\langle x|\psi_k^x\rangle| > \frac{1}{2}$, then k must be of the order of $\sqrt{N}$. In particular, we find upper and lower bounds for $D_k = \sum_x \|\psi_k^x - \psi_k\|^2$.

We begin with establishing the upper bound. First, note that

$$D_{k+1} = \sum_x \|U_x \psi_k^x - \psi_k\|^2 = \sum_x \|U_x(\psi_k^x - \psi_k) + (U_x - I)\psi_k\|^2 , \tag{7.21}$$

and

$$
\begin{aligned}
D_{k+1} &\leq \sum_x (\|\psi_k^x - \psi_k\| + \|(U_x - I)\psi_k\|)^2 \\
&= \sum_x (\|\psi_k^x - \psi_k\|^2 + 4\|\psi_k^x - \psi_k\||\langle x|\psi_k\rangle| + |\langle x|\psi_k\rangle|^2) \\
&\leq D_k + 4(\sum_x \|\psi_k^x - \psi_k\|^2)^{1/2}(\sum_x |\langle x|\psi_k\rangle|^2)^{1/2} + 4 \\
&\leq D_k + 4\sqrt{D_k} + 4 .
\end{aligned}
\tag{7.22}
$$

Now, we will use this and induction to show $D_k \leq 4k^2$. First, we have $D_0 = 0$. Now

$$D_1 = \sum_x \|U_1 U_x \psi_{in} - U_1 \psi_{in}\|^2 = \sum_x \|U_x \psi_{in} - \psi_{in}\|^2, \tag{7.23}$$

but $(U_x - I)|\psi_{in}\rangle = -2|x\rangle\langle x|\psi_{in}\rangle$, so we have that

$$D_1 = 4 \sum_x |\langle x|\psi_{in}\rangle|^2 = 4. \tag{7.24}$$

So, it is clearly true that $D_k \leq 4k^2$ for $k = 0, 1$. Now, assuming it is true for k

$$D_{k+1} \leq D_k + 4\sqrt{D_k} + 4 \leq 4k^2 + 8k + 4 = 4(k+1)^2. \tag{7.25}$$

Therefore, $D_k \leq 4k^2$.

Next we establish the lower bound. To this end, let us define $Q_x = I - |x\rangle\langle x|$, then we have that

$$
\begin{aligned}
\|\psi_k^x - \psi_k\|^2 &= \| |x\rangle(\langle x|\psi_x^k\rangle - \langle x|\psi_k\rangle) + Q_x(\psi_k^x - \psi_k)\|^2 \\
&= |\langle x|\psi_x^k\rangle - \langle x|\psi_k\rangle|^2 + \|Q_x(\psi_k^x - \psi_k)\|^2 \\
&\geq |\langle x|\psi_x^k\rangle|^2 + |\langle x|\psi_k\rangle|^2 - 2|\langle x|\psi_x^k\rangle| \cdot |\langle x|\psi_k\rangle| \\
&\quad + \|Q_x\psi_k^x\|^2 + \|Q_x\psi_k)\|^2 - 2|\langle Q_x\psi_k^x|\psi_k\rangle| \\
&\geq 2 - 2|\langle x|\psi_k\rangle| - 2\|Q_x\psi_k^x\|. \tag{7.26}
\end{aligned}
$$

Now suppose we assume that after k steps, when we measure the state $|\psi_k^x\rangle$, our probability of finding x is greater than $1/2$, i.e. $|\langle x|\psi_k^x\rangle|^2 > 1/2$. We also have that $|\langle x|\psi_k^x\rangle|^2 + \|Q_x\psi_k^x\|^2 = 1$ from which $\|Q_x\psi_k^x\|^2 \leq 1/2$ follows, so $\|\psi_k^x - \psi_k\|^2 \geq 2 - 2|\langle x|\psi_k\rangle| - \sqrt{2}$ and

$$
\begin{aligned}
D_k &\geq \sum_x (2 - 2|\langle x|\psi_k\rangle| - \sqrt{2}) \\
&\geq N(2 - \sqrt{2}) - 2(\sum 1^2)^{1/2}(\sum_x |\langle x|\psi_k\rangle|^2)^{1/2} \\
&\geq N(2 - \sqrt{2}) - 2\sqrt{N}. \tag{7.27}
\end{aligned}
$$

Putting the bounds together gives

$$4k^2 \geq N(2 - \sqrt{2}) - 2\sqrt{N} \tag{7.28}$$

and from here

$$k \geq \frac{(2 - \sqrt{2})^{1/2}}{2}\sqrt{N}\left(1 - \frac{2}{\sqrt{N}}\frac{1}{2 - \sqrt{2}}\right)^{1/2} \tag{7.29}$$

follows. So, reducing the number of function calls from N to $\sqrt{N}$ is indeed the best we can do and the Grover search algorithm is optimal.

7.4 Period Finding: Simon's Algorithm

Now we take a look at Simon's algorithm which is a simple period-finding algorithm. A more sophisticated version is one of the major components of Shor's factoring algorithm.

Consider a function $F : \mathbb{Z}_2^{\otimes n} \to \mathbb{Z}_2^{\otimes n}$ which is $2 \to 1$. In particular

$$f(x) = f(y) \quad \text{iff } y = x \oplus \xi \text{ where } x, y, \xi \in \mathbb{Z}_2^{\otimes n} . \tag{7.30}$$

Here $\oplus$ stands for component-wise mod 2 addition, i.e. for $w, z \in \mathbb{Z}_2^{\otimes n}$ we have that $w \oplus z = (w_1 + z_1 \ (mod\ 2), \ldots, w_n + z_n \ (mod\ 2))$ and ξ is fixed. The object is to find ξ with only $poly(n)$ function evaluations.

Start with $|0 \ldots 0\rangle$ and apply a Hadamard gate to each qubit to get $2^{-n/2} \sum_x |x\rangle$. Now apply U_f, which has the following action

$$U_f|x\rangle|y\rangle = |x\rangle|y \oplus f(x)\rangle . \tag{7.31}$$

So

$$U_f \left(\frac{1}{2^{n/2}} \sum_x |x\rangle|0\rangle \right) = \frac{1}{2^{n/2}} \sum_x |x\rangle|f(x)\rangle . \tag{7.32}$$

Now measure the second register. This gives some result, x_0, and leaves the first in the state $\frac{1}{\sqrt{2}} (|x_0\rangle + |x_0 \oplus \xi\rangle)$, where x_0 is random. This randomness makes measuring the above state useless, if we want to determine ξ. Instead, apply $H^{\otimes n}$ to the state. This gives us

$$\frac{1}{2^{(n+1)/2}} \sum_y \left[(-1)^{x_0 \cdot y} + (-1)^{(x_0 \oplus \xi) \cdot y} \right] |y\rangle$$

$$= \frac{1}{2^{(n+1)/2}} \sum_y (-1)^{x_0 \cdot y} \left[1 + (-1)^{\xi \cdot y} \right] |y\rangle$$

$$= \frac{1}{2^{(n+1)/2}} \sum_{\{y|y\cdot\xi=0\}} (-1)^{x_0 \cdot y} |y\rangle . \tag{7.33}$$

Now measure this state. We get some value of y, call it y_1, such that $y_1 \cdot \xi = 0$. With $\mathcal{O}(n)$ iterations of this procedure, we obtain n independent equations of the form $y_j \cdot \xi = 0$, $j = 1, \ldots, n$, and we can solve this linear system to determine ξ.

7.5   Quantum Fourier Transform and Phase Estimation

The quantum Fourier transform is a component of a number of quantum algorithms, the Shor factoring algorithm in particular. We will not treat the Shor algorithm in this book, as it has been covered extensively in many other places. We will show instead how the quantum Fourier transform can be used to find an unknown eigenvalue of a unitary transformation.

Let $|a\rangle$ be a member of the computational basis in the m-qubit Hilbert space. The m-bit binary number a can be expressed as $a = 2^{m-1}a_1 + 2^{m-1}a_2 + \ldots + 2^0 a_m$, where each of the a_j are either 0 or 1. The quantum Fourier transform, U_F, takes $|a\rangle$ into the state

$$U_F|a\rangle = \frac{1}{2^m} \sum_{y=0}^{2^m-1} e^{2\pi i a \cdot y/2^m} |y\rangle. \tag{7.34}$$

The inverse transformation is given by

$$U_F^{-1}|a\rangle = \frac{1}{2^m} \sum_{y=0}^{2^m-1} e^{-2\pi i a \cdot y/2^m} |y\rangle. \tag{7.35}$$

This transformation can implemented efficiently using only one and two-qubit gates.

Now let us see how we can use the quantum Fourier transform to estimate an unknown eigenvalue. The circuit is shown in Fig. 7.6.

Suppose that we have the unitary operator U, where $U|\psi\rangle = \exp(2\pi i \phi)|\psi\rangle$ and $0 \le \phi < 1$. We are given one copy of $|\psi\rangle$ and gates that perform Controlled-U^k operations for $k = 1, 2, 2^2, \ldots 2^{m-1}$. We want to find ϕ, which we do not know, to m-bit accuracy. We start with each of the qubits in the m control lines in the state $(|0\rangle + |1\rangle)/\sqrt{2}$, so that the initial state of our computation is

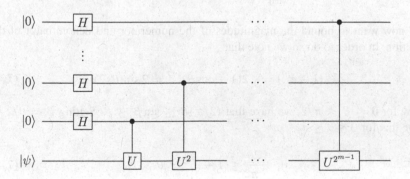

Fig. 7.6 Quantum circuit for phase estimation

$$2^{-m/2}[\prod_{j=0}^{m-1}(|0\rangle_j + |1\rangle_j)] \otimes |\psi\rangle. \tag{7.36}$$

We now apply the Controlled-U^{2^j} gates, the control being the jth qubit and the target being the system in the state $|\psi\rangle$. This results in the state

$$2^{-m/2}[\prod_{j=0}^{m-1}(|0\rangle_j + e^{2\pi i 2^j \phi}|1\rangle_j)] \otimes |\psi\rangle = 2^{-m/2}\sum_{y=0}^{2^m-1} e^{2\pi i \phi y}|y\rangle \otimes |\psi\rangle, \tag{7.37}$$

where $|y\rangle$ is an m-qubit computational basis state. Now if ϕ is of the form $a/2^m$, where a is an m-digit binary number, we can simply apply the inverse quantum Fourier transform to the above state, and the result will be $|a\rangle$. We will then have learned ϕ.

Now let us see what happens if ϕ is not of the form $a/2^m$. Let $\phi = (a/2^m) + \delta$, where a is the closest m-bit binary number to $2^m\phi$. This implies that $0 < |\delta| \le 2^{-(m+1)}$. We now apply the inverse Fourier transform to the state in Eq. (7.37) yielding

$$2^{-m}\sum_{y=0}^{2^m-1}\sum_{x=0}^{2^m-1} e^{-2\pi i x \cdot y/2^m} e^{2\pi i \phi y}|x\rangle = 2^{-m}\sum_{y=0}^{2^m-1}\sum_{x=0}^{2^m-1} e^{2\pi i (a-x)\cdot y/2^m} e^{2\pi i \delta y}|x\rangle, \tag{7.38}$$

where we have dropped $|\psi\rangle$ since it is not entangled with the rest of the state and plays no further role. Now let us look at the coefficient of the state $|a\rangle$ in the above equation. It is given by

$$2^{-m}\sum_{y=0}^{2^m-1} e^{2\pi i \delta y} = 2^{-m}\left(\frac{1 - e^{2\pi i \delta 2^m}}{1 - e^{2\pi i \delta}}\right). \tag{7.39}$$

We now want to bound the magnitudes of the numerator and denominator of this fraction. In order to do so we note that

$$|1 - e^{i\theta}| = \sqrt{2}(1 - \cos\theta)^{1/2} = 2\sin(\theta/2). \tag{7.40}$$

Now for $0 \le \beta \le \pi/2$, we have that $(2/\pi)\beta \le \sin\beta \le \beta$. Setting $\beta = \theta/2$ we have that for $0 \le \theta \le \pi$,

$$\frac{2\theta}{\pi} \le |1 - e^{i\theta}| \le \theta. \tag{7.41}$$

Note that because $|1 - e^{i\theta}| = |1 - e^{-i\theta}|$, the above inequalities can be modified to hold in the range $-\pi \leq \theta \leq \pi$ by inserting absolute value signs appropriately

$$\frac{2|\theta|}{\pi} \leq |1 - e^{i\theta}| \leq |\theta|. \tag{7.42}$$

Now because $|\delta| \leq 1/2^{m+1}$, we have that $2\pi\delta 2^m \leq \pi$ and, therefore, $|1 - e^{2\pi i \delta 2^m}| \geq 4\delta 2^m$, and we also have that $|1 - e^{2\pi i \delta}| \leq 2\pi\delta$. This implies that the probability of obtaining the state $|a\rangle$ when measuring the output state of the circuit is

$$2^{-2m} \left| \frac{1 - e^{2\pi i \delta 2^m}}{1 - e^{2\pi i \delta}} \right|^2 \geq 2^{-2m} \left(\frac{4\delta 2^m}{2\pi\delta} \right)^2 = \frac{4}{\pi^2}. \tag{7.43}$$

Therefore, the probability of obtaining the best m-bit approximation to ϕ is $(4/\pi^2) = 0.4$. A more detailed analysis shows that the probability of getting an error greater than $k/2^m$ is less than $1/(2k - 1)$.

One possible use for this algorithm is related to the Grover search. Suppose we are given a black box Boolean function that is of one of two types. There is either one input, x_0, which we do not know, for which $f(x_0) = 1$, with all other inputs, $x \neq x_0$ yielding $f(x) = 0$, or all inputs yield $f(x) = 0$. We would like to find which type of black box function we have. One approach is to run the Grover algorithm and see if we get the same answer almost all of the time. If so, they we have the first kind of black box. If we get different answers each time, then we have the second type. A second approach is to use the phase estimation algorithm. The operator $Q = U_{w_0^\perp} U_f$ has different eigenvalues for the two different types of oracles. In the case that all inputs yield $f(x) = 0$, we have that $U_f = I$, which implies that $Q = U_{w_0^\perp}$. In that case, Q is just a reflection, so that its eigenvalues are just ± 1. In particular, the state $|w_0\rangle$ is an eigenstate with eigenvalue 1. If one of the inputs yields $f(x_0) = 1$, then in the subspace S', Q can be expressed as a 2×2 matrix in the $\{|w_0\rangle, |w_0^\perp\rangle\}$ basis

$$Q = \begin{pmatrix} \cos 2\alpha & -\sin 2\alpha \\ \sin 2\alpha & \cos 2\alpha \end{pmatrix}, \tag{7.44}$$

where α is the angle between $|w_0\rangle$ and $|x_0^\perp\rangle$ and is $O(N^{-1/2})$. This matrix has eigenvalues $e^{\pm 2i\alpha}$, and the eigenstates are $|\alpha_\pm\rangle = (|w_0\rangle \mp i|w_0^\perp\rangle)/\sqrt{2}$. Now suppose that N, the number of possible inputs to our Boolean function is $N = 2^n$. In order to discriminate between the two types of oracles, we need to determine the eigenvalues of Q to $O(2^{-n/2})$, because $1 - e^{2i\alpha}$ is of this order. We then make use of the phase estimation algorithm with $m > n/2$ and an input state into the target qubits of the Controlled-Q^{2^j} gates of $|w_0\rangle$. Now $|w_0\rangle$ is not an eigenstate of Q, but it is the sum of two eigenstates $|w_0\rangle = (|\alpha_+\rangle + |\alpha_-\rangle)/\sqrt{2}$. The output of the phase estimation circuit will be approximately of the form $(|a_+\rangle|\alpha_+\rangle + |a_-\rangle|\alpha_-\rangle)/\sqrt{2}$ where $a_+/2^m$

is a good estimate of $\alpha/2\pi$ and $a_-/2^m$ is a good estimate of $(2\pi - \alpha)/2\pi$. If we simply measure the first m qubits of the output state in the computational basis, we will obtain, with equal probability an estimate of either a_+ or a_-. If either one of these is different from zero, then the we know that there is an x_0 such that $f(x_0) = 1$.

The procedure we have just outlined is most useful when there is more than one value of x such that $f(x) = 1$, and we want to find out how many values of x satisfying this condition there are. This is a procedure known as quantum counting. In that case the eigenvalues of Q depend on the number of solutions, and by estimating the eigenvalues we can determine that number.

7.6 Quantum Walks

Finding new quantum algorithms has not been easy, and one approach one might try to find new ones is to see if there are particular mathematical structures that have proved useful in classical algorithms and then try to generalize them to the quantum realm. One area in which this approach has been fruitful is in algorithms based on random walks. There are a number of classical algorithms based on random walks, and we shall present an example of one shortly. It has been possible to define a quantum version of a random walk, known as a quantum walk, and there are now new quantum algorithms that are based on quantum walks. In this section we will describe what a quantum walk is and some of the things they can do.

The simplest example of a classical random walk is one on a line. The walk starts at a point, which we shall call the origin. The walker then flips an unbiased coin. If it comes up heads, he takes one step to the right, if tails, one step to the left (all steps are the same length). This process is repeated for the desired number of steps, n. The result can be described by a probability distribution, $p(x; n)$, which is the probability of being at position x after n steps. The position is measured in units of step length, and is positive to the right of the origin (which is $x = 0$) and negative to the left. For example, for a walk of two steps, the only possible final positions are $x = -2, 0, 2$ and we find that $p(-2; 2) = p(2; 2) = 1/4$ and $p(0; 2) = 1/2$.

It is also possible to perform random walks on more general structures known as graphs. A graph consists of a set of vertices, V, and a set of edges, E. Each edge connects two of the vertices, and an edge is labelled by an unordered pair of vertices, which are just the vertices connected by that edge. In general, not all of the vertices will be connected by an edge. A graph in which each pair of vertices is connected by an edge is known as a complete graph, and if there are N vertices, there will be $N(N - 1)/2$ edges in a complete graph. In order to perform a random walk on a graph, we choose one vertex on which to start. For the first step, we see which vertices are connected to the vertex we are on by an edge, and then we randomly choose one of them, each having the same probability, and then move to that vertex. So, for example, if our starting vertex is connected to three other vertices, then we would end up on each of those vertices with a probability of $1/3$. We then repeat this process for the new vertex in order to make the second step, and keep repeating it for as many steps as we wish.

A simple example of an algorithm based on a random walk is one that determines whether two vertices in a graph are connected or not. In order to determine whether there is a a path connecting a specified vertex u to another specified vertex v, we can start a walker at u, execute a random walk for a certain number of steps, and see after each step whether we have reached v. It can be shown that if the graph has N vertices, and we run the walk for $2N^3$ steps, then the probability of not reaching v if there is a path from u to v is less than one half. So if we run a walk of this length m times, and do not reach v during any of these walks, the probability of this occurring if there is a path from u to v is less than $1/2^m$. Therefore, we shall say that if during one of these walks we find v, then there is a path from u to v, and if after m walks of length $2N^3$ during which we do not reach v, then there is no path from u to v. Our probability of making a mistake is less than 2^{-m}. This gives us a probabilistic algorithm for determining whether there is a path from u to v.

There are a number of different ways to define a quantum walk, but we shall only explore one of them, known as the scattering quantum walk. In this walk, the particle resides on the edges and can be though of as scattering when it goes through a vertex. In particular, suppose an edge connects vertices v_1 and v_2. There are two states corresponding to this edge, and these states are assumed to be orthogonal. There is the state $|v_1 v_2\rangle$ which corresponds to the particle being on the edge and going from vertex v_1 to v_2, and the state $|v_2, v_1\rangle$, which corresponds to the particle being on the edge and going from v_2 to v_1. The set of these states for all of the edges form an orthonormal basis for the Hilbert space of the walking particle.

Next we need a unitary operator that will advance the walk one time step. We obtain this operator by combining the action of local unitaries that describe what happens at the individual vertices. Let us consider a vertex v, and let ω_v be the linear span of the set of edge states entering v and Ω_v be the span of the set of edge states leaving v. Because each edge attached to v has two states, one entering and one leaving v, ω_v and Ω_v have the same dimension. The local unitary, U_v at v maps ω_v to Ω_v. We are going to require that the action of U_v be completely symmetric, that is we want it to act on all of the edges in the same way. In particular, suppose there are n edges attached to v. We want the amplitude for the particle to be reflected back onto the edge from which it entered v to be $-r$ and the amplitude for it to be transmitted through the vertex and leave by a different edge to be t. That is, if we denote the vertices attached to v by $1, 2, \ldots n$, and if the particle enters v from vertex j, then

$$U_v|j, v\rangle = -r|v, j\rangle + t \sum_{k=1, k \neq j}^{n} |v, k\rangle. \tag{7.45}$$

In order for U_v to be unitary, we must have that the state on the right-hand side of this equation be normalized

$$|r|^2 + (n-1)|t|^2 = 1, \quad .$$ (7.46)

and that output states resulting from orthogonal input states be orthogonal

$$-r^*t - rt^* + (n-2)|t|^2 = 0.$$ (7.47)

If, for convenience, we also require that r and t be real, we find that

$$r = \frac{n-2}{n} \quad t = \frac{2}{n}.$$ (7.48)

Note that with this choice, $r + t = 1$. The action of the unitary operator U that advances the walk one step, is given by the combined action of all of the operators U_v at the different vertices.

Let us look at a walk on a simple graph known as a star graph, shown in Fig. 7.7. It consists of a central vertex with N edges attached to it and N vertices attached to the other ends of these edges. We shall denote the central vertex by 0 and the outer vertices by $1, 2, \ldots N$. The local unitary corresponding to the central vertex is described by the operator U_v above with $r = (N-2)/N$ and $t = 2/N$. The outer vertices reflect the particle except for one, which we shall assume is vertex 1, that reflects the particle and flips the phase of the state as well. That is the marked vertex, the one that is different from the others, that we are trying to find. Therefore, we have $U|0, j\rangle = |j, 0\rangle$ for $j \geq 2$ and $U|0, 1\rangle = -|1, 0\rangle$. We shall start the walk in the state

$$|\psi_{init}\rangle = \frac{1}{\sqrt{N}} \sum_{j=1}^{N} |0, j\rangle.$$ (7.49)

Because of the symmetry of the problem the walk takes place in a only a subspace of the entire Hilbert space, and the dimension of this subspace is small. In particular, if we define

Fig. 7.7 A star graph consists of a central vertex 0 and N outer vertices. The outer vertices are connected to the central vertex by N edges. For the figure $N = 8$

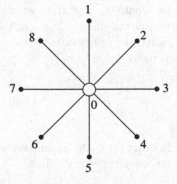

$$|\psi_1\rangle = |0, 1\rangle$$

$$|\psi_2\rangle = |1, 0\rangle$$

$$|\psi_3\rangle = \frac{1}{\sqrt{N-1}} \sum_{j=2}^{N} |0, j\rangle$$

$$|\psi_4\rangle = \frac{1}{\sqrt{N-1}} \sum_{j=2}^{N} |j, 0\rangle \tag{7.50}$$

then the action of U on these states is given by

$$U|\psi_1\rangle = -|\psi_2\rangle$$

$$U|\psi_2\rangle = -r|\psi_1\rangle + t\sqrt{N-1}|\psi_3\rangle$$

$$U|\psi_3\rangle = |\psi_4\rangle$$

$$U|\psi_4\rangle = r|\psi_3\rangle + t\sqrt{N-1}|\psi_1\rangle. \tag{7.51}$$

From this we see that the four-dimensional subspace spanned by these vectors is invariant under U. Our initial state, which can be expressed as

$$|\psi_{init}\rangle = \frac{1}{\sqrt{N}}|\psi_1\rangle + \sqrt{\frac{N-1}{N}}|\psi_3\rangle, \tag{7.52}$$

is also in this subspace, and so the entire quantum walk will take place in the four-dimensional invariant subspace. This drastically simplifies finding the state of the particle after n steps.

Now from the way this walk has been set up, you might suspect that it will simply mimic the action of the Grover algorithm. If that is the case, you are right. In order to see this, we first note that the action of U in the invariant subspace can be described by a 4×4 matrix

$$M = \begin{pmatrix} 0 & -r & 0 & t\sqrt{N-1} \\ -1 & 0 & 0 & 0 \\ 0 & t\sqrt{N-1} & 0 & r \\ 0 & 0 & 1 & 0 \end{pmatrix}, \tag{7.53}$$

where the matrix elements of M are given by $M_{jk} = \langle\psi_j|U|\psi_k\rangle$. In order to find out how the walk behaves, we first find the eigenvalues and eigenvectors of U. The characteristic equation for the eigenvalues, λ, of M is

$$\lambda^4 - 2r\lambda^2 + 1 = 0. \tag{7.54}$$

We will solve the equation in the large N limit. In that case, we express the equation as

$$\lambda^4 - 2\lambda^2 + 1 + 2t\lambda^2 = 0. \tag{7.55}$$

We ignore the last term on the left-hand side, which is small when N is large, in order to find zeroth order solutions, λ_0. This gives $\lambda_0 = \pm 1$. We now set $\lambda = \lambda_0 + \delta\lambda$, and substitute it back into the equation. Keeping terms of up to second order in small quantities we find that for $\lambda_0 = 1$

$$\delta\lambda^2 + \frac{1}{2}t(1 + 2\delta\lambda) = 0, \tag{7.56}$$

and for $\lambda_0 = -1$ we find

$$\delta\lambda^2 + \frac{1}{2}t(1 - 2\delta\lambda) = 0. \tag{7.57}$$

In both cases, the solutions are, to lowest order in $1/N$

$$\delta\lambda = \pm i\sqrt{\frac{t}{2}}, \tag{7.58}$$

which is of order $N^{-1/2}$.

It is also necessary to find the eigenstates of M. Setting $\Delta = \sqrt{t/2}$, we have that for $\lambda = 1 + i\Delta$ and $\lambda = 1 - i\Delta$, the eigenstates are, respectively,

$$|u_1\rangle = \frac{1}{2}\begin{pmatrix} -1 \\ 1 \\ -i \\ -i \end{pmatrix} \qquad |u_2\rangle = \frac{1}{2}\begin{pmatrix} -1 \\ 1 \\ i \\ i \end{pmatrix}, \tag{7.59}$$

and for $\lambda = -1 + i\Delta$ and $\lambda = -1 - i\Delta$, the eigenstates are, respectively,

$$|u_3\rangle = \frac{1}{2}\begin{pmatrix} 1 \\ 1 \\ -i \\ i \end{pmatrix} \qquad |u_4\rangle = \frac{1}{2}\begin{pmatrix} 1 \\ 1 \\ i \\ -i \end{pmatrix}. \tag{7.60}$$

In terms of the eigenstates, we see that

$$|\psi_{init}\rangle = \frac{i}{2}(|u_1\rangle - |u_2\rangle + |u_3\rangle - |u_4\rangle) + O(N^{-1/2}). \tag{7.61}$$

Noting that $1 \pm i\Delta \cong e^{\pm i\Delta}$ and $-1 \pm i\Delta \cong -e^{\mp i\Delta}$, we have that

$$U^n|\psi_{init}\rangle = \frac{i}{2}[e^{in\Delta}|u_1\rangle - e^{-in\Delta}|u_2\rangle$$

$$+(-1)^n(e^{-in\Delta}|u_3\rangle - e^{in\Delta}|u_4\rangle)] + O(N^{-1/2}), \qquad (7.62)$$

or

$$U^n|\psi_{init}\rangle = \frac{1}{2}\begin{pmatrix} \sin(n\Delta) \\ -\sin(n\Delta) \\ \cos(n\Delta) \\ \cos(n\Delta) \end{pmatrix} + \frac{1}{2}(-1)^n\begin{pmatrix} \sin(n\Delta) \\ \sin(n\Delta) \\ \cos(n\Delta) \\ -\cos(n\Delta) \end{pmatrix}, \qquad (7.63)$$

up to order $N^{-1/2}$.

From this result, we see that when $n\Delta$ is close to $\pi/2$, the particle will be located on the edge connected to the marked vertex. In n is even it will be in the state $|0, 1\rangle$ and if n is odd it will be in the state $-|1, 0\rangle$. By simply measuring the location of the particle, in particular, which edge it is on, we will find which vertex is the marked one. Note that if $n\Delta$ is close to $\pi/2$, then n is of order $\sqrt{N}$. Classically, in order to find the marked vertex, we would have to check each vertex, which would require $O(N)$ operations, whereas if we run a quantum walk, we can find the marked vertex in $O(\sqrt{N})$ steps. Therefore, we obtain a quadratic speedup.

So far, we have only used a quantum walk to do something we already knew how to do, find a marked element in a list. Let us see if we can use it to do something else. Suppose that instead of a marked vertex, our star graph has an extra edge. That is, there is an edge between two of the outer vertices, and we would like to find out where it is. A quantum walk can provide a quadratic speedup for this type of search as well. The graph is depicted in Fig. 7.8.

Let's assume the extra edge is between vertices 1 and 2. That means that besides the states $|0, j\rangle$ and $|j, 0\rangle$, for $j = 1, 2, \ldots N$, we also have the states $|1, 2\rangle$ and

Fig. 7.8 A star graph with an extra edge between two outer vertices

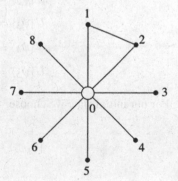

$|2, 1\rangle$. For simplicity we shall assume that vertices 1 and 2 just transmit the particle. Our unitary operator will now act as $U|0, j\rangle = |j, 0\rangle$ for $j > 2$, and

$$U|0, 1\rangle = |1, 2\rangle \quad U|0, 2\rangle = |2, 1\rangle$$
$$U|1, 2\rangle = |2, 0\rangle \quad U|2, 1\rangle = |1, 0\rangle. \tag{7.64}$$

Its action on the states $|j, 0\rangle$ is as before. The walk resulting from this choice of U can also be analyzed easily, because it stays within a five-dimensional subspace of the entire Hilbert space. Define the states

$$|\psi_1\rangle = \frac{1}{\sqrt{2}}(|0, 1\rangle + 0, 2\rangle)$$

$$|\psi_2\rangle = \frac{1}{\sqrt{2}}(|1, 0\rangle + |2, 0\rangle)$$

$$|\psi_3\rangle = \frac{1}{\sqrt{N-2}} \sum_{j=3}^{N} |0, j\rangle$$

$$|\psi_4\rangle = \frac{1}{\sqrt{N-2}} \sum_{j=3}^{N} |j, 0\rangle$$

$$|\psi_5\rangle = \frac{1}{\sqrt{2}}(|1, 2\rangle + |2, 1\rangle). \tag{7.65}$$

These states span a five-dimensional space we shall call S. The unitary transformation, U, that advances the walk one step acts on these states as follows:

$$U|\psi_1\rangle = |\psi_5\rangle$$
$$U|\psi_2\rangle = -(r - t)|\psi_1\rangle + 2\sqrt{rt}|\psi_3\rangle$$
$$U|\psi_3\rangle = |\psi_4\rangle$$
$$U|\psi_4\rangle = (r - t)|\psi_3\rangle + 2\sqrt{rt}|\psi_1\rangle$$
$$U|\psi_5\rangle = |\psi_2\rangle. \tag{7.66}$$

For our initial state we choose

$$|\psi_{init}\rangle = \frac{1}{\sqrt{2N}} \sum_{j=1}^{N}(|0, j\rangle - |j, 0\rangle)$$

$$= \frac{1}{\sqrt{N}}(|\psi_1\rangle - |\psi_2\rangle)$$

$$+\sqrt{\frac{N-2}{2N}}(|\psi_3\rangle - |\psi_4\rangle)), \tag{7.67}$$

which is in S. Since the initial state is in S, and S is an invariant subspace of U, the entire walk will remain in S, and so we find ourselves in a situation similar to the previous one This search, however, is more sensitive to the choice of initial state than the previous one. While we didn't mention it before, in the previous search we could also have taken a superposition of all ingoing states instead of all outgoing ones as our initial state. In the present case, the minus sign in the first expression for initial state is essential; if it is replaced by a plus sign, the search will fail.

In order to find the evolution of the quantum state for this walk, we proceed as before and find the eigenvalues and eigenstates of U restricted to S. The matrix that describes the action of U on S is given by

$$M = \begin{pmatrix} 0 & -(r-t) & 0 & 2\sqrt{rt} & 0 \\ 0 & 0 & 0 & 0 & 1 \\ 0 & 2\sqrt{rt} & 0 & (r-t) & 0 \\ 0 & 0 & 1 & 0 & 0 \\ 1 & 0 & 0 & 0 & 0 \end{pmatrix}. \tag{7.68}$$

The characteristic equation for this matrix is

$$\lambda^5 - (r-t)\lambda^3 + (r-t)\lambda^2 - 1 = 0. \tag{7.69}$$

One root of this equation is $\lambda = 1$, and if we factor out $(\lambda - 1)$ from the above equation, we are left with

$$\lambda^4 + \lambda^3 + 2t\lambda^2 + \lambda + 1 = 0. \tag{7.70}$$

As before, we will use a perturbation expansion to find the roots of this equation with the transmission amplitude, t, as the small parameter. The zeroth order solutions are found by setting $t = 0$, which gives us the large N limit, and we find

$$\lambda^4 + \lambda^3 + \lambda + 1 = (\lambda + 1)(\lambda^3 + 1) = 0, \tag{7.71}$$

so the zeroth order roots are -1 twice, $e^{i\pi/3}$, and $e^{-i\pi/3}$. Setting $\lambda = -1 + \delta\lambda$, substituting into the above equation and keeping terms of up to $(\delta\lambda)^2$ gives

$$3(\delta\lambda)^2 - 4t\delta\lambda + 2t = 0, \tag{7.72}$$

whose solution, keeping lowest order terms is

$$\delta\lambda = \pm i\sqrt{\frac{2t}{3}} = O(N^{-1/2}). \tag{7.73}$$

If we set $\lambda = e^{\pm i\pi/3} + \delta\lambda$, we find that $\delta\lambda = O(N^{-1})$, so these roots and their corresponding eigenvalues are not of interest, because they will not yield a quadratic speedup.

We now need to find the eigenvectors. If the components of the eigenvectors are denoted by x_j, where $j = 1, \ldots 5$, the eigenvector equations are

$$- (r - t)x_2 + 2\sqrt{rt}x_4 = (-1 \pm i\Delta)x_1$$

$$x_5 = (-1 \pm i\Delta)x_2$$

$$2\sqrt{rt}x_2 + (r - t)x_4 = (-1 \pm i\Delta)x_3$$

$$x_3 = (-1 \pm i\Delta)x_4$$

$$x_1 = (-1 \pm i\Delta)x_5, \tag{7.74}$$

where now $\Delta = (2t/3)^{1/2}$. To lowest order (the terms that were dropped are of order $1/\sqrt{N}$ or lower) the eigenvector corresponding to the eigenvalue $-1 + i\Delta$ is

$$|v_1\rangle = \frac{1}{\sqrt{6}} \begin{pmatrix} 1 \\ 1 \\ -i\sqrt{3/2} \\ i\sqrt{3/2} \\ -1 \end{pmatrix}, \tag{7.75}$$

and the eigenvector corresponding to the eigenvalue $-1 - i\Delta$ is

$$|v_2\rangle = \frac{1}{\sqrt{6}} \begin{pmatrix} 1 \\ 1 \\ i\sqrt{3/2} \\ -i\sqrt{3/2} \\ -1 \end{pmatrix}. \tag{7.76}$$

We find that, up to terms of order $N^{-1/2}$, our initial state can be expressed as

$$|\psi_{init}\rangle = \frac{i}{\sqrt{2}}(|v_1\rangle - |v_2\rangle). \tag{7.77}$$

Expressing the eigenvalues corresponding to $|v_1\rangle$ and $|v_2\rangle$ as

$$-1 + i\Delta \cong -e^{-i\Delta} \qquad -1 - i\Delta \cong -e^{i\Delta} \tag{7.78}$$

we find that the state after n steps is

$$U^n|\psi_{init}\rangle = \frac{(-1)^n}{\sqrt{3}} \begin{pmatrix} \sin(n\Delta) \\ \sin(n\Delta) \\ \sqrt{3/2}\cos(n\Delta) \\ -\sqrt{3/2}\cos(n\Delta) \\ -\sin(n\Delta) \end{pmatrix}. \tag{7.79}$$

From this equation, we can see that when $n\Delta = \pi/2$, the particle is located on one of the edges leading to the extra edge or on the extra edge itself. This will happen when $n = O(\sqrt{N})$.

We now need to discuss how to interpret this result. It is reasonable to assume that if we are given a graph with an extra edge in an unknown location, we only have access to the edges connecting the central vertex to the outer ones, and not to the extra edge itself (if we had access to the extra edge, then we would have to know where it is). That is, in making a measurement, we can only determine which of the edges connecting central vertex to to the outer ones the particle is on. If it is on the extra edge, we will not detect it. So, after n steps, where $n\Delta = \pi/2$, we measure the edges to which we have access to find out where the particle is. With probability $2/3$ it will be on an edge connected to the extra edge, and with probability $1/3$ it will be on the extra edge itself, in which case we won't detect it.

In comparing this procedure to a classical search for the extra edge, we shall assume that classically the graph is specified by an adjacency list, which is an efficient specification for sparse graphs. For each vertex of the graph, one lists the vertices that are connected to it by an edge. In our case, the central vertex is connected to all of the other vertices, the vertices not connected to the extra edge are connected only to the central vertex, and two of the outer vertices are connected to the central vertex and to each other. Searching this list classically would require $O(N)$ steps to find the extra edge, while the quantum procedure will succeed in $O(\sqrt{N})$ steps. Therefore, we again obtain a quadratic speedup by using a quantum walk. Finally, let us mention that it is possible to use a quantum walk to find not just an extra edge, but a path through a simple maze, with a quantum speedup.

We have just examined the use of quantum walks in search problems, but they have been useful in developing other types of algorithms as well. One example is element distinctness. One has a function in the form of a black box, that is one puts in an input x and the output is $f(x)$, but we have no knowledge about the function. We can only send in inputs and obtain outputs. Our task is to find two inputs, if they exist, that give the same output. This can be accomplished by using a kind of quantum walk, which requires fewer queries to the black box than is necessary on a classical computer. It is also possible to use quantum walks to evaluate certain types of Boolean formulas with fewer queries than are possible classically.

7.7   Quantum Simulation

Suppose we would like to calculate the dynamics of a of a spin chain consisting of spin-1/2 particles interacting with nearest neighbor interactions. An example would be the Heisenberg spin chain, with Hamiltonian

$$H = -J \sum_{n=1}^{N-1} \mathbf{S}_n \cdot \mathbf{S}_{n+1}, \tag{7.80}$$

where J is the coupling constant ($J > 0$ for a ferromagnetic chain and $J < 0$ for an anti-ferromagnetic one) and $\mathbf{S}_n$ is the spin operator for the nth spin. What we would like to find is $U(t) = \exp(-itH/\hbar)$ acting on an initial state. In order to do so we would expand the state in terms of a basis, which means we would have to keep track of 2^N amplitudes. This exponential dependence on the number of spins presents a huge problem. We will only be able to treat systems consisting of a small number of spins before the calculations become intractable.

An alternative would be to simulate the system on a quantum computer. Instead of an exponential number of classical bits, we would only need N qubits, one for each spin in the chain. We then need a way of breaking down the operator $U(t)$ into a series of quantum gates. We first note that each term in our Hamiltonian acts on only two qubits, so we should be able to implement $U_n(t) = \exp(-itH_n/\hbar)$, where $H_n = -J\mathbf{S}_n \cdot \mathbf{S}_{n+1}$, with one and two-qubit gates. However, because the operators H_n do not all commute, it is not the case that $U(t)$ is just the product of the operators $U_n(t)$. The solution to this problem is provided by the Trotter product formula, which implies that for Hermitian operators A and B,

$$e^{i(A+B)\Delta t} = e^{iA\Delta t}e^{iB\Delta t} + O((\Delta t)^2), \tag{7.81}$$

and for our Hamiltonian,

$$U(\Delta t) = \prod_{n=1}^{N-1} U_n(\Delta t) + O((\Delta t)^2). \tag{7.82}$$

This dependence for the error of each piece implies that the total error will go to zero when $\Delta t \to 0$. Suppose that $\Delta t = t/M$, for some large integer M. When we multiply the terms for each of the M intervals together, the leading term in the error is $(\Delta t)^2$ times the sum of M operators. Roughly speaking, we can think of the size of this term as of order $M(\Delta t)^2 = t(\Delta t)$, which does go to zero as Δt goes to 0. Consequently, the total error can be made small.

In order to simulate the time development generated by our Hamiltonian, we break the time interval $(0, t)$ into small parts of size Δt, and implement the time evolution for each part with one and two-qubit gates. The interval Δt is chosen sufficiently small so that the total error is below a specified limit. This, of course, is

just the basic strategy, and to actually simulate a quantum system requires a much
more detailed analysis. A more in-depth presentation of quantum simulation is given
by the final paper in the references.

7.8 QAOA

For the foreseeable future, we will not have a large quantum computer with error-
corrected qubits. What we will have are machines with roughly 100 somewhat
noisy qubits. This period has been referred to by John Preskill of Caltech as the
NISQ (Noisy Intermediate Scale Quantum) era. One would like to find tasks that
machines of this type can usefully tackle. One possibility is a quantum-classical
hybrid calculation that uses a quantum computer for part of the computation and a
classical computer for the remaining part. A number of procedures of this type have
been proposed, and we will discuss one of them.

The quantum approximate optimization algorithm (QAOA), proposed by
E. Farhi, J. Goldstone, and S. Gutmann, is a hybrid quantum-classical procedure for
tackling optimization problems with constraints. In this approach, the information
about the problem is encoded into the phases of amplitudes of the computational
basis states, and then a mixing procedure is applied. In order to see how this works,
it is useful to look at an example, in particular, the MaxCut problem for a graph. In
this problem, one is given a graph, and the object is to divide the vertices into two
sets in order to maximize the number of edges between the two sets. One represents
each vertex as a qubit and the state $|0\rangle$ corresponds to being in one set and the state
$|1\rangle$ corresponds to being in the other. We define a Hamitonian

$$H_C = \sum_{j,k \in E} \frac{1}{2}(1 - \sigma_j^z \sigma_k^z), \qquad (7.83)$$

where E is the set of edges of the graph and σ_j^z is the σ^z operator for the jth qubit.
Note that if the vertices j and k are in the same set, the term corresponding to that
edge is zero, while if they are in different sets, the term is one. The eigenstates of
H_C are just the computational basis states, and what we would like to do is find
computational basis states that correspond to large eigenvalues of H_C. Our goal is
not to find the optimal solution, which would be the eigenstate of H_C with the largest
eigenvalue, but to find states that are close to the optimal solution.

We now apply $U_C(\gamma) = \exp(-i\gamma H_C)$ to the state

$$|\psi_{in}\rangle = \frac{1}{2^{N/2}} \sum_{x=0}^{2^N-1} |x\rangle, \qquad (7.84)$$

where N is the number of vertices in the graph and $|x\rangle$ is an N qubit computational basis state corresponding to the N-digit binary number x. The parameter γ is a variational parameter. Next, one applies the operator $U_B(\beta) = \exp(-i\beta H_B)$, where

$$H_B = \sum_{j=1}^{N} \sigma_j^x, \tag{7.85}$$

and β is a second variational parameter. This is a mixing operator. This procedure, the application of $U_B U_C$, can be iterated, with different values of γ and β at each step, but we will just look at the simplest case. After the application of $U_B U_C$, we are left with the state $|\gamma, \beta\rangle = U_B(\beta)U_C(\gamma)|\psi_{in}\rangle$.

We now come to the question of how the parameters γ and β should be chosen. One option is to compute the expectation value

$$F(\gamma, \beta) = \langle \gamma, \beta|H_C|\gamma, \beta\rangle \tag{7.86}$$

on a classical or quantum computer and find the values of γ and β that give the maximum value of $F(\gamma, \beta)$. Let's call these values γ_m and β_m. One then prepares the state $|\gamma_m, \beta_m\rangle$ as above, and measures it in the computational basis. If for $F(\gamma_m, \beta_m)$ the largest contribution is from computational basis states with large eigenvalues, then this measurement should yield one of these states and provide a configuration with a large number of edges between the sets. Whether this procedure will yield a quantum advantage is a subject of active investigation.

For up-to-date information on quantum algorithms, the Quantum Algorithm Zoo is a good place to look: https://quantumalgorithmzoo.org.

7.9 Problems

1. Suppose we have a star graph with a loop on one of its outer vertices, say vertex 1. The other vertices simply reflect the particle, $U|0, j\rangle = |j, 0\rangle$ for $j > 1$. The loop has one quantum state, which we shall denote by $|l_1\rangle$. The unitary operator has the action $U|0, 1\rangle = |l_1\rangle$ and $U|l_1\rangle = |1, 0\rangle$. Show that starting with the initial state in Eq. (7.67) the particle making the walk will become localized on the loop and the edge connected to the loop in $O(\sqrt{N})$ steps.

2. We have a Controlled-U gate, which acts on two qubits. Qubit a is the control qubit and qubit b is the target qubit, so that $|0\rangle_a|j\rangle_b \rightarrow |0\rangle_a|j\rangle_b$ and $|1\rangle_a|j\rangle_b \rightarrow |1\rangle_a U|j\rangle_b$ for $j = 0, 1$. Suppose that the eigenvalues of U are ± 1. Our task is to generate two qubits one in the $+1$ eigenstate, $|u_+\rangle$, and one in the -1 eigenstate, $|u_-\rangle$ with one use of the gate. Show, making use of the rotational invariance of the singlet state

$$|\phi_s\rangle = \frac{1}{\sqrt{2}}(|0\rangle|1\rangle - |1\rangle|0\rangle),$$

that if we start with the three-qubit state

$$|\Psi_{in}\rangle_{abc} = |+x\rangle_a|\phi_s\rangle_{bc}$$

and send qubits a and b through the Controlled-U gate and make the proper measurement of qubit a, then one of the remaining qubits will be in the state $|u_+\rangle$ and the other will be in the state $|u_-\rangle$, and we will know which qubit is in which state.

3. Suppose we have a black box that evaluates the Boolean function $f(x)$, where x is the n-bit string $x_1, x_2, \ldots x_n$. This function is a sum of linear and quadratic terms in the variables x_j and each variable appears in only one term. By considering the function $f(x) + f(\bar{x})$, where $\bar{x}$ is the n-bit string $x_1 + 1, x_2 + 1, \ldots x_n + 1$, show that we can use the Bernstein-Vazirani algorithm to determine which variables appear in quadratic terms with two function evaluations. How many evaluations would be required classically?

Further Reading

1. D. Aharonov, A. Ambainis, J. Kempe, U. Vazirani, Quantum walks on graphs, in *Proceedings of the 33rd Symposium on the Theory of Computing (STOC01)* (ACM Press, New York, 2001), pp. 50–59. quant-ph/0012090
2. E. Bernstein, U. Vazirani, Quantum complexity theory. SIAM J. Comput. **26**, 1411 (1997)
3. A.M. Childs, W. van Dam, Quantum algorithms for algebraic problems. Rev. Mod. Phys. **82**, 1(2010)
4. R. Cleve, A. Ekert, C. Macchiavello, M. Mosca, Quantum algorithms revisited. Proc. R. Soc. Lond. A **454**, 339 (1998)
5. E. Farhi, J. Goldstone, and S. Gutmann, A quantum approximate approximation algorithm, arXiv: 1411.4028 (2014)
6. E. Feldman, M. Hillery, H.-W. Lee, D. Reitzner, H. Zheng, V. Bužek, Finding structural anomalies in graphs by means of quantum walks. Phys. Rev. A **82**, 040302R (2010)
7. I.M. Georgescu, S. Ashhab, F. Nori, Quantum simulation. Rev. Mod. Phys. **86**, 154 (2014)
8. L.K. Grover, Quantum mechanics helps in searching for a needle in a haystack. Phys. Rev. Lett. **79**, 325 (1997)
9. D. Reitzner, M. Hillery, D. Koch, Finding paths with quantum walks, or quantum walking through a maze. Phys. Rev. A **96**, 032323 (2017)

Chapter 8
Quantum Machines

8.1 Introduction

If we are given quantum information in the form of qubits in a particular state, we have seen that we can process that information by sending the qubits through different sequences of gates. A particular collection of gates constitutes a quantum machine that manipulates the information encoded in the qubits in a particular way. A quantum machine can either perform a single task, or, if it is programmable, a number of different tasks, the exact task depending on the program.

In this chapter we want to look at several different quantum machines. The first two, a quantum cloner and a Universal-NOT (or, more concisely, U-NOT) gate, are single task machines. They both perform, approximately, tasks that cannot be performed exactly. We then move on to programmable machines. We will first prove a general result that shows there is no deterministic, universal programmable quantum processor. We will then examine two different probabilistic programmable machines. The first is based on the same circuit as the cloner. We will show, as an example, how it can be used to implement a quantum phase gate in any basis, the basis being determined by the program. The second unambiguously discriminates between two states, but the two states it is discriminating between are given in the form of a program and not hard-wired into the machine.

8.2 Cloners and U-NOT Gates

As we have seen, a device that perfectly clones a quantum state is impossible to construct. However, if we relax the requirement that the copies be perfect, it is possible to copy quantum information. A second operation that is not possible to perform exactly in the U-NOT operation. This ideally would take a qubit in an

© Springer Nature Switzerland AG 2021

J. A. Bergou et al., *Quantum Information Processing*, Graduate Texts in Physics,
https://doi.org/10.1007/978-3-030-75436-5_8

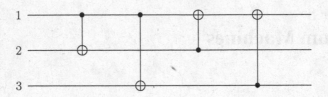

Fig. 8.1 Quantum circuit for the cloning machine

arbitrary state $|\psi\rangle = \alpha|0\rangle + \beta|1\rangle$ and send it into the orthogonal state $|\psi_\perp\rangle = \beta^*|0\rangle - \alpha^*|1\rangle$. The indication that this is an impossible operation is the appearance of complex conjugates. The perfect U-NOT operation is an anti-unitary one, but quantum operations must be unitary. It is, nonetheless, possible to construct an approximate U-NOT gate, and, it turns out, approximate cloners and approximate U-NOT gates are closely related.

Let us begin with the cloner. Consider the circuit shown in Fig. 8.1.

The circuit consists of three qubits being acted upon by four Controlled-NOT gates. The input qubit is qubit number 1, and it is its state we wish to copy. In order to see how this works, let us consider what happens with different input states for the remaining two qubits. Define the two-qubit states

$$|\Xi_{00}\rangle = \frac{1}{\sqrt{2}}(|0\rangle|0\rangle + |1\rangle|1\rangle)$$

$$|\Xi_{0x}\rangle = \frac{1}{\sqrt{2}}|0\rangle(|0\rangle + |1\rangle). \tag{8.1}$$

Now if qubit 1 is in the state $|\psi\rangle_1$ and qubits 2 and 3 are in one of the two states above, then the cloning circuit will implement the following transformations

$$|\psi\rangle_1|\Xi_{00}\rangle_{23} \to |\psi\rangle_1|\Xi_{00}\rangle_{23}$$

$$|\psi\rangle_1|\Xi_{0x}\rangle_{23} \to |\psi\rangle_2|\Xi_{00}\rangle_{13}. \tag{8.2}$$

Examining these equations, we see that in the first the quantum information from the first qubit appears in output 1, and in the second it appears in output 2, so what this circuit does is move the information from the first qubit around, and the location to which it gets moved is determined by the state sent into inputs 2 and 3. This suggests that if instead of sending either $|\Xi_{00}\rangle$ or $|\Xi_{0x}\rangle$ into inputs 2 and 3, we send in a linear combination of them, some of the quantum information from qubit 1 will appear in output 1 and some of it will appear in output 2, thereby cloning the state. This is, in fact, exactly what happens. If we choose

$$|\Psi\rangle_{23} = c_0|\Xi_{00}\rangle_{23} + c_1|\Xi_{0x}\rangle_{23}, \tag{8.3}$$

as the input state for qubits 2 and 3, with c_0 and c_1 real for simplicity, and $c_0^2 + c_1^2 + c_0 c_1 = 1$ so that the state is normalized, the reduced density matrices for outputs 1 and 2 are

$$\rho_1^{(out)} = (c_0^2 + c_0 c_1)|\psi\rangle\langle\psi| + \frac{c_1^2}{2}I$$

$$\rho_2^{(out)} = (c_1^2 + c_0 c_1)|\psi\rangle\langle\psi| + \frac{c_0^2}{2}I. \tag{8.4}$$

Note that by choosing c_0 and c_1 we can control how much information about $|\psi\rangle$ goes to which output. In particular, if we choose $c_0 = c_1 = 1/\sqrt{3}$, then the information is divided equally, and we find that

$$\rho_1^{(out)} = \rho_2^{(out)} = \frac{5}{6}|\psi\rangle\langle\psi| + \frac{1}{6}|\psi_\perp\rangle\langle\psi_\perp|, \tag{8.5}$$

where $|\psi_\perp\rangle$ is the qubit state orthogonal to $|\psi\rangle$. Therefore, the fidelity of the cloner output $\rho_1^{(out)}$ (or $\rho_2^{(out)}$ since they are the same in this case) to the ideal output, $|\psi\rangle$, which is given by $\langle\psi|\rho_1^{(out)}|\psi\rangle$, is 5/6. A fidelity of one would imply perfect cloning, so what we have here is a device that produces two copies of the input qubit that are pretty good approximations to it. Note that the fidelity does not depend on the input state, that is all states are cloned equally well. This feature of this cloning machine is known as universality.

Note that the cloner employs three qubits, and we have only discussed the final state of two of them. One might wonder if the output state of the third qubit is of interest. This is, in fact, where the connection with the U-NOT gate enters. The output state of the third qubit is given by

$$\rho_3^{(out)} = c_0 c_1|\psi^*\rangle\langle\psi^*| + \frac{1}{2}(1 - c_0 c_1)I, \tag{8.6}$$

where $|\psi^*\rangle = \alpha^*|0\rangle + \beta^*|1\rangle$ and I is the two-by-two identity matrix. If we now apply the unitary operator $U_0 = -i\sigma_y$, which has the effect $U_0|0\rangle = -|1\rangle$ and $U_0|1\rangle = |0\rangle$, to this density matrix, and make use of the fact that $I = |\psi\rangle\langle\psi| + |\psi_\perp\rangle\langle\psi_\perp|$, we find

$$U_0 \rho_3^{(out)} U_0^{-1} = \frac{1}{2}(1 + c_0 c_1)|\psi_\perp\rangle\langle\psi_\perp| + \frac{1}{2}(1 - c_0 c_1)|\psi\rangle\langle\psi|. \tag{8.7}$$

In the case that $c_0 = c_1 = 1/\sqrt{3}$ this becomes

$$U_0 \rho_3^{(out)} U_0^{-1} = \frac{2}{3}|\psi_\perp\rangle\langle\psi_\perp| + \frac{1}{3}|\psi\rangle\langle\psi|. \tag{8.8}$$

This is, in fact, the best approximation to the state orthogonal to that of the input qubit that can be realized, a fact we will not prove here. Note that the fidelity of the output to the ideal output state, $|\psi_\perp\rangle$, is 2/3. Therefore, the cloner with the addition of a U_0 gate to the third output not only clones states, but it also realizes the best possible approximate U-NOT gate.

The same result for the U-NOT operation can be achieved by measuring the original qubit. We measure $|\psi\rangle$ along a random direction in our two-dimensional Hilbert space

$$|\eta\rangle = \cos(\theta/2)|0\rangle + e^{i\phi}\sin(\theta/2)|1\rangle, \tag{8.9}$$

that is, we measure the projection $|\eta\rangle\langle\eta|$. If we obtain the result 1, we produce the state $|\eta_\perp\rangle$, where

$$|\eta_\perp\rangle = e^{-i\phi}\sin(\theta/2)|0\rangle - \cos(\theta/2)|1\rangle, \tag{8.10}$$

and if we get 0, we produce the state $|\eta\rangle$. The density matrix resulting from this procedure is

$$\rho^{(out)}(\eta) = |\langle\psi|\eta\rangle|^2|\eta_\perp\rangle\langle\eta_\perp| + |\langle\psi|\eta_\perp\rangle|^2|\eta\rangle\langle\eta|. \tag{8.11}$$

If we now average this over η we find

$$\rho^{(out)} = \frac{1}{4\pi}\int_0^{2\pi} d\phi \int_0^\pi d\theta \sin(\theta)\rho^{(out)}(\eta)$$

$$= \frac{2}{3}|\psi_\perp\rangle\langle\psi_\perp| + \frac{1}{3}|\psi\rangle\langle\psi|. \tag{8.12}$$

Therefore, the best approximate U-NOT gate can be achieved in two different ways. One is to use the cloning circuit, and the second is to the measure the qubit in a random direction and then produce a qubit whose direction is opposite to that indicated by our measurement result.

One might wonder if a similar strategy can be applied to cloning. That is one measures the original qubit in a random direction $|\eta\rangle$, and if one gets 1 produces two qubits in the state $|\eta\rangle|\eta\rangle$, and if one gets 0 produces two qubits in the state $|\eta_\perp\rangle|\eta_\perp\rangle$. This procedure does work, but, unlike in the case of the U-NOT, it is not optimal. One finds after averaging over η that the fidelity of the output state to the ideal output state, $|\psi\rangle|\psi\rangle$ is 2/3, which is less than the 5/6 achieved by the cloning circuit.

8.3 Programmable Machines: A General Result

We now want to consider programmable quantum machines, which we shall often refer to as quantum processors. Programmable machines have a number of advantages over machines that perform a single function. First, they are much more flexible. In order to change what they do, you just change the program rather than rewiring the entire quantum circuit. Second, they offer the possibility of performing several operations on the data in parallel, by using superpositions of program states, where each element of the superposition corresponds to a different operation. Programmable machines have two inputs, one for the data, which is to be acted upon, and one for the program, which will specify the operation to be performed on the data. Both the data and the program are quantum states. In particular, the processor is a unitary operator acting on the Hilbert space $\mathcal{H}_d \otimes \mathcal{H}_p$, where $\mathcal{H}_d$ is the data Hilbert space and $\mathcal{H}_p$ is the program Hilbert space. Ideally, we would like to be able to program any unitary operator acting on the data. For example, if our data space is two-dimensional, we would like to be able to have a program for each element of $SU(2)$. Such a processor would be universal, that is, it could be used to deterministically perform any unitary operation on a qubit. Unfortunately, as shown by Nielsen and Chuang, it is impossible to construct such a processor.

In order to show this, we need to examine the resources that are necessary in order to implement a given set of operations on the data. What Nielsen and Chuang showed is that if the program $|\Xi_1\rangle_p \in \mathcal{H}_p$ implements the unitary operator U_1 on the data state, and $|\Xi_2\rangle_p \in \mathcal{H}_p$ implements the unitary operator U_2, then $_p\langle\Xi_1|\Xi_2\rangle_p = 0$. This implies that for every unitary operator that the processor can implement on the data state, we need an extra dimension in the program space. Since the number of operations in $SU(2)$ is uncountably infinite, a program space that is finite, or even countably finite, would not be big enough to account for every operation.

Let us now prove the no-go theorem for deterministic programmable quantum processors. We assume the processor is represented by a unitary operator, G, acting $\mathcal{H}_d \otimes \mathcal{H}_p$, where $\mathcal{H}_d$ is the data space and $\mathcal{H}_p$ is the program space. We suppose that we have a program $|\Xi_1\rangle_p \in \mathcal{H}_p$ that implements the unitary operator U_1 on $\mathcal{H}_d$, in particular

$$G(|\psi\rangle_d \otimes |\Xi_1\rangle_p) = U_1|\psi\rangle_d \otimes |\Xi_1'\rangle_p. \tag{8.13}$$

Now it could be the case that the output in the program space depends on the state $|\psi\rangle_d$ that is sent into the data input. In order to show that this is not the case, assume that

$$G(|\psi_1\rangle_d \otimes |\Xi_1\rangle_p) = U_1|\psi\rangle_d \otimes |\Xi_1'\rangle_p$$
$$G(|\psi_2\rangle_d \otimes |\Xi_1\rangle_p) = U_1|\psi\rangle_d \otimes |\Xi_1''\rangle_p. \tag{8.14}$$

Taking the inner products of the left-hand sides of the above equations and equating that to the inner product of the right-hand sides, and assuming that $_d\langle\psi_1|\psi_2\rangle_d \neq$

0, gives us $_p\langle \Xi_1' | \Xi_1'' \rangle_p = 1$, thereby implying that the program state outputs are identical.

Now suppose that the program state $|\Xi_1\rangle_p$ implements the operator U_1 and the program state $|\Xi_2\rangle_p$ implements U_2. We then have that

$$G(|\psi\rangle_d \otimes |\Xi_1\rangle_p) = U_1|\psi_1\rangle_d \otimes |\Xi_1'\rangle_p$$
$$G(|\psi\rangle_d \otimes |\Xi_2\rangle_p) = U_2|\psi_2\rangle_d \otimes |\Xi_1'\rangle_p. \qquad (8.15)$$

Taking inner products we find

$$_p\langle \Xi_2 | \Xi_1 \rangle_p = {}_d\langle \psi | U_2^{-1} U_1 | \psi \rangle_d \; _p\langle \Xi_2' | \Xi_1' \rangle_p. \qquad (8.16)$$

We will examine both the case $_p\langle \Xi_2' | \Xi_1' \rangle_p \neq 0$ and the case $_p\langle \Xi_2' | \Xi_1' \rangle_p = 0$. If $_p\langle \Xi_2' | \Xi_1' \rangle_p \neq 0$, we have

$$\frac{_p\langle \Xi_2 | \Xi_1 \rangle_p}{_p\langle \Xi_2' | \Xi_1' \rangle_p} = {}_d\langle \psi | U_2^{-1} U_1 | \psi \rangle_d, \qquad (8.17)$$

and we note that the left-hand side does not depend on $|\psi\rangle_d$, so the right-hand side cannot either. That implies that $U_2^{-1} U_1$ is a multiple of the identity, and since both of the operators are unitary, we must have $U_2 = e^{i\theta} U_1$ for some θ between 0 and 2π. Now if, on the other hand, $_p\langle \Xi_2' | \Xi_1' \rangle_p = 0$, then we see that we must also have that $_p\langle \Xi_2 | \Xi_1 \rangle_p = 0$. Summarizing, what we have found is that if U_1 and U_2 are different, that is, they are not multiples of each other, then they must correspond to orthogonal program states. Therefore, the dimension of the program space must be at least as great as the number of unitary operators that the processor can perform,

Similar reasoning can be employed to show that a deterministic scheme employing measurement is also impossible. We can call this a measure-and-correct scheme. Suppose that we send a program and data into our processor, and at the output measure the program state in a fixed basis. Each measurement outcome corresponds to a different unitary operator being applied to the data state, but for each program state the resulting operators are related to each other in the same way. That means that for any program state, if we do not obtain the desired measurement result, we can correct the resulting output state by applying an operator that does not depend on the program state.

Let us look at a simple example. Suppose that both the data and program spaces are two-dimensional, and that our processor acts as follows:

$$G(|\psi\rangle_d \otimes |\Xi_1\rangle_p) = \frac{1}{\sqrt{2}}(U_1|\psi\rangle_d \otimes |0\rangle_p + VU_1|\psi\rangle_d \otimes |1\rangle_p)$$

$$G(|\psi\rangle_d \otimes |\Xi_2\rangle_p) = \frac{1}{\sqrt{2}}(U_2|\psi\rangle_d \otimes |0\rangle_p + VU_2|\psi\rangle_d \otimes |1\rangle_p). \qquad (8.18)$$

Here, V is a fixed unitary operator. Such a processor is capable of deterministically applying four different unitary operators to the data state, U_1, VU_1, U_2 and VU_2. For example, suppose we want to apply U_1. We use the program $|\Xi_1\rangle$, and then measure the program state in the basis $\{|0\rangle, |1\rangle\}$. If we obtain $|0\rangle$ we are done, and if we obtain $|1\rangle$, then we can apply V^{-1} to the data state. In either case, we obtain the output state $U_1|\psi\rangle_d$. We will also be able to deterministically obtain the superpositions $c_1 U_1 + c_2 U_2$ and $c_1 V U_1 + c_2 V U_2$, where c_1 and c_2 are complex numbers. It appears that we have beaten the no-go theorem, because we are able to deterministically realize four unitary operators with a two-dimensional program space. Unfortunately, it will not work. If we take the inner products of the two equations above, we find that

$$_p\langle\Xi_1|\Xi_2\rangle_p = {}_d\langle\psi|U_1^{-1}U_2|\psi\rangle_d. \tag{8.19}$$

The left-hand side does not depend on $|\psi\rangle_d$, which, as before, implies that U_1 and U_2 are related by a phase factor, and that the program states are multiples of each other. Therefore, we can only realize two operators in this way, U_1 and VU_1, and we have not gained anything.

8.4 Probabilistic Processors

The no-go result we proved in the previous section only applies to deterministic processors. If the processor in probabilistic, its limitations no longer apply. Let us first illustrate this with a simple example, and then proceed to a more complicated one. Suppose our data system is a qubit, and we want to implement the one parameter group of transformations, $U(\alpha) = \exp(i\alpha\sigma_z)$, where $0 \leq \alpha < 2\pi$. This can be accomplished with a success probability of $1/2$ by using a qubit program and a Controlled-NOT gate. The Controlled-NOT gate has two inputs, a control input and a target input, and in the case we wish to consider here, the target qubit is the program and the control qubit is the data. The program states are

$$|\Xi(\alpha)\rangle = \frac{1}{\sqrt{2}}(e^{i\alpha}|0\rangle + e^{-i\alpha}|1\rangle). \tag{8.20}$$

If the data state input is $|\psi\rangle$, the output of this processor is then

$$|\Psi_{out}\rangle = \frac{1}{\sqrt{2}}(U(\alpha)|\psi\rangle|0\rangle + U^{-1}(\alpha)|\psi\rangle|1\rangle). \tag{8.21}$$

By measuring the program state output in the basis $\{|0\rangle, |1\rangle)\}$, and keeping the result only if we get $|0\rangle$, which happens with a probability of $1/2$, we obtain the data state output $U(\alpha)|\psi\rangle$, which is the desired result. Note that in this case, a single processor is able to realize, with a one-qubit program space, a continuous group of transformations. The cost is that in each application, the desired transformation is only realized with a probability of $1/2$

Now let us look at a more complicated example. We will begin by going back and considering the three-qubit circuit for the approximate cloner. Qubit 1 will now be our data state, and qubits 2 and 3 will be our program. We will denote the data state by $|\psi\rangle_1$ and the program state by $|\Xi\rangle_{23}$. Define the two-qubit Bell states to be

$$|\Psi_\pm\rangle = \frac{1}{\sqrt{2}}(|00\rangle \pm |11\rangle)$$

$$|\Phi_\pm\rangle = \frac{1}{\sqrt{2}}(|01\rangle \pm |10\rangle). \tag{8.22}$$

If these states are used as programs in our processor, we find that

$$|\psi\rangle_1|\Psi_+\rangle_{23} \rightarrow |\psi\rangle_1|\Psi_+\rangle_{23}$$
$$|\psi\rangle_1|\Psi_-\rangle_{23} \rightarrow \sigma_z|\psi\rangle_1|\Psi_-\rangle_{23}$$
$$|\psi\rangle_1|\Phi_+\rangle_{23} \rightarrow \sigma_x|\psi\rangle_1|\Phi_+\rangle_{23}$$
$$|\psi\rangle_1|\Psi_-\rangle_{23} \rightarrow (-i\sigma_y)|\psi\rangle_1|\Psi_-\rangle_{23}, \tag{8.23}$$

where σ_x, σ_y, and σ_z are the Pauli matrices. Suppose we want to implement the operator

$$U_\phi = |\phi_\perp\rangle\langle\phi_\perp| - |\phi\rangle\langle\phi| = I - 2|\phi\rangle\langle\phi| \tag{8.24}$$

on the data state, where $|\phi\rangle$ and $|\phi_\perp\rangle$ are specified, orthogonal one-qubit states. The operator U_ϕ is similar to σ_z, but instead of flipping the phase of the state $|1\rangle$ and leaving $|0\rangle$ unchanged, it flips the phase of the state $|\phi\rangle$ and leave the phase of $|\phi_\perp\rangle$ unchanged. In order to find a program state that will implement this operator, we first express it in terms of the Pauli martrices. Setting $|\phi\rangle = \mu|0\rangle + v|1\rangle$, we find

$$U_\phi = -(\mu v^* + \mu^* v)\sigma_x + (\mu v^* - \mu^* v)(-i\sigma_y)$$
$$+ (|v|^2 - |\mu|^2)\sigma_z. \tag{8.25}$$

We can now apply the operation U_ϕ to $|\psi\rangle_1$ by sending in the program state

$$|\Xi_\phi\rangle = -(\mu v^* + \mu^* v)|\Phi_+\rangle_{23} + (\mu v^* - \mu^* v)|\Phi_-\rangle_{23}$$
$$+ (|v|^2 - |\mu|^2)|\Psi_-\rangle_{23}, \tag{8.26}$$

and measuring the program outputs to see if they are in the state $(|\Phi_+\rangle_{23} + |\Phi_-\rangle_{23} + |\Phi_-\rangle_{23})/\sqrt{3}$. This will occur with a probability of $1/3$. When we do obtain this result, the output of the data state is $U_\phi|\psi\rangle_1$. Note that both the measurement we make and its probability of success do not depend on the state $|\phi\rangle$. We can express the program vector in a neater form if we introduce the operator, U_{in}, defined by

$$U_{in}|00\rangle = -|10\rangle \quad U_{in}|10\rangle = -|11\rangle$$

$$U_{in}|01\rangle = |00\rangle \quad U_{in}|11\rangle = |01\rangle. \tag{8.27}$$

The program state can then be expressed as

$$|\Xi\rangle_{23} = \frac{1}{\sqrt{2}} U_{in}(|\phi\rangle_2|\phi_\perp\rangle_3 + |\phi_\perp\rangle_2|\phi\rangle_3). \tag{8.28}$$

Summarizing, with this device we can implement a phase flip in any basis, and the basis itself is specified by the program state.

Now let us return to our simple Controlled-NOT processor. Suppose that we want to increase the probability of a successful outcome. One possibility is to try again if get the wrong result of our measurement on the program state. If we obtained the result $|1\rangle$ from our measurement, then the data qubit is in the state $U^{-1}(\alpha)|\psi\rangle$. We can take this qubit and run it through the processor again, but this time use the program $|\Xi(2\alpha)\rangle$. If we do so, the output state is

$$|\Psi'_{out}\rangle = \frac{1}{\sqrt{2}}(U(\alpha)|\psi\rangle|0\rangle + U^{-1}(3\alpha)|\psi\rangle|1\rangle). \tag{8.29}$$

We again measure the program state and keep the result if we get $|0\rangle$. This again happens with a probability of $1/2$. Adding this second step has increased our overall success probability to $3/4$, and the procedure can be repeated to bring the success probability as close to one as we wish. What we need to do this, however, is a collection of qubits in the proper program states, that is, besides a qubit in the state $|\Xi(\alpha)\rangle$, we need an additional one in the state $|\Xi(2\alpha)\rangle$.

We can also accomplish the same thing by enlarging our program space. Our data space still consists of one qubit, but the program space now contains two qubits. Let us label the three inputs, input 1 being the data input, input 2 the first program input and input 3 the second program input. The processor now consists of two gates. The first is a controlled-NOT gate whose control qubit is qubit 1 and whose target qubit is qubit 2. The second gate is a Toffoli gate. This gate has two control qubits and one target qubit. The states of the control qubits are not changed, and if they are in the states $|0\rangle|0\rangle$, $|0\rangle|1\rangle$, or $|1\rangle|0\rangle$, neither is the state of the target qubit. However, if they are in the state $|1\rangle|1\rangle$, then σ_x is applied to the target qubit. In our processor, qubits 1 and 2 are the control qubits and qubit 3 is the target qubit. The input state is $|\psi\rangle_1|\Xi(\alpha)\rangle_2|\Xi(2\alpha)\rangle_3$, and the output state is

$$|\Psi''_{out}\rangle = \frac{1}{2}[U(\alpha)|\psi\rangle_1(|0\rangle_2|0\rangle_3 + |0\rangle_2|1\rangle_3 + |1\rangle_2|0\rangle_3) + U^{-1}(3\alpha)|\psi\rangle|1\rangle_2|1\rangle_3]. \tag{8.30}$$

At the output we measure the program qubits in the computational basis and keep the data state output if we get $|0\rangle|0\rangle$, $|0\rangle|1\rangle$, or $|1\rangle|0\rangle$. If we do, the data output is in the state $U(\alpha)|\psi\rangle$, and we have achieved our goal. This happens with a probability of $3/4$. By increasing the dimension of the program space further, we can increase our probability of success. We have, therefore, two strategies for increasing the success probability for a probabilistic processor.

8.5   A Programmable State Discriminator

In a previous chapter, we discussed unambiguous state discrimination. One is given a qubit, which is in one of two known states, $|\psi_1\rangle$ or $|\psi_2\rangle$, and one's task is to determine which of the two states the qubit is in. In the case of unambiguous discrimination, one cannot make a mistake, but the procedure is allowed to fail. We found a POVM that optimally accomplishes this task. It has three outcomes, the state is $|\psi_1\rangle$, the state is $|\psi_2\rangle$, and failure. The optimal POVM is the one that minimizes the probability of failure.

The actual state-distinguishing device, a realization of the optimal POVM, for two *known* states depends on the two states, $|\psi_1\rangle$ and $|\psi_2\rangle$, i. e. these two states are "hard wired" into the machine. What we now wish to do is to see if we can construct a machine in which the information about $|\psi_1\rangle$ and $|\psi_2\rangle$ is supplied in the form of a program. In particular, we want the program to consist of the two qubit states that we wish to distinguish. In other words, we are given two qubits, one in the state $|\psi_1\rangle$ and another in the state $|\psi_2\rangle$. We have no knowledge of the states $|\psi_1\rangle$ and $|\psi_2\rangle$. Then we are given a third qubit that is guaranteed to be in one of these two program states, and our task is to determine, as best we can, in which one. We are allowed to fail, but not to make a mistake.

In order to solve this problem, what we need to do is to find a POVM, and our task is then reduced to the following measurement optimization problem. One has two input states

$$|\Psi_1^{in}\rangle = |\psi_1\rangle_A |\psi_2\rangle_B |\psi_1\rangle_C \,,$$

$$|\Psi_2^{in}\rangle = |\psi_1\rangle_A |\psi_2\rangle_B |\psi_2\rangle_C \,, \tag{8.31}$$

where the subscripts A and B refer to the program registers (A contains $|\psi_1\rangle$ and B contains $|\psi_2\rangle$), and the subscript C refers to the data register. Our goal is to unambiguously distinguish between these inputs, keeping in mind that one has no knowledge of $|\psi_1\rangle$ and $|\psi_2\rangle$. In particular, one wants to find a POVM that will accomplish this.

Let the elements of our POVM be Π_1, corresponding to unambiguously detecting $|\Psi_1^{in}\rangle$, Π_2, corresponding to unambiguously detecting $|\Psi_2^{in}\rangle$, and Π_0, corresponding to failure. The probabilities of successfully identifying the two possible input states are given by

$$\langle\Psi_1^{in}|\Pi_1|\Psi_1^{in}\rangle = p_1 \qquad \langle\Psi_2^{in}|\Pi_2|\Psi_2^{in}\rangle = p_2 \,, \tag{8.32}$$

and the condition of no errors implies that

$$\Pi_2|\Psi_1^{in}\rangle = 0 \qquad \Pi_1|\Psi_2^{in}\rangle = 0 \,. \tag{8.33}$$

In addition, because the alternatives represented by the POVM exhaust all possibilities, we have that

$$I = \Pi_1 + \Pi_2 + \Pi_0 \,. \tag{8.34}$$

The fact that we know nothing about $|\psi_1\rangle$ and $|\psi_2\rangle$ means that the only way we can guarantee satisfying the above conditions is to take advantage of the symmetry properties of the states, i.e. that $|\Psi_1^{in}\rangle$ is invariant under interchange of the first and third qubits, and $|\Psi_2^{in}\rangle$ is invariant under interchange of the second and third qubits. That means that Π_1 should give zero when acting on states that are symmetric in qubits B and C, while Π_2 should give zero when acting on states that are symmetric in qubits A and C. Defining the antisymmetric states for the corresponding pairs of qubits

$$|\psi_{BC}^{(-)}\rangle = \frac{1}{\sqrt{2}}(|0\rangle_B|1\rangle_C - |1\rangle_B|0\rangle_C) \,,$$

$$|\psi_{AC}^{(-)}\rangle = \frac{1}{\sqrt{2}}(|0\rangle_A|1\rangle_C - |1\rangle_A|0\rangle_C) \,, \tag{8.35}$$

we introduce the projectors onto the antisymmetric subspaces of the corresponding qubits as

$$P_{BC}^{as} = |\psi_{BC}^{(-)}\rangle\langle\psi_{BC}^{(-)}| \,,$$

$$P_{AC}^{as} = |\psi_{AC}^{(-)}\rangle\langle\psi_{AC}^{(-)}| \,. \tag{8.36}$$

We can now take for Π_1 and Π_2 the operators

$$\Pi_1 = c_1 I_A \otimes P_{BC}^{as} \,,$$

$$\Pi_2 = c_2 I_B \otimes P_{AC}^{as} \,, \tag{8.37}$$

where I_A and I_B are the identity operators on the spaces of qubits A and B, respectively, and c_1 and c_2 are as yet undetermined nonnegative real numbers. Using the above expressions for Π_j, where $j = 1, 2$ in Eq. (8.32), we find that

$$p_j = \langle \Psi_j^{in} | \Pi_j | \Psi_j^{in} \rangle = c_j \frac{1}{2}(1 - |\langle \psi_1 | \psi_2 \rangle|^2) \,. \tag{8.38}$$

The average probability, P, of successfully determining which state we have, assuming that the input states occur with equal probability is given by

$$P = \frac{1}{2}(p_1 + p_2) = \frac{1}{4}(c_1 + c_2)(1 - |\langle \psi_1 | \psi_2 \rangle|^2), \tag{8.39}$$

and we want to maximize this expression subject to the constraint that $\Pi_0 = I - \Pi_1 - \Pi_2$ is a positive operator.

Let S be the four-dimensional subspace of the entire eight-dimensional Hilbert space of the three qubits, A, B, and C, that is spanned by the vectors $|0\rangle_A |\psi_{BC}^{(-)}\rangle$, $|1\rangle_A |\psi_{BC}^{(-)}\rangle$, $|0\rangle_B |\psi_{AC}^{(-)}\rangle$, and $|1\rangle_B |\psi_{AC}^{(-)}\rangle$. In the orthogonal complement of S, $S^\perp$, the operator Π_0 acts as the identity, so that in $S^\perp$, Π_0 is positive. Therefore, we need to investigate its action in S. First, let us construct an orthonormal basis for S. Applying the Gram-Schmidt process to the four vectors, given above, that span S, we obtain the orthonormal basis

$$|\Phi_1\rangle = |0\rangle_A |\psi_{BC}^{(-)}\rangle \,,$$

$$|\Phi_2\rangle = \frac{1}{\sqrt{3}}(2|0\rangle_B |\psi_{AC}^{(-)}\rangle - |0\rangle_A |\psi_{BC}^{(-)}\rangle) \,,$$

$$|\Phi_3\rangle = |1\rangle_A |\psi_{BC}^{(-)}\rangle \,,$$

$$|\Phi_4\rangle = \frac{1}{\sqrt{3}}(2|1\rangle_B |\psi_{AC}^{(-)}\rangle - |1\rangle_A |\psi_{BC}^{(-)}\rangle). \tag{8.40}$$

In this basis, the operator Π_0, restricted to the subspace S, is given by the 4×4 matrix

$$\Pi_0 = \begin{pmatrix} 1 - c_1 - \frac{1}{4}c_2 & -\frac{\sqrt{3}}{4}c_2 & 0 & 0 \\ -\frac{\sqrt{3}}{4}c_2 & 1 - \frac{3}{4}c_2 & 0 & 0 \\ 0 & 0 & 1 - c_1 - \frac{1}{4}c_2 & -\frac{\sqrt{3}}{4}c_2 \\ 0 & 0 & -\frac{\sqrt{3}}{4}c_2 & 1 - \frac{3}{4}c_2 \end{pmatrix} \tag{8.41}$$

Because of the block diagonal nature of Π_0, the characteristic equation for its eigenvalues, λ, is given by the biquadratic equation

$$[\lambda^2 - (2 - c_1 - c_2)\lambda + 1 - (2 - c_1 - c_2) + \frac{3}{4}c_1 c_2]^2 = 0 \,. \tag{8.42}$$

It is easy to obtain the eigenvalues explicitly, but for our purposes, the conditions that guarantee that they are nonnegative are more useful. These can be read out from the above equation, yielding

$$2 - c_1 - c_2 \geq 0 \,,$$
$$1 - (2 - c_1 - c_2) + \frac{3}{4}c_1 c_2 \geq 0 \,. \tag{8.43}$$

The second is the stronger of the two conditions. When it is satisfied the first one is always met but the first one can still be used to eliminate nonphysical solutions. We can use the second condition to express c_2 in terms of c_1,

$$c_2 \leq \frac{2 - 2c_1}{2 - (3/2)c_1} \,. \tag{8.44}$$

For the maximum probability of success, we chose the equal sign. Inserting the resulting expression into (8.39) gives

$$P = \frac{1}{4}(c_1 + \frac{2 - 2c_1}{2 - (3/2)c_1})(1 - |\langle \psi_1 | \psi_2 \rangle|^2) \,. \tag{8.45}$$

We can easily find $c_1 = c_{1,opt}$ for which the right-hand side of this expression is maximum and using this together with Eq. (8.44) we obtain

$$c_{1,opt} = c_{2,opt} = \frac{2}{3}. \tag{8.46}$$

These values, in conjunction with Eqs. (8.37) completely specify the POVM. Inserting these optimal values into (8.39) gives

$$P_{POVM} = \frac{1}{3}(1 - |\langle \psi_1 | \psi_2 \rangle|^2) \,. \tag{8.47}$$

If we know the states $|\psi_1\rangle$ and $|\psi_2\rangle$, the probability of successfully determining the state is $1 - |\langle \psi_1 | \psi_2 \rangle|$. This is always greater than or equal to the probability in the previous equation, but this is to be expected. Knowledge of the states $|\psi_1\rangle$ and $|\psi_2\rangle$ corresponds to being given an infinite number of examples of each state, which we can then measure to determine exactly what the states are. Our programmable device only has access to one example of each state. It is, however, a very flexible device. Note that the POVM elements do not in any way depend on the states $|\psi_1\rangle$ and $|\psi_2\rangle$, which means that it will work for any two program states.

8.6 Problems

1. If we restrict the class of state that we would like to clone, we can achieve higher
   fidelities for the clones than is possible with a device that is designed to clone
   all states optimally. An example of this is phase-covariant cloning. Suppose we
   want to clone only states of the form

$$|\psi(\theta)\rangle = \frac{1}{\sqrt{2}}(|0\rangle + e^{i\theta}|1\rangle).$$

   Consider the following cloning transformation, U, acting on two qubits

   $$U|0\rangle_1|0\rangle_2 = |0\rangle_1|0\rangle_2$$
   $$U|1\rangle_1|0\rangle_2 = \cos\eta|1\rangle_1|0\rangle_2 + \sin\eta|0\rangle_1|1\rangle_2.$$

   The input state to this cloner is $|\psi(\theta)\rangle_1|0\rangle_2$, and the angle η controls how the
   information about the input state is split between outputs 1 and 2. Find the
   reduced density matrices of the outputs 1 and 2, the fidelities of these outputs
   to the input state, $|\psi(\theta)\rangle$, and show that in the case $\eta = \pi/4$ these fidelities
   exceed $5/6$.
2. Consider the cloner discussed in Sect. 8.2. Show that the fidelities of the output
   states (see Eq. (8.4)) satisfy the relation

$$\sqrt{(1 - F_1)(1 - F_2)} = F_1 + F_2 - \frac{3}{2},$$

   where F_1 is the fidelity of the output state in output 1 to the input state, and F_2 is
   the fidelity of the output state in output 2 to the input state.
3. Let us again consider our cloning circuit composed of four C-NOT gates. Show
   that it performs the following transformations

$$|\psi\rangle_1|0\rangle_2|-x\rangle_3 \rightarrow (\sigma_z|\psi\rangle_2)|\Psi_-\rangle_{13}$$
$$|\psi\rangle_1|+x\rangle_2|1\rangle_3 \rightarrow (\sigma_x|\psi\rangle_3)|\Phi_+\rangle_{12}$$

   Now show that if the input state is $|\psi\rangle_1(\alpha|\Psi_+\rangle_{23} + \beta|0\rangle_2|-x\rangle_3 + \gamma|+x\rangle_2|1\rangle_3)$,
   with the normalization condition

$$|\alpha + \beta|^2 + |\alpha + \gamma|^2 + |\beta - \gamma|^2 = 2,$$

the reduced density matrices of the outputs are

$$\rho_1 = \left[\left| \alpha + \frac{\beta + \gamma}{2} \right|^2 - \frac{|\beta - \gamma|^2}{4} \right] \rho_{in} + \frac{|\beta - \gamma|^2}{2} I$$

$$\rho_2 = \left[\left| \beta + \frac{(\alpha - \gamma)}{2} \right|^2 - \frac{|\alpha + \gamma|^2}{4} \right] \sigma_z \rho_{in} \sigma_z + \frac{|\alpha + \gamma|^2}{2} I$$

$$\rho_3 = \left[\left| \gamma + \frac{(\alpha - \beta)}{2} \right|^2 - \frac{|\alpha + \beta|^2}{4} \right] \sigma_x \rho_{in} \sigma_x + \frac{|\alpha + \beta|^2}{2} I,$$

where $\rho_{in} = |\psi\rangle\langle\psi|$. This implies that the cloner can not only split quantum information, but it can split it and then cause operations to be performed on the parts.

4. Suppose we want to use the probabilistic processor composed of four C-NOT gates to implement the operation $V_\phi = |\phi\rangle\langle\phi_\perp| + |\phi_\perp\rangle\langle\phi|$ on the data state. Find a program state that will cause this to happen with a probability of $1/3$, and show that if the program state is expressed in the form $U_{in}|\Xi'\rangle_{23}$, then the state $|\Xi'\rangle_{23}$ can be expressed very simply in terms of $|\phi\rangle$ and $|\phi_\perp\rangle$.

Further Reading

1. V. Bužek, M. Hillery, Quantum copying: beyond the no-cloning theorem. Phys. Rev. A **54**, 1844 (1996)
2. V. Bužek, M. Hillery, R.F. Werner, Optimal manipulations with qubits: universal NOT Gate. Phys. Rev. A **60**, R2626 (1999)
3. For reviews of quantum cloning see V. Scarani, S. Iblisdir, N. Gisin, A. Acin, Quantum cloning. Rev. Mod. Phys. **77**, 1225 (2005) and N. J. Cerf, J. Fiurašek, Optical quantum cloning—a review. Progr. Optics **49**, 455 (2006)
4. M.A. Nielsen, I.L. Chuang, Programmable quantum gate arrays. Phys. Rev. Lett. **79**, 321 (1997)
5. M. Hillery, V. Bužek, M. Ziman, Probabilistic implementation of universal quantum processors. Phys. Rev. A **65**, 022301 (2002)
6. J. Preskill, Proc. Roy. Soc. Lond. A **454**, 385 (1998)
7. G. Vidal, L. Masanes, J.I. Cirac, Phys. Rev. Lett. **88**, 047905 (2002)
8. J. Bergou, M. Hillery, A universal programmable quantum state discriminator that is optimal for unambiguously distinguishing between unknown states. Phys. Rev. Lett. **94**, 160501 (2005)

Chapter 9
Decoherence and Quantum Error Correction

One of the biggest problems in building a quantum computer is noise or decoherence. Qubits are coupled to other systems whether we want them to be or not, e.g. atoms couple to the electromagnetic field and spins couple to other spins via dipole-dipole interactions. These unwanted couplings can cause errors, and we need to protect quantum information against these errors.

In most of this chapter we will study quantum error-correcting codes. These allow us to protect quantum information from the effects of decoherence. We will begin with a discussion of the general theory of quantum error-correcting codes and then discuss in detail one particular class of these codes, the CSS codes. We will conclude with a very short introduction to another technique for protecting quantum information from decoherence, decoherence-free subspaces.

9.1 General Theory of Quantum Error-Correcting Codes

Classically, to protect against errors, we can just repeat the bit. We can encode one bit in three as $0 \rightarrow 000$ and $1 \rightarrow 111$. Errors can flip bits, that is change a 0 to a 1 or vice versa. To decode the bit, we use majority voting; if there are more 0's than 1's we call it 0, and if there are more 1's than 0's, we call it 1. This will protect against one bit-flip error.

Let's look at this in terms of probabilities. Suppose that the probability of one bit-flip error is p, and that the occurrence of errors in the different bits is independent. Then the probability of no errors is $(1-p)^3$, of one error $3p(1-p)^2$, of two errors $3p^2(1-p)$ and of three errors p^3. The probability that the error correction fails is just the sum of the probabilities that two or three errors occur, or $p^2(3-2p)$. This will be smaller that the probability of an error in an unencoded bit if $p^2(3-2p) < p$, which is true if $p < 1/2$. If this condition is satisfied, then it is better to encode the bit than not.

© Springer Nature Switzerland AG 2021
J. A. Bergou et al., *Quantum Information Processing*, Graduate Texts in Physics,
https://doi.org/10.1007/978-3-030-75436-5_9

We would like to do something similar for qubits, but we face several problems in doing so. The task we would like to accomplish is harder, because we do not just want to protect $|0\rangle$ and $|1\rangle$, but any state of the form $a|0\rangle + b|1\rangle$. Some of the problems we face in protecting qubit states are:

1. Qubits are susceptible to more kinds of errors than are classical bits. There are phase errors that send $|0\rangle \rightarrow |0\rangle$ and $|1\rangle \rightarrow -|1\rangle$, which has the effect of changing $a|0\rangle + b|1\rangle$ to $a|0\rangle - b|1\rangle$. In addition, there are general small errors that have the effect $a|0\rangle + b|1\rangle \rightarrow (a + O(\epsilon))|0\rangle + (b + O(\epsilon))|1\rangle$, where $\epsilon \ll 1$ is a parameter that characterizes the size of the error.
2. We have to be very careful about how we look at a qubit to detect the error, because by looking at a state, we mean measuring it, and measuring a state can change it.
3. We cannot just copy the qubit state, because of the no-cloning theorem.

What this means is that we have to be careful and clever.

The first quantum error-correcting code was due to Peter Shor, and we will examine it in detail. We start by analogy with the classical case and encode $|0\rangle$ by $|000\rangle$ and $|1\rangle$ by $|111\rangle$, which means that the state $a|0\rangle + b|1\rangle$ will be encoded as $a|000\rangle + b|111\rangle$. We would like to see if this encoding will help detect and correct bit-flip errors. Note that any single-qubit state is mapped into the subspace of three-qubit states spanned by $|000\rangle$ and $|111\rangle$.

If we just measure each qubit in the $\{|0\rangle, |1\rangle\}$ basis, to detect a bit-flip, we will destroy any superpositions, so something else is required. Notice that in the states $|000\rangle$ and $|111\rangle$, all of the qubits are in the same state, in particular, qubits 1 and 2 are in the same state and qubits 2 and 3 are in the same state. Denoting σ_z by Z (we will also denote σ_x by X), let us measure $Z_1 Z_2$ and $Z_2 Z_3$ and see what happens. Acting on either $|000\rangle$ or $|111\rangle$ we have summarized the results in Table 9.1.

So, by looking at the result, we can tell which bit flipped. In addition, any state of the form $a|000\rangle + b|111\rangle$, or this state with a single bit flipped, is an eigenstate of $Z_1 Z_2$ and $Z_2 Z_3$, so measuring them does not change the state. Therefore, if one bit flips, we can determine which one it is by measuring these two observables, and we will not change the state. We can then correct the error by flipping that bit back. For example, if bit 2 flipped we would have $a|000\rangle + b|111\rangle \rightarrow a|010\rangle + b|101\rangle$, measuring $Z_1 Z_2$ and $Z_2 Z_3$ would give us -1 and -1, telling is that it was bit 2 that flipped and not altering the state. We could then apply X_2 to the state to flip bit 2 back to its proper value.

Table 9.1 Truth table for the operations $Z_1 Z_2$ and $Z_2 Z_3$ on either $|000\rangle$ or $|111\rangle$		$Z_1 Z_2$	$Z_2 Z_3$
	No flips	1	1
	Bit 1 flipped	-1	1
	Bit 2 flipped	-1	-1
	Bit 3 flipped	1	-1

This procedure also works if there is only some amplitude for one bit to flip. Suppose

$$|000\rangle \rightarrow (1 - \epsilon^2)^{1/2}|000\rangle + \epsilon|010\rangle$$
$$|111\rangle \rightarrow (1 - \epsilon^2)^{1/2}|111\rangle + \epsilon|101\rangle, \qquad (9.1)$$

which implies that

$$a|000\rangle + b|111\rangle \rightarrow (1 - \epsilon^2)^{1/2}(a|000\rangle + b|111\rangle) + \epsilon(a|010\rangle + b|101\rangle). \qquad (9.2)$$

Now let us see what happens when we make our measurements. Measuring $Z_1 Z_2$, we obtain 1 with probability $1 - \epsilon^2$ and -1 with probability ϵ^2. If we obtain 1, the state is restored and becomes $a|000\rangle + b|111\rangle$. If we obtain -1 the state is $a|010\rangle + b|101\rangle$. Now let's measure $Z_2 Z_3$. If we obtained 1 for the first measurement, we will also obtain one for the second, since the state is now restored to what it should be. In that case we have obtained 1 for both measurements, so we do nothing. If we obtained -1 for the first measurement, we will obtain -1 for the second, since bit 2 is definitely flipped. Having obtained 1 for both measurements, we apply X_2 to correct the error.

At this point we can correct one bit-flip error, but now we need to worry about phase-flip errors. Phase-flip errors behave like bit-flip errors if we look at them in a different basis. Note that a phase-flip error turns the state $|+x\rangle$ into $|-x\rangle$ and $|-x\rangle$ into $|+x\rangle$, which is the same effect a bit-flip error has in the basis $\{|0\rangle, |1\rangle\}$. If we encode $|0\rangle \rightarrow |+x, +x, +x\rangle$ and $|1\rangle \rightarrow |-x, -x, -x\rangle$, then we can detect single-bit phase-flip errors. To detect the error, we measure $X_1 X_2$ and $X_2 X_3$, which tells us in which bit the error occurred. We then correct the error by applying Z to the appropriate bit.

We now want to combine the bit and phase-flip codes, so that we can protect against both kinds of errors. Think of starting with the phase-flip code and encoding each of the qubits in it with the the bit-flip code. This gives us a nine qubit code, which is the Shor code. In detail, the encoding is given by

$$|0\rangle \rightarrow \frac{1}{2\sqrt{2}}(|000\rangle + |111\rangle)(|000\rangle + |111\rangle)(|000\rangle + |111\rangle)$$

$$|1\rangle \rightarrow \frac{1}{2\sqrt{2}}(|000\rangle - |111\rangle)(|000\rangle - |111\rangle)(|000\rangle - |111\rangle). \qquad (9.3)$$

We can find bit flip errors by measuring products of Z operators. In particular, measuring $Z_1 Z_2$ and $Z_2 Z_3$ will detect bit flip errors in the first three-qubit cluster, measuring $Z_4 Z_5$ and $Z_5 Z_6$ will detect bit flip errors in the second three-qubit cluster, and $Z_7 Z_8$ and $Z_8 Z_9$ will detect bit flip errors in the third three-qubit cluster. Once the bit-flip has been detected, we can apply an X operator to the appropriate qubit to flip it back.

A phase-flip error in any qubit will cause the sign in one of the clusters to flip. We can find which cluster by measuring $\prod_{j=1}^{6} X_j$ and $\prod_{j=4}^{9} X_j$. If both give 1, then there is no error, if the first gives 1 and the second -1, the error is in the first cluster, if both give -1, the error is in the second cluster, and if the first gives 1 and the second -1, the error is in the third cluster. Once we have determined in which cluster the error occurred, we can apply a Z operator to any of the qubits in that cluster to correct the error. Note that this code will correct an error in one qubit, but not more.

So far we have only considered bit-flip and phase-flip errors. It doesn't seem as though this would be sufficient, but it is. To see why we must take a more general look at quantum error correction.

Start by considering a single qubit interacting with its environment. Let the qubit Hilbert space be $\mathcal{H}_A$ and the environment Hilbert space be $\mathcal{H}_E$. We shall call the initial state of the environment $|0\rangle_E$ and the operator that describes the evolution of the qubit and the environement U_{AE}. We have that

$$U_{AE}(|0\rangle_A \otimes |0\rangle_E) = |0\rangle_A \otimes |e_{00}\rangle_E + |1\rangle_A \otimes |e_{01}\rangle_E$$

$$U_{AE}(|1\rangle_A \otimes |0\rangle_E) = |0\rangle_A \otimes |e_{10}\rangle_E + |1\rangle_A \otimes |e_{11}\rangle_E. \tag{9.4}$$

The states $|e_{jk}\rangle_E$ are not necessarily orthogonal or normalized, but they must obey the constraints imposed by the unitarity of U_{AE}. For example, we must have that $\|e_{00}\|^2 + \|e_{01}\|^2 = 1$ and $\|e_{10}\|^2 + \|e_{11}\|^2 = 1$. We now want to see the effect of U_{AE} acting on a general qubit state, i.e. on $|\psi\rangle_A \otimes |0\rangle_E$, where $|\psi\rangle_A = a|0\rangle_A + b|1\rangle_A$. After some work we find that

$$U_{AE}(|\psi\rangle_A \otimes |0\rangle_E) = a(|0\rangle_A \otimes |e_{00}\rangle + |1\rangle_A \otimes |e_{01}\rangle_E)$$

$$+ b(|0\rangle_A \otimes |e_{10}\rangle_E + |1\rangle_A \otimes |e_{11}\rangle_E)$$

$$= I|\psi\rangle_A \otimes |e_I\rangle_E + X|\psi\rangle_A \otimes |e_X\rangle_E$$

$$+ Y|\psi\rangle_A \otimes |e_Y\rangle_E + Z|\psi\rangle_A \otimes |e_Z\rangle_Z, \tag{9.5}$$

where I is the identity operator, $Y = iXZ$, and

$$|e_I\rangle_E = \frac{1}{2}(|e_{00}\rangle + |e_{11}\rangle) \qquad |e_X\rangle_E = \frac{1}{2}(|e_{01}\rangle + |e_{10}\rangle)$$

$$|e_Y\rangle_E = \frac{i}{2}(|e_{10}\rangle - |e_{01}\rangle) \qquad |e_Z\rangle_E = \frac{1}{2}(|e_{00}\rangle - |e_{11}\rangle). \tag{9.6}$$

Therefore, we can expand the action of U_{AE} on the qubit in terms of the Pauli matrices. This is a consequence of the fact that these matrices plus the identity form a basis for 2×2 matrices. Note that the vectors $|e_I\rangle_E$, $|e_X\rangle_E$, $|e_Y\rangle_E$, and $|e_Z\rangle_E$ are not necessarily normalized or orthogonal. For n qubits we can expand the unitary evolution operator that mixes the qubits and the environment in terms of $\{I, X, Y, Z\}^{\otimes n}$. Let us call the members of this set E_a so that

$$U_{AE}(|\psi\rangle_A \otimes |0\rangle_E) = \sum_a E_a |\psi\rangle_A \otimes |e_a\rangle_E. \tag{9.7}$$

Note that E_a is unitary and that $\mathcal{H}_A$ is now the n-qubit Hilbert space.

When designing a code, we choose a subset $\mathcal{E} \subseteq \{I, X, Y, Z\}^{\otimes n}$; these are the errors we want to be able to correct. Typically, $\mathcal{E}$ is chosen to be all E_a of weight t or less. The weight of E_a is the number of operators it contains that are not the identity. Next, we choose a code subspace, $\mathcal{H}_c \subseteq \mathcal{H}_A$, which will contain the code words, and suppose $\{|\bar{j}\rangle_A\}$ is an orthonormal basis of that space. Suppose we had for $E_a, E_b \in \mathcal{E}$

$$_A\langle \bar{j}|E_b^\dagger E_a|\bar{k}\rangle_A = \delta_{ab}\delta_{\bar{j}\bar{k}}. \tag{9.8}$$

This implies that each error in $\mathcal{E}$ maps the code space into a different subspace, and that all of these subspaces are orthogonal, i.e. $E_a \mathcal{H}_C$ is orthogonal to $E_b \mathcal{H}_C$ for $a \neq b$, and hence these subspaces are distinguishable. Within one of these subspaces, errors map code words (the basis elements $|\bar{j}\rangle_A$) onto orthogonal states, that is $E_a|\bar{j}\rangle_A$ is orthogonal to $E_a|\bar{k}\rangle_A$ for $\bar{j} \neq \bar{k}$.

This means that we can find which error occurred (into which orthogonal subspace it mapped the code word), and we can correct it. If we found that E_a occurred, we just apply $E_a^\dagger$. In fact, we can correct any error that is a combination of the elements of $\mathcal{E}$. If

$$|\psi\rangle_A \otimes |0\rangle_E \rightarrow \sum_a E_a|\psi\rangle_A \otimes |e_a\rangle_E, \tag{9.9}$$

then we can measure the observable $\sum_a \lambda_a P_a$, where the λ_a are distinct, and P_a projects onto $E_a \mathcal{H}_C$. If we obtain $\lambda_{a'}$, then the state becomes $E_{a'}|\psi_A\rangle \otimes |e_{a'}\rangle_E$, and we can apply $E_{a'}^\dagger$ to correct the error. Therefore, by being able to correct a finite number of errors, in particular the elements of $\mathcal{E}$, we are able to correct an infinite number of them, i.e. combinations of the errors in $\mathcal{E}$.

It turns out that the condition in Eq. (9.8) is too strong. The Shor code does not obey it, and it still works. In that code, different phase-flip errors in the same cluster lead to identical states. A code satisfying Eq. (9.8) is called a nondegenerate code. Codes that do not satisfy it are called degenerate.

Before discussing the general condition for a quantum code to correct a set of errors, let us show that both errors and the recovery process can be represented as superoperators. Let $\{|\mu\rangle_E\}$ be an orthonormal basis for $\mathcal{H}_E$. We can expand the states $|e_a\rangle_E$ appearing in Eq. (9.9) in this basis, and this allows us to express Eq. (9.9) as

$$U_{AE}|\psi\rangle_A \otimes |0\rangle_E = \sum_\mu M_\mu |\psi\rangle_A \otimes |\mu\rangle_E, \tag{9.10}$$

where

$$M_\mu = \sum_a {}_E\langle\mu|e_a\rangle_E E_a. \tag{9.11}$$

The unitarity of U_{AE} implies that $\sum_\mu M_\mu^\dagger M_\mu = I$. Tracing out the environment, we see that the error takes the density matrix in the code subspace, ρ_A, to

$$T_E(\rho_A) = \sum_\mu M_\mu \rho_A M_\mu^\dagger. \tag{9.12}$$

We see, then, that errors can be represented as superoperators.

Now let us look at the recovery process. Let ρ_A' be the state of n qubits after the error. We measure ρ_A', the measurement being described by a POVM with operators $\tilde{R}_\nu$, and if we get result ν we apply the operator U_ν to correct the error. Therefore, with probability $p_\nu = \mathrm{Tr}(\tilde{R}_\nu^\dagger \tilde{R}_\nu \rho_A')$, we obtain the state

$$\rho_{A\nu} = \frac{1}{p_\nu} U_\nu \tilde{R}_\nu \rho_A' \tilde{R}_\nu^\dagger U_\nu^\dagger. \tag{9.13}$$

Defining $R_\nu = U_\nu \tilde{R}_\nu$, we have that the entire density matrix after the correction procedure has been applied is

$$R(\rho_A') = \sum_\nu p_\nu \rho_{A\nu} = \sum_\nu R_\nu \rho_A' R_\nu^\dagger. \tag{9.14}$$

Note that

$$\sum_\nu R_\nu^\dagger R_\nu = \sum_\nu \tilde{R}_\nu^\dagger \tilde{R}_\nu = I, \tag{9.15}$$

because $\{\tilde{R}_\nu\}$ is a POVM, and therefore R is a superoperator.

We are now in a position to show that the condition for a quantum code to be able to correct an error described by the superoperator T_E, which has Kraus operators M_μ, is

$$_A\langle\bar{j}|M_{\mu'}^\dagger M_\mu|\bar{k}\rangle_A = C_{\mu'\mu}\delta_{\bar{j}\bar{k}}, \tag{9.16}$$

for all M_μ and $M_{\mu'}$, where $C_{\mu'\mu}$ is an arbitrary hermitian matrix. In order to analyze this claim, we will work on an extended space $\mathcal{H}_A \otimes \mathcal{H}_E \otimes \mathcal{H}_B$, which will allow us to use state vectors instead of density matrices. On this space T_E can be represented as $U_{AE} \otimes I_B$, that is a unitary operator that acts on $\mathcal{H}_A \otimes \mathcal{H}_E$ and the identity on

$\mathcal{H}_B$, and R can be represented as $U_{AB} \otimes I_E$, that is a unitary operator that acts on $\mathcal{H}_A \otimes \mathcal{H}_B$ and the identity on $\mathcal{H}_E$. In detail we have

$$T_E : |\bar{j}\rangle_A \otimes |0\rangle_E \otimes |v\rangle_B \rightarrow \sum_\mu M_\mu |\bar{j}\rangle_A \otimes |\mu\rangle_E \otimes |v\rangle_B$$

$$R : |\bar{j}\rangle_A \otimes |\mu\rangle_E \otimes |0\rangle_B \rightarrow \sum_\nu R_\nu |\bar{j}\rangle_A \otimes |\mu\rangle_E \otimes |v\rangle_B. \tag{9.17}$$

If the recovery operation is to correct the error on the code subspace, we must have

$$R \circ T_E : |\bar{j}\rangle_A \otimes |0\rangle_E \otimes |0\rangle_B \rightarrow \sum_{\mu,\nu} R_\nu M_\mu |\bar{j}\rangle_A \otimes |\mu\rangle_E \otimes |v\rangle_B = |\bar{j}\rangle_A \otimes |\Psi\rangle_{EB},$$

$$\tag{9.18}$$

where $|\Psi\rangle_{EB}$ is independent of $\bar{j}$. Taking the inner product of both sides with $_E\langle \mu'|_B\langle v'|$, we have that

$$R_{\nu'} M_{\mu'} |\bar{j}\rangle_A = \lambda_{\mu'\nu'} |\bar{j}\rangle_A, \tag{9.19}$$

where $\lambda_{\mu'\nu'} = {}_E\langle \mu'|_B\langle v'|\Psi\rangle_{EB}$ is independent of $\bar{j}$. This implies that for any $|\psi\rangle_A$ in the code space that $R_\nu M_\mu |\psi\rangle_A = \lambda_{\mu\nu} |\psi\rangle_A$, so that for $|\phi\rangle_A$ in the code space

$$_A\langle \phi|R_\nu M_\mu|\psi\rangle_A = \lambda_{\mu\nu} \, _A\langle \phi|\psi\rangle_A = \, _A\langle (R_\nu M_\mu)^\dagger \phi|\psi\rangle_A. \tag{9.20}$$

This further implies that $(R_\nu M_\mu)^\dagger |\phi\rangle_A = \lambda_{\mu\nu}^* |\phi\rangle_A$ for $|\phi\rangle_A$ in the code space. We now have

$$M_\sigma^\dagger M_\mu |\bar{j}\rangle_A = M_\sigma^\dagger (\sum_\nu R_\nu^\dagger R_\nu) M_\mu |\bar{j}\rangle_A = (\sum_\nu \lambda_{\sigma\nu}^* \lambda_{\mu\nu}) |\bar{j}\rangle_A. \tag{9.21}$$

so that, setting $C_{\sigma\mu} = \sum_\nu \lambda_{\sigma\nu}^* \lambda_{\mu\nu}$,

$$_A\langle \bar{k}|M_\sigma^\dagger M_\mu|\bar{j}\rangle_A = C_{\sigma\mu}\delta_{\bar{k}\bar{j}}. \tag{9.22}$$

Therefore, we have shown that if the recovery operation is able to correct the error, the condition in Eq. (9.16) must be satisfied.

Now let us show the reverse, if the condition in Eq. (9.16) is satisfied, then we can recover from the error produced by T_E. First, let us define a new Kraus representation for T_E by

$$\tilde{M}_\mu = \sum_{\mu'} u_{\mu\mu'} M_{\mu'}, \tag{9.23}$$

where $u_{\mu\mu'}$ is a unitary matrix. This gives us

$$_A\langle\bar{k}|\tilde{M}_\sigma^\dagger\tilde{M}_\mu|\bar{j}\rangle_A = \delta_{\bar{k}\bar{j}}\sum_{\sigma'\mu'}u_{\sigma\sigma'}^*C_{\sigma'\mu'}u_{\mu\mu'} = \delta_{\bar{k}\bar{j}}\sum_{\sigma'\mu'}u_{\sigma\sigma'}^*C_{\sigma'\mu'}(u^*)_{\mu'\mu}^\dagger. \quad (9.24)$$

We can now choose u^* to diagonlize C, so that the above equation becomes

$$_A\langle\bar{k}|\tilde{M}_\sigma^\dagger\tilde{M}_\mu|\bar{j}\rangle_A = \delta_{\bar{k}\bar{j}}\tilde{C}_\mu\delta_{\sigma\mu}. \quad (9.25)$$

Note that because $\sum_\mu\tilde{M}_\mu^\dagger\tilde{M}_\mu = I$, we have that $\sum_\mu\tilde{C}_\mu = 1$. For each $\tilde{C}_\nu \neq 0$ define

$$R_\nu = \frac{1}{\sqrt{\tilde{C}_\nu}}\sum_{\bar{k}}|\bar{k}\rangle_A\langle\bar{k}|\tilde{M}_\nu^\dagger. \quad (9.26)$$

First we note that

$$R_\nu\tilde{M}_\mu|\bar{j}\rangle_A = \frac{1}{\sqrt{\tilde{C}_\nu}}\sum_{\bar{k}}|\bar{k}\rangle_A\langle\bar{k}|\tilde{M}_\nu^\dagger\tilde{M}_\mu|\bar{j}\rangle_A = \sqrt{\tilde{C}_\nu}\delta_{\mu\nu}|\bar{j}\rangle_A. \quad (9.27)$$

Going back to our representation of the superoperators on $\mathcal{H}_A \otimes \mathcal{H}_E \otimes \mathcal{H}_B$, we have that

$$\sum_{\mu\nu}R_\nu\tilde{M}_\mu|\bar{j}\rangle_A \otimes |\mu\rangle_E \otimes |\nu\rangle_B = |\bar{j}\rangle_A \otimes \sum_\mu\sqrt{\tilde{C}_\nu}|\mu\rangle_E \otimes |\mu\rangle_B = |\bar{j}\rangle_A|\Psi\rangle_{EB},$$

$$(9.28)$$

so that it does recover the original state in the code space. Finally, we need to verify that $\sum_\nu R_\nu^\dagger R_\nu = I$. We begin by noting that

$$\sum_\nu R_\nu^\dagger R_\nu = \sum_\nu\sum_{\bar{j}}\frac{1}{\tilde{C}_\nu}\tilde{M}_\nu|\bar{j}\rangle_A\langle\bar{j}|\tilde{M}_\nu^\dagger. \quad (9.29)$$

Now let us apply this operator to any vector of the form $\tilde{M}_\sigma|\bar{k}\rangle_A$,

$$\sum_\nu R_\nu^\dagger R_\nu\tilde{M}_\sigma|\bar{k}\rangle_A = \sum_\nu\sum_{\bar{j}}\frac{1}{\tilde{C}_\nu}\tilde{M}_\nu|\bar{j}\rangle_A\langle\bar{j}|\tilde{M}_\nu^\dagger\tilde{M}_\sigma|\bar{k}\rangle_A$$

$$= \sum_\nu\sum_{\bar{j}}\frac{1}{\tilde{C}_\nu}\tilde{M}_\nu|\bar{j}\rangle_A\tilde{C}_\nu\delta_{\nu\sigma}\delta_{\bar{j}\bar{k}} = \tilde{M}_\sigma|\bar{k}\rangle_A. \quad (9.30)$$

Defining $\mathcal{H}_{\tilde{M}} = \text{span}\{\tilde{M}_\sigma |\psi\rangle_A\}$ for all $\tilde{M}_\sigma$ and $|\psi\rangle_A \in \mathcal{H}_c$, we see that $\sum_\nu R_\nu^\dagger R_\nu$ is just the projection onto $\mathcal{H}_{\tilde{M}}$. To complete the recovery operation, we just add to it $P_{\tilde{M}}^\perp$ the projection onto the orthogonal complement of $\mathcal{H}_{\tilde{M}}$. Adding this operator does not affect our recovery operation, because this operation takes place in $\mathcal{H}_{\tilde{M}}$, and $P_{\tilde{M}}^\perp$ maps any state in this space to zero.

Summarizing, what we have shown is that we can recover from an error T_E with Kraus operators M_μ, if and only if Eq. (9.16) is satisfied. Now this does not appear to be too impressive; we can recover from one error. However, the situation is better than it seems. The same recovery procedure will work for any error whose Kraus operators are linear combinations of the M_μ. In order to see this consider an error T_F with Kraus operators

$$F_\sigma = \sum_\mu m'_{\sigma\mu} M_\mu = \sum_\mu m_{\sigma\mu} \tilde{M}_\mu. \tag{9.31}$$

Applying our recovery operator to a code word affected by F_σ gives us

$$R_\nu F_\sigma |\bar{j}\rangle_A = \frac{1}{\sqrt{\tilde{C}_\nu}} \sum_{\bar{k}} \sum_\mu m_{\sigma\mu} |\bar{k}\rangle_A \langle \bar{k}| \tilde{M}_\nu^\dagger \tilde{M}_\mu |\bar{j}\rangle_A = \sqrt{\tilde{C}_\nu} m_{\sigma\nu} |\bar{j}\rangle_A. \tag{9.32}$$

Going back to our description of the error and recovery operations on the extended space $\mathcal{H}_A \otimes \mathcal{H}_E \otimes \mathcal{H}_B$, we have

$$\sum_{\nu\sigma} R_\nu F_\sigma |\bar{j}\rangle_A \otimes |\sigma\rangle_E \otimes |\nu\rangle_B = |\bar{j}\rangle_A \otimes \sum_{\nu\sigma} \sqrt{\tilde{C}_\nu} m_{\sigma\nu} |\sigma\rangle_E \otimes |\nu\rangle_B = |\bar{j}\rangle_A |\Psi\rangle_{EB}, \tag{9.33}$$

so the error is corrected.

Now that we know what is necessary to correct errors, let us go back and consider the basic errors $E_a \in \mathcal{E}$ from which we built up all of the others. Define $T_\mathcal{E}$ to have Kraus operators $\sqrt{p_a} E_a$, where $0 \leq p_a \leq$ and $\sum_a p_a = 1$. Then, if and only if our code space satisfies

$$_A\langle \bar{j}| E_b^\dagger E_a |\bar{k}\rangle_A = C_{ba} \delta_{\bar{j}\bar{k}}, \tag{9.34}$$

can we recover from $T_\mathcal{E}$. But if we can recover from $T_\mathcal{E}$, then we can recover from any error whose Kraus operators are liner combinations of the $E_a \in \mathcal{E}$. For example, if $\mathcal{E}$ consists of bit flips, phase flips or both on t qubits or fewer, then we can recover from all errors on t qubits of fewer if our code satisfies the above condition.

9.2   An Example: CSS Codes

Now we want to look at a particular class of quantum codes, the CSS (Calderbank-Shor-Steane) codes. Before we do, however, it is necessary to learn something about classical linear codes. A linear code that encodes k bits of information into n bits is called an $[n, k]$ code. It can be described by an $n \times k$ matrix (n rows and k columns) whose elements are 0 or 1. This matrix, G, is known as the generator matrix for the code. A k-digit binary number is encoded into an n-digit code word by writing it as a column vector of length k and then multiplying this vector by the generator matrix to give a column vector of length n, which is the code word. All of the operations here are modulo 2, so the elements of the vectors and matrix are members of the field F_2 that contains the elements 0 and 1, and whose operations, addition and multiplication, are done modulo 2. As an example, consider the $[6, 2]$ code with generator matrix

$$G = \begin{pmatrix} 1 & 0 \\ 1 & 0 \\ 1 & 0 \\ 0 & 1 \\ 0 & 1 \\ 0 & 1 \end{pmatrix}. \tag{9.35}$$

This encodes two bits into six as follows

$$\begin{pmatrix} 0 \\ 0 \end{pmatrix} \rightarrow \begin{pmatrix} 0 \\ 0 \\ 0 \\ 0 \\ 0 \\ 0 \end{pmatrix}, \quad \begin{pmatrix} 1 \\ 0 \end{pmatrix} \rightarrow \begin{pmatrix} 1 \\ 1 \\ 1 \\ 0 \\ 0 \\ 0 \end{pmatrix}, \tag{9.36}$$

etc. The space of code words, C, is spanned by the columns of G. These should be linearly independent over F_2 so that the encoding is unique. Every vector in C will be a code word. .

Another way to specify the code subspace is by means of constraints. Our code subspace had dimension k and lies in an n-dimensional space, so we can specify it by imposing $n - k$ constraints. This can be done by means of an $n - k$ by n matrix H. The code subspace is the set of n-component vectors that is mapped to 0 by H. If the $n - k$ constraints are independent, then the rows of H will be independent. H is called the parity check matrix, and, as we shall see, it is useful in correcting errors.

Clearly G and H are related. Since the columns of G are in the code subspace, we have that $HG = 0$. Let's find H for the $[6, 2]$ code. It will be a 4×6 matrix, and its rows need to be orthogonal to the columns of G. That means we need four linearly independent six-component vectors that are orthogonal to the two columns of G. We start by noticing that $(1\ 1\ 0)^T$ and $(1\ 0\ 1)^T$ are linearly independent and are orthogonal to $(1\ 1\ 1)^T$, where T denotes transpose. We can, therefore, choose

$$H = \begin{pmatrix} 1\ 1\ 0\ 0\ 0\ 0 \\ 1\ 0\ 1\ 0\ 0\ 0 \\ 0\ 0\ 0\ 1\ 1\ 0 \\ 0\ 0\ 0\ 1\ 0\ 1 \end{pmatrix}. \tag{9.37}$$

The parity check matrix is useful for detecting and correcting errors, because an error will usually send the code word out of the code subspace, and we can detect this by acting on the corrupted code word with H. To see how this works, we first define the weight of an n-component vector consisting of 0's and 1's to be the number of 1's. We can represent a corrupted code word x by $x + e$, where e is an n-component vector representing the error. Each 1 in e causes a bit-flip error in x, so that the number of bit-flip errors is equal to the weight of e. Note that because $Hx = 0$, we have that $H(x + e) = He$, and we call He the syndrome of error e. Define the distance of a code to be the minimum weight of any nonzero code word, i.e. of any nonzero $x \in C$. The Hamming distance, which we shall just call the distance, between two code words x and y is just the number of places in which they differ, which is the same as the weight of $x + y$. We shall denote this distance by $d(x, y)$. As a result of these definitions, we see that for $x \neq y$, $d(x, y)$ will be greater than or equal to the weight of the code, because since $x + y \in C$, its weight must be greater than or equal to the weight of the code. Therefore, if a code C has a distance of $2t + 1$, then errors of weight t will not change one code word into another. Each error will produce a unique syndrome so that we can correct it. To see this, note that if $e_1 \neq e_2$ but $He_1 = He_2$, then $H(e_1 + e_2) = 0$ so that we would have $e_1 + e_2 \in C$. This, however, is not possible, because the weight of $e_1 + e_2$ is less than or equal to $2t$, but the weight of the code is $2t + 1$. Therefore, $He_1 \neq He_2$, and the error syndromes are unique. Once we know which error has occurred, say e, we can correct it by adding e to the corrupted code word, because $(x + e) + e = x$.

For each code C, there is a dual code $C^\perp$. This comes from the observation that $HG = 0$ implies that $G^T H^T = 0$, so that we can interpret H^T as a generator matrix for an $[n, n - k]$ code and G^T as its parity check matrix. This equation implies that each code word in $C^\perp$ is orthogonal to all of the columns of G, so that each code word in $C^\perp$ is orthogonal to all of the code words in C. Because vectors in F_2^n can be orthogonal to themselves, C and $C^\perp$ can intersect. A code is called weakly self-dual if $C \subseteq C^\perp$ and self-dual if $C = C^\perp$. For an $[n, k]$ code to be self-dual, we must have $n = 2k$.

In concluding our brief introduction to classical linear codes, we want to prove an identity relating C and $C^\perp$, which will come in useful shortly. The identity is

$$\sum_{x \in C}(-1)^{x \cdot y} = \begin{cases} 2^k & y \text{ in } C^\perp \\ 0 & y \text{ not in } C^\perp \end{cases}. \tag{9.38}$$

The first part is easy. If $y \in C^{\perp}$, and $x \in C$, then $x \cdot y = 0$. Now, using the fact that C has 2^k code words we get the first result. The second part follows from the identity, where $w \in \{0, 1\}^k$,

$$\sum_{v \in \{0,1\}^k} (-1)^{v \cdot w} = 0, \tag{9.39}$$

for $w \neq 0$. We can express $x \in C$ as $x = Gv$ for some $v \in \{0, 1\}^k$, so we have that

$$\sum_{x \in C} (-1)^{x \cdot y} = \sum_{v \in \{0,1\}^k} (-1)^{(Gv) \cdot y} = \sum_{v \in \{0,1\}^k} (-1)^{v \cdot (G^T y)} = 0, \tag{9.40}$$

if $G^T y \neq 0$. But, $G^T y \neq 0$ implies that y is not in $C^{\perp}$.

Now we can use these classical codes to define a quantum code. Let C_1 be an $[n, k_1]$ classical code and C_2 be an $[n, k_2]$ classical code, where $k_1 > k_2$ and $C_2 \subset C_1$. We further suppose that C_1 has a distance d_1 and $C_2^{\perp}$ has a distance $d_2^{\perp}$. We define two elements of C_1, x and y, to be equivalent if and only if $x + y \in C_2$. This breaks C_1 up into $|C_1|/|C_2| = 2^{k_1 - k_2}$ equivalence classes, or cosets. We define an n-qubit quantum state for each coset as

$$|x + C_2\rangle = \frac{1}{\sqrt{|C_2|}} \sum_{y \in C_2} |x + y\rangle. \tag{9.41}$$

The fact that the cosets are disjoint means that these states are orthogonal for x and x' in different cosets. These states span a $2^{k_1 - k_2}$ dimensional subspace of the n-qubit space, so this is an $[n, k_1 - k_2]$ quantum code; it encodes $k_1 - k_2$ qubits in n qubits.

Let us see what happens when we apply $H^{\otimes n}$, a Hadamard gate to each qubit, to this state. Remember that

$$H^{\otimes n}|x\rangle = \frac{1}{2^{n/2}} \sum_{y=0}^{2^n - 1} (-1)^{x \cdot y} |y\rangle \tag{9.42}$$

so

$$H^{\otimes n}|x + C_2\rangle = \frac{1}{\sqrt{|C_2|}} \sum_{y \in C_2} \frac{1}{2^{n/2}} \sum_{u=0}^{2^n - 1} (-1)^{(x+y) \cdot u}$$

$$= \frac{1}{2^{(n+k_2)/2}} \sum_{u=0}^{2^n - 1} (-1)^{x \cdot u} \sum_{y \in C_2} (-1)^{y \cdot u} |u\rangle$$

$$= \frac{1}{2^{(n-k_2)/2}} \sum_{u \in C_2^{\perp}} (-1)^{x \cdot u} |u\rangle. \tag{9.43}$$

What we get is a superposition, with phases, of code words in $C_2^\perp$. As we shall see, it will be possible to correct bit-flip errors in the original code and phase-flip errors in $C_2^\perp$.

Now suppose $d_1 > 2t_f + 1$ and $d_2^\perp > 2t_p + 1$. Our code will then be able to correct t_f bit-flip errors and t_p phase-flip errors. Let e_1 be a vector with weight less than t_f, and e_2 be a vector with weight less than t_p. The ones in e_1 correspond to bit flips and the ones in e_2 correspond to phase flips. The errors have the effect

$$|x + C_2\rangle \rightarrow \frac{1}{\sqrt{|C_2|}} \sum_{y \in C_2} (-1)^{(x+y)\cdot e_2} |x + y + e_1\rangle. \tag{9.44}$$

To correct the bit-flip errors we append an n-qubit ancilla and apply a unitary operator that takes

$$U_f |v\rangle |0\rangle = |v\rangle |H_1 v\rangle, \tag{9.45}$$

where H_1 is the parity-check matrix for C_1. We then have that

$$U_f \left(\frac{1}{\sqrt{|C_2|}} \sum_{y \in C_2} (-1)^{(x+y)\cdot e_2} |x + y + e_1\rangle \right) |0\rangle$$

$$= \left(\frac{1}{\sqrt{|C_2|}} \sum_{y \in C_2} (-1)^{(x+y)\cdot e_2} |x + y + e_1\rangle \right) |H_1 e_1\rangle. \tag{9.46}$$

Now measure the ancilla in the computational basis. The result tells us which qubits have been flipped, as C_1 can correct up to t_f bit-flip errors. Apply X to these bits to flip them back, and throw away the ancilla. Our state is now

$$\frac{1}{\sqrt{|C_2|}} \sum_{y \in C_2} (-1)^{(x+y)\cdot e_2} |x + y\rangle. \tag{9.47}$$

Now apply $H^{\otimes n}$ to this state

$$H^{\otimes n} \frac{1}{\sqrt{|C_2|}} \sum_{y \in C_2} (-1)^{(x+y)\cdot e_2} |x + y\rangle$$

$$= \frac{1}{\sqrt{|C_2|}} \sum_{y \in C_2} \frac{1}{2^{n/2}} \sum_{u=0}^{2^n - 1} (-1)^{(x+y)\cdot(e_2 + u)} |u\rangle$$

$$= \frac{1}{2^{(n-k_2)/2}} \sum_{u+e_2 \in C_2^\perp} (-1)^{x\cdot(u+e_2)} |u\rangle$$

$$= \frac{1}{2^{(n-k_2)/2}} \sum_{u' \in C_2^\perp} (-1)^{x\cdot u'} |u' + e_2\rangle, \tag{9.48}$$

where $u' = u + e_2$. Now append an n-qubit ancilla and apply a unitary operator that takes

$$U_p|v\rangle|0\rangle = |v\rangle|G_2^T v\rangle, \tag{9.49}$$

where G_2 is the generator for C_2 so that G_2^T is the parity-check matrix for $C_2^\perp$. Doing so gives us

$$U_p\left(\frac{1}{2^{(n-k_2)/2}}\sum_{u'\in C_2^\perp}(-1)^{x\cdot u'}|u' + e_2\rangle\right)|0\rangle$$

$$= \left(\frac{1}{2^{(n-k_2)/2}}\sum_{u'\in C_2^\perp}(-1)^{x\cdot u'}|u' + e_2\rangle\right)|G_2^T e_2\rangle. \tag{9.50}$$

We again measure the ancilla in the computational basis, and this tells us which bits have flipped. Apply X to these bits to flip them back, and discard the ancilla. We how have the state

$$\frac{1}{2^{(n-k_2)/2}}\sum_{u'\in C_2^\perp}(-1)^{x\cdot u'}|u'\rangle. \tag{9.51}$$

Now apply $H^{\otimes n}$, and from Eq. (9.43) and the fact that $H^2 = I$ we have

$$H^{\otimes n}\frac{1}{2^{(n-k_2)/2}}\sum_{u'\in C_2^\perp}(-1)^{x\cdot u'}|u'\rangle = \frac{1}{\sqrt{|C_2|}}\sum_{y\in C_2}|x + y\rangle, \tag{9.52}$$

and all of the errors have been corrected. Note that if we assume that $t_f = t_p = t$, what we have shown is that we can correct t bit flips, t phase flips, and t products of bit flips and phase flips. This implies that we can correct all errors on t qubits or fewer.

An example of a CSS code is the 7-qubit Steane code, which can correct errors in one qubit. It is based on the classical [7, 4] Hamming code. A Hamming code is derived by choosing an integer $r \geq 2$, and then taking for the parity-check matrix, H, the matrix whose columns are the $2^r - 1$ bit strings of length r, excluding the string with all zeroes. This gives an r by $2^r - 1$ parity-check matrix, which implies the generator is a $2^r - 1$ by $2^r - r - 1$ matrix, so we have a $[2^r - 1, 2^r - r - 1]$ code. If $r = 3$, this gives a [7, 4] code. The parity-check matrix for this code is

$$H = \begin{pmatrix} 1\,0\,1\,0\,1\,0\,1 \\ 0\,1\,1\,0\,0\,1\,1 \\ 0\,0\,0\,1\,1\,1\,1 \end{pmatrix}. \tag{9.53}$$

This code has distance 3. To see this first note that the string $x_3 = (1110000)^T$, which has weight 3, satisfies $Hx_3 = 0$, so it is in the code. If x_1 were a code word of weight 1, then $Hx_1 = 0$ would imply that one of the columns of H would have to be zero, which is not the case. Therefore, there are no weight-one code words. If x_2 is a code word of weight 2, we could express it as $x_2 = x_1 + x_1'$, where x_1 and x_1' are both of weight one and $x_1 \neq x_1'$. Then $H(x_1 + x_1') = Hx_1 + Hx_1' = 0$ implies that two columns of H must be identical, which is also not the case. Therefore, there are no weight-two code words, and the distance of the code is 3. The generator matrix for this code is

$$G = \begin{pmatrix} 1 & 0 & 0 & 1 \\ 0 & 1 & 0 & 1 \\ 1 & 1 & 0 & 1 \\ 0 & 0 & 1 & 0 \\ 1 & 0 & 1 & 0 \\ 0 & 1 & 1 & 0 \\ 1 & 1 & 1 & 0 \end{pmatrix}. \tag{9.54}$$

Note that the rows of H are in the code, and they are just the first three columns of G. These vectors are orthogonal to themselves.

The matrix H^T is the generator of the dual code, which is a [7, 3] code. In this case $C^\perp \subset C$, and $C^\perp$ consists of all of the code words in C with an even weight. This code also has a distance of 3.

To construct the CSS code, we take $C_1 = C$ and $C_2 = C^\perp$, so that $C_2^\perp = C$ has weight 3. This implies that the Steane code can correct one-qubit errors. It is a [7, 1] quantum code, $(k_1 - k_2 = 4 - 3 = 1)$. There are only two cosets in this case, each with eight members. Letting y_j, for $j = 1, 2, 3, 4$ be the columns of G, we have that the members of the coset containing the identity are given by $c_1 y_1 + c_2 y_2 + c_3 y_3$, where $c_j \in \{0, 1\}$, and the members of the other coset are given by $c_1 y_1 + c_2 y_2 + c_3 y_3 + y_4$. Note that the members of the first coset have even weight, while the members of the second have odd weight.

9.3 Decoherence-Free Subspaces

So far we have focussed on error-correcting codes as a way of defeating the effects of decoherence. There are a number of other approaches, and we will give here a very brief introduction to one of them. This method takes advantage of the fact that if qubits are subject to the same errors, there are subspaces that remain free from the effects of decoherence.

Let us start with a single qubit and suppose it is subject to random-phase errors. In particular, we have that

$$|0\rangle \rightarrow |0\rangle \quad |1\rangle \rightarrow e^{i\phi}|1\rangle, \tag{9.55}$$

where ϕ is distributed according the the probability distribution $p(\phi)$. Let us suppose that the qubit is initially in the state $|\psi\rangle = a|0\rangle + b|1\rangle$ and see what happens to it under the action of this type of decoherence (phase decoherence). Defining the operator $R(\phi) = \exp[i\phi(I - \sigma_z)/2]$, which has the action $R(\phi)|0\rangle = |0\rangle$ and $R(\phi)|1\rangle = e^{i\phi}|1\rangle$, we have that after the effects of the decoherence, the density matrix of the qubit is given by

$$\rho = \int_0^{2\pi} d\phi \, p(\phi)R(\phi)|\psi\rangle\langle\psi|R^\dagger(\phi)$$

$$= |a|^2|0\rangle\langle 0| + |b|^2|1\rangle\langle 1| + a^*bz|1\rangle\langle 0| + ab^*z^*|0\rangle\langle 0|, \tag{9.56}$$

where

$$z = \int_0^{2\pi} d\phi \, p(\phi)e^{i\phi}. \tag{9.57}$$

Since $|z| \leq 1$, and is usually less than one, phase decoherence causes the magnitude of the off-diagonal elements of the initial density matrix to decrease. An extreme case is when the phase is uniformly distributed ($p(\phi) = 1/2\pi$), in which case $z = 0$, and the off-diagonal elements will vanish. This would lead to a complete destruction of the phase information in the initial state $|\psi\rangle$.

Now let us consider two qubits, and we shall assume that they are subject to the same random-phase errors. That means that after the phase decoherence has taken place, the two qubit state $|\Psi\rangle$ will become

$$\rho = \int_0^{2\pi} d\phi \, p(\phi)R(\phi) \otimes R(\phi)|\Psi\rangle\langle\Psi|R^\dagger(\phi) \otimes R^\dagger(\phi). \tag{9.58}$$

Note than under the action of this kind of decoherence, the states $|0\rangle|1\rangle \rightarrow e^{i\phi}|0\rangle|1\rangle$ and $|1\rangle|0\rangle \rightarrow e^{i\phi}|1\rangle|0\rangle$ are affected in the same way. Furthermore, we see that any superposition of them

$$R(\phi) \otimes R(\phi)(a|0\rangle|1\rangle + b|1\rangle|0\rangle) = e^{i\phi}(a|0\rangle|1\rangle + b|1\rangle|0\rangle), \tag{9.59}$$

is just multiplied by an overall phase. When a state of this type is inserted into Eq. (9.58), the overall phase simply cancels out and the state is unchanged. Therefore, states of the form $(a|0\rangle|1\rangle + b|1\rangle|0\rangle)$ are not affected by phase decoherence

We can take advantage of this fact and protect the state of a single qubit from phase decoherence by encoding it into two qubits in the subspace spanned by $|0\rangle|1\rangle$

and $|1\rangle|0\rangle$. In particular we can encode the single qubit state $|0\rangle$ as $|0\rangle|1\rangle$, and the single qubit state $|1\rangle$ as $|1\rangle|0\rangle$. As long as the phase decoherence affects both qubits in the same way, this encoding will ensure that any state of a single qubit will be free from the effects of phase decoherence.

9.4 Problems

1. Show that the three-qubit bit-flip code, $|0\rangle \rightarrow |0\rangle^{\otimes 3}$ and $|1\rangle \rightarrow |1\rangle^{\otimes 3}$ satisfies the quantum error correction condition for the error sets $\{I, X_1, X_2, X_3\}$ and $\{I, Y_1, Y_2, Y_3\}$, but not for the combined set $\{I, X_1, X_2, X_3, Y_1, Y_2, Y_3\}$.

2. For a nondegenerate $[n, k]$ quantum code, each error maps the code space into a different subspace, and all of those subspaces are orthogonal. Suppose we want to be able to correct up to t single-qubot errors, X, Y, or Z.

   (a) Show that

   $$\sum_{j=0}^{t} \binom{n}{j} 3^j \leq 2^{n-k}.$$

   This is the quantum Hamming bound.

   (b) Find the smallest value of n allowed by this bound for $k = 1$ and $t = 1, 2$.

3. Find the decoherence-free subspaces for three qubits all undergoing the same random phase noise errors.

4. One type of code we did not discuss is one that protects against erasure errors. A one-qubit erasure error is equivalent to losing one qubit. We are going to look at the case of qutrits. Consider the following encoding for a qutrit

   $$|0\rangle \rightarrow \frac{1}{\sqrt{3}}(|000\rangle + |111\rangle + |222\rangle)$$

   $$|1\rangle \rightarrow \frac{1}{\sqrt{3}}(|012\rangle + |120\rangle + |201\rangle)$$

   $$|2\rangle \rightarrow \frac{1}{\sqrt{3}}(|021\rangle + |102\rangle + |210\rangle).$$

   We now use this encoding to encode a general one-qutrit state $|\psi\rangle = \alpha|0\rangle + \beta|1\rangle + \gamma|2\rangle$ into a three qutrit state.

   (a) Show that if we lose two of the qutrits, there is no information about the state $|\psi\rangle$ remaining.

   (b) Show that if we only lose one of the qutrits, we can perfectly recover the state $|\psi\rangle$.

Further Reading

1. J. Preskill, *Lecture notes for Physics 219, Quantum Computation*.
2. D.A. Lidar, K.B. Whaley, Decoherence-free subspaces, in *Irreversible Quantum Dynamics*, edited by F. Benatti, R. Floreanini. Springer Lecture Notes in Physics vol. 622 (Berlin 2003), p. 83 and quant-ph/0301032

Chapter 10
The Stabilizer Formalism
and the Gottesman–Knill Theorem

The stabilizer formalism provides another method of constructing quantum error-correcting codes. However, we are going to be interested in it primarily for a different reason; it allows us to prove the Gottesman–Knill theorem. That theorem serves as a useful warning. Just because you are manipulating qubits with quantum gates does not guarantee that what you are doing cannot be simulated efficiently on a classical computer.

10.1 Stabilizer Formalism

The idea behind the stabilizer formalism is that we can describe a state by the operators that stabilize it, that is, the operators for which the state is an eigenstate with eigenvalue one. For example, the state $|0\rangle$ is stabilized by Z because $Z|0\rangle = |0\rangle$ and the state $|1\rangle$ is stabilized by $-Z$, as $(-Z)|1\rangle = |1\rangle$. Similarly $|+x\rangle$ is stabilized by X and $|-x\rangle$ is stabilized by $-X$. We will be interested in operators that are members of the Pauli group, which for one qubit is given by

$$G_1 = \{\pm I, \pm iI, \pm X, \pm iX, \pm Y, \pm iY, \pm Z, \pm iZ\}, \tag{10.1}$$

and for n qubits is $G_n = G_1^{\otimes n}$. Note that the members of G_n either commute or anti-commute.

Now let S be a subgroup of G_n, and define V_S to be the subspace of n-qubit states that are invariant under the action of all of the elements of S. Then S is said to be the stabilizer of V_S. It is convenient to describe S in terms of its generators. A group, G is generated by a set of elements $g_1, g_2, \ldots g_m$ if every element of G can be written as a product of elements in that set, and we express this as $G = \langle g_1, g_2, \ldots g_m \rangle$. For example, consider the subgroup of G_3 given by $S = \{I, Z_1 Z_2, Z_1 Z_3, Z_2 Z_3\}$. We find that $S = \langle Z_1 Z_2, Z_2 Z_3 \rangle$, because $Z_1 Z_3 = (Z_1 Z_2)(Z_2 Z_3)$, and $I = (Z_1 Z_2)^2$.

In order to find V_S, we only need to find the subspace stabilized by the generators, because if a space is stabilized by the generators, it is stabilized by the group. The eigenstates of $Z_1 Z_2$ with eigenvalue one are $|0\rangle |0\rangle |\psi_0\rangle$ and $|1\rangle |1\rangle |\psi_1\rangle$, where $|\psi_0\rangle$ and $|\psi_1\rangle$ are arbitrary. If these states are to be eigenstates of $Z_2 Z_3$ with eigenvalue one, we must have $|\psi_0\rangle = |0\rangle$ and $|\psi_1\rangle = |1\rangle$, so that V_S is the linear span of the states $|0\rangle |0\rangle |0\rangle$ and $|1\rangle |1\rangle |1\rangle$.

There are two conditions that a subgroup, S has to satisfy if the subspace that it stabilizes is to be nontrivial. The first is that $-I$ is not a member of S. If it is, then for $|\psi\rangle$ to be in V_S, we must have $(-I)|\psi\rangle = |\psi\rangle$, which implies that $|\psi\rangle = 0$. The second is that S must be abelian, that is, all of its elements must commute. To see why, suppose g_1 and g_2 are in S, and that they anti-commute. Now if $|\psi\rangle \in V_S$, then $g_1 g_2 |\psi\rangle = |\psi\rangle$. But $g_1 g_2 |\psi\rangle = -g_2 g_1 |\psi\rangle$, and since $|\psi\rangle$ is in V_S, it must be the case that $g_2 g_1 |\psi\rangle = |\psi\rangle$. Therefore, we again have that $|\psi\rangle = -|\psi\rangle$, and $|\psi\rangle = 0$.

When specifying an subgroup by its generators, we want the generators to be independent. That means that if we remove any of the generators, we change the subgroup,

$$\langle g_1, \ldots g_{j-1}, g_{j+1}, \ldots g_m \rangle \neq \langle g_1, \ldots g_m \rangle, \tag{10.2}$$

for any g_j. One way of testing whether a set of generators is independent is by using something called the check matrix. This is an $m \times 2n$ matrix, where each row corresponds to a generator of S, and all of the elements are 0 or 1. The elements of the check matrix are constructed as follows. If the generator g_j has an I in the kth place, then $M_{jk} = M_{j,k+n} = 0$. If it has an X in the kth place, then $M_{jk} = 1$ and $M_{j,k+n} = 0$, a Y, then $M_{jk} = M_{j,k+n} = 1$, and a Z, then $M_{jk} = 0$ and $M_{j,k+n} = 1$. Factors of ± 1 and $\pm i$ are ignored. This matrix has the useful property that if $r(g)$ is the row corresponding to the generator g, and $r(g')$ is the row corresponding to the generator g', then

$$r(g) + r(g') = r(gg'), \tag{10.3}$$

where the addition is bitwise, modulo 2. This follows from looking at the correspondence

$$X \leftrightarrow (1,0) \quad Y \leftrightarrow (1,1) \quad Z \leftrightarrow (0,1), \tag{10.4}$$

where the numbers in parentheses are $(M_{jk}, M_{j,k+n})$. For example, note that $XY = iZ$ and $(1,0) + (1,1) = (0,1)$. One can check that the correspondences hold for the other products.

We can now state our theorem.

Theorem 1 *If $S = \langle g_1, \ldots g_m \rangle$, where $-I$ is not in S and $g_1, \ldots g_m$ all commute (which implies that S is abelian), then the generators are independent if and only if the rows of the check matrix are linearly independent.*

Proof If $g \in G_n$, then either $g^2 = I$ or $g^2 = -I$. However, since $-I$ is not in S, we must have $g^2 = I$ for $g \in S$, and this implies that for the generators of S, $g_j^2 = I$, for $j = 1, 2, \ldots m$. Note that this means that each generator is its own inverse. What we are going to show is that the generators are not independent if and only if the rows of the check matrix are linearly dependent. This implies our result, that the generators are independent if and only if the rows are linearly independent. Now let us suppose that the rows are linearly dependent. This means that we have

$$\sum_{j=1}^{m} a_j r(g_j) = 0, \tag{10.5}$$

where $a_j \in \{0, 1\}$, not all of the a_j are 0, and additions are bitwise modulo 2. This and the result from the previous paragraph imply that

$$\prod_{j=1}^{m} g_j^{a_j} = \pm I \text{ or } \pm iI, \tag{10.6}$$

and since $-I$ is not in S, which implies that $\pm iI$ are also not in S, then the right-hand side of the above equation is just I. Now suppose that $a_k = 1$, which gives us

$$g_k = g_k^{-1} = \prod_{j=1, j \neq k}^{m} g_j^{a_j}, \tag{10.7}$$

so the set of generators is not independent. The reverse argument, assuming the generators are not independent and showing that the rows of the check matrix are linearly dependent, makes use of essentially the same steps. Hence, our result is proved.

This result can be used to prove the following theorem, which we shall need shortly.

Theorem 2 If $S = \langle g_1, \ldots g_{n-k} \rangle$ is generated by $n - k$ independent, commuting elements from G_n, and $-I$ is not in S, then V_S is a 2^k-dimensional vector space.

Since the proof is a bit long, we have put it at the end of the chapter, where those who are interested can look at it. Note that this theorem implies that a state, which is a subspace of dimension one, will have a stabilizer generated by n elements.

10.2 Description of Gates and Measurements

Now that we know how to describe a state in terms of the n generators of its stabilizer, we would like to see how the stabilizer changes under the action of a number of different kinds of gates. This is a way of keeping track of how a state changes as it goes through a quantum circuit. Before looking at specific gates, let us start with a general consideration. Suppose that $|\psi\rangle \in V_S$, $g \in S$, and we send $|\psi\rangle$ through a quantum gate described by the unitary operator U. We then have that

$$U|\psi\rangle = Ug|\psi\rangle = UgU^{\dagger}U|\psi\rangle, \tag{10.8}$$

which implies that $U|\psi\rangle$ is stabilized by $UgU^{\dagger}$. This means that UV_S is stabilized by $USU^{\dagger} = \{UgU^{\dagger} \mid g \in S\}$.

If we are going to describe the progress of a state through a quantum circuit by means of its stabilizer, we want the gates in that circuit to map G_n into itself. The set of all unitary operators that do this is called the normalizer of G_n. Luckily, there are a number of common gates that have this property. Both the Hadamard gate, H and the phase gate, F, where $F|0\rangle = |0\rangle$ and $F|1\rangle = i|1\rangle$, have that property

$$HXH^{\dagger} = Z \qquad FXF^{\dagger} = Y$$
$$HYH^{\dagger} = -Y \qquad FYF^{\dagger} = -X$$
$$HZH^{\dagger} = X \qquad FYF^{\dagger} = Z. \tag{10.9}$$

So do the one qubit rotations, X, Y, and Z (Pauli gates). For example, $XXX^{\dagger} = X$, $XYX^{\dagger} = -Y$, and $XZX^{\dagger} = -Z$. The C-NOT gate, a two-qubit gate that can create entanglement, also maps G_n into itself. The unitary corresponding to the C-NOT gate can be expressed as

$$U_{C-NOT} = \frac{1}{2}(I_1 + Z_1) \otimes I_2 + \frac{1}{2}(I_1 - Z_1) \otimes X_2, \tag{10.10}$$

and we find, for example, that

$$U_{C-NOT}(X_1 \otimes I_2)U_{C-NOT}^{\dagger} = X_1 \otimes X_2$$
$$U_{C-NOT}(I_1 \otimes Z_2)U_{C-NOT}^{\dagger} = Z_1 \otimes Z_2. \tag{10.11}$$

Summarizing, we see that if an n-qubit quantum state goes through a series of gates, all of which are in the normalizer of G_n, rather than describe the state directly at each stage of the circuit, we can describe how its stabilizer changes.

Now let us incorporate measurement into this picture. The members of G_n that are just products of Pauli matrices are hermitian, and therefore, they are observables, and we can talk about measuring them. Suppose $g \in G_n$ is just a product of Pauli matrices, that is, it has no -1 or $\pm i$ factors. In addition, assume that g commutes with the generators of the stabilizer of the state $|\psi\rangle$, $\langle g_1, g_2, \ldots g_n\rangle$. Note that in general a group element either commutes or anti-commutes with each of the generators, and we will consider the case when g anti-commutes with some of them later. Now $g_j g|\psi\rangle = g g_j|\psi\rangle = g|\psi\rangle$, and, since this holds for all of the generators, we have that $g|\psi\rangle \in V_S$, where V_S is the space left invariant by the group $\langle g_1, g_2, \ldots g_n \rangle$. In this case, V_S is one-dimensional, it is just multiples of the state $|\psi\rangle$, so that $g|\psi\rangle = \lambda|\psi\rangle$. We also have that $g^2|\psi\rangle = \lambda^2|\psi\rangle$ and that $g^2 = I$, which implies that $\lambda = \pm 1$. Therefore, $|\psi\rangle$ is an eigenstate of g with eigenvalue either 1 or -1, and measuring g in this state will not change the state. This kind of measurement will, as a result, not change the stabilizer group.

Now suppose that g anti-commutes with some of the g_j, and that one of these is g_1. Note that if we replace g_j by $g_1 g_j$ we do not change the stabilizer group, that is

$$\langle g_1, \ldots g_j, \ldots g_n \rangle = \langle g_1, \ldots g_1 g_j, \ldots g_n \rangle, \tag{10.12}$$

and that g commutes with $g_1 g_j$. What we can now do is to replace each generator with which g anti-commutes, g_j, by $g_1 g_j$, except for $j = 1$, without changing the stabilizer group. Thus we can assume that g commutes with all of the generators except for g_1. Now let us see what happens to the stabilizer group if we measure g. The observable g has eigenvalues ± 1, so after the measurement we have, up to normalization, one of the two possible states

$$\text{result} = 1 \qquad |\psi^+\rangle = \frac{1}{2}(I + g)|\psi\rangle$$

$$\text{result} = -1 \qquad |\psi^-\rangle = \frac{1}{2}(I - g)|\psi\rangle. \tag{10.13}$$

The vector space stabilized by $\langle g_2, g_3, \ldots g_n \rangle$ is two-dimensional and is spanned by $|\psi^+\rangle$ and $|\psi^-\rangle$. Therefore, $|\psi^+\rangle$ is stabilized by $\langle g, g_2, \ldots g_n \rangle$ and $|\psi^-\rangle$ is stabilized by $\langle -g, g_2, \ldots g_n \rangle$. Summarizing, if we measure g and get 1, the state after the measurement is stabilized by $\langle g, g_2, \ldots g_n \rangle$, and if we get -1, the post-measurement state is stabilized by $\langle -g, g_2, \ldots g_n \rangle$. So, we can keep track of how a state changes by going through a quantum circuit consisting of the gates mentioned in this section and measurements of observables in G_n by seeing how the generators of the stabilizer group of the state change.

10.3 The Gottesman–Knill Theorem

We now have all the machinery necessary to state and prove the Gottesman–Knill theorem. In fact, we have essentially already proved it. We start with an n-qubit state in the computational basis, $|j_1\rangle \otimes |j_2\rangle \ldots \otimes |j_n\rangle$, where $j_k = 0$ or 1. The state is described by the stabilizer group $\langle (-1)^{j_1} Z_1, (-1)^{j_2} Z_2, \ldots (-1)^{j_n} Z_n \rangle$. We now send the state trough a circuit consisting of Hadamard, phase, Pauli, and C-NOT gates. In addition we make measurements of observables in G_n. At each step we can describe the state by n stabilizer group elements, and each of these elements has n entries, e. g.

$$g_1 = Z_1 \otimes I_2 \otimes I_3 \ldots \otimes X_n$$

$$g_2 = I_2 \otimes (-iY_2) \otimes Z_3 \ldots \otimes I_n$$

$$\vdots \tag{10.14}$$

This means that each step in the computation requires updating n^2 elements, and if the computation has m steps (gates or measurements), the total number of updates is mn^2. However, all of these updates could be done on a classical computer in $O(mn^2)$ steps. Therefore, the entire quantum computation can be simulated efficiently on a classical computer. This can all be summarized in the following theorem.

Gottesman–Knill Theorem *Suppose a quantum computation is performed that involves only: state preparation in the computational basis, Hadamard gates, phase gates, Pauli gates, C-NOT gates, and the measurement of observables in the Pauli group together with the possibility of classical control conditioned on these measurements. Such a computation can be efficiently simulated on a classical computer.*

This result tells us that if we want to really get a quantum advantage in a computation, we will have to go beyond the limits specified in the statement of the theorem. Different kinds of gates, measurements, or state preparations, other than those mentioned in the theorem, would have to be used.

10.4   Proof of Theorem 2

We now want to prove Theorem 2, which relates the dimension of the subspace stabilized by a subgroup, S, of G_n to the number of independent generators of S. In order to do so, we first have to discuss an additional property of the check matrix. This is that the rows of the check matrix tell us whether the corresponding elements of G_n commute or anti-commute. Let us now see how this works.

We begin by defining the $2n \times 2n$ matrix

$$\Lambda = \begin{pmatrix} 0 & I_n \\ I_n & 0 \end{pmatrix}. \tag{10.15}$$

It is then the case that if $r(g_1)$ and $r(g_2)$ are the rows of the check matrix corresponding to group elements g_1 and g_2, respectively, then g_1 and g_2 commute if and only if $r(g_1)^T \Lambda r(g_2) = 0$, where T denotes the transpose. Note that since the only two values that $r(g_1)^T \Lambda r(g_2)$ can have are 0 or 1, since all operations are modulo 2, that means if it is 1, then g_1 and g_2 anti-commute. To see how this works, let us look at the jth and $(n+j)$th elements of $r(g_1)$ and $r(g_2)$, which correspond to the jth elements of g_1 and g_2, respectively. Each of those elements is one of the pairs $(0,0)$, $(0,1)$, $(1,0)$, and $(1,1)$. Now the jth elements of g_1 and g_2 commute only if at least one of them is the identity, or if they are both the same. Otherwise they anti-commute. Notice that if we take the inner product of the two-component vector corresponding to the jth and $(n+j)$th elements of $r(g_1)$ with the swapped version (jth and $(n+j)$th elements swapped) of the one for $r(g_2)$, we get exactly what we want, 0 if the jth elements commute, and 1 if they anti-commute. If we now sum over j, which gives us $r(g_1)^T \Lambda r(g_2)$, we get 0 if the number of anti-commuting elements is even, which implies that g_1 and g_2 commute, and 1 if it is odd, which implies that g_1 and g_2 anti-commute.

Now let us prove the theorem. Let $S = \langle g_1, \ldots g_{n-k} \rangle$ be generated by $n-k$ commuting elements of G_n, and $-I$ is not in S. We want to show that V_S is a 2^k dimensional subspace of $\mathcal{H}_n$, the space of n qubits. The strategy is to break $\mathcal{H}_n$ into 2^{n-k} equal size blocks, one of which is V_S. In order to do so, we will define a set of projection operators. Let x be a vector with $n-k$ components each of which is either 0 or 1, and define the projection operator

$$P_x = \frac{1}{2^{n-k}} \prod_{j=1}^{n-k} [I + (-1)^{x_j} g_j]. \tag{10.16}$$

Note that the operators $(I \pm g_j)/2$ project onto the states such that $g_j|\psi\rangle = \pm|\psi\rangle$, respectively. P_0 projects onto V_S, because in the subspace onto which it projects, all of the the g_j have the eigenvalue $+1$. It is also the case that $P_x P_{x'} = 0$ if $x \neq x'$, because for some j we will have the product $(I + g_j)(I - g_j) = 0$.

It is also the case that

$$\sum_{x \in \{0,1\}^{n-k}} P_x = I. \tag{10.17}$$

This can be proved by induction. Define

$$Q_m = \frac{1}{2^m} \sum_{x \in \{0,1\}^m} \prod_{j=1}^{m} [I + (-1)^{x_j} g_j], \tag{10.18}$$

and note that $Q_1 = I$. We have that

$$Q_{m+1} = \frac{1}{2^{m+1}} \sum_{x_{m+1} \in \{0,1\}} \sum_{x \in \{0,1\}^m} [I + (-1)^{x_{m+1}} g_{m+1}]$$

$$\prod_{j=1}^{m} [I + (-1)^{x_j} g_j],$$

$$= \frac{1}{2}[(I + g_{m+1})Q_m + (I - g_{m+1})Q_m] = Q_m. \tag{10.19}$$

Therefore, $Q_m = I$ and Eq. (10.17) is proved.

What we have so far is that the projections break $\mathcal{H}_n$ into orthogonal subspaces, $P_x \mathcal{H}_n$, and this implies that

$$\dim(\mathcal{H}_n) = 2^n = \sum_{x \in \{0,1\}^{n-k}} \dim(P_x \mathcal{H}_n). \tag{10.20}$$

If we can show that each of the subspaces $V_x = P_x \mathcal{H}_n$ has the same dimension, we will be done. We can do this by showing that there are unitary operators that transform one subspace into another. In order to do this, we will start with the check matrix, G for $g_1, \ldots g_{n-k}$. It is a theorem from linear algebra that the row rank of a matrix is equal to its column rank, and since the elements $g_1, \ldots g_{n-k}$ are assumed to be independent, the row, and hence the column, rank of G is $n - k$. Therefore any $n - k$ component vector, x, can be written as a linear combination of the columns of G, and, in particular, for any x there is a $2n$ component vector y such that

$$G \Lambda y = x. \tag{10.21}$$

Now y corresponds to a member $g \in G_n$ in the check matrix representation with the property that $gg_j = (-1)^{x_j} g_j g$, so that, since $gg^\dagger = I$,

$$g P_0 g^\dagger = \frac{1}{2^{n-k}} \prod_{j=1}^{n-k} g(I + g_j)g^\dagger$$

$$= \frac{1}{2^{n-k}} \prod_{j=1}^{n-k} [I + (-1)^{x_j} g_j] = P_x. \tag{10.22}$$

The operator g is unitary, and this implies that $\dim(V_x) = \dim(V_0)$, which further implies that $2^n = 2^{n-k} \dim(V_0)$, and finally that $\dim(V_0) = 2^k$.

10.5 Problems

1. (a) Find independent generators for the stabilizer group in G_3 for the GHZ state $(1/\sqrt{2})(|000\rangle + |111\rangle)$.
   (b) Find the check matrix for the generators you found in part (a) and use it to show that the generators are independent.
   (c) Show that there is no subgroup in G_3 that stabilizes the W state, $(1/\sqrt{3})(|001\rangle + |010\rangle + |100\rangle)$.

2. The Toffoli gate is a three qubit gate that acts as follows. The first two qubits are control qubits, and their state is not changed by the action of the gate. If the control qubits are in the state $|00\rangle$, $|01\rangle$, or $|10\rangle$, then the state of the third qubit is not affected. If the control qubits are in the state $|11\rangle$, then a bit flip operation, X, is applied to the third qubit. By finding an expression for $U(X_1 \otimes I_2 \otimes I_3)U^\dagger$, where U is the Toffoli gate, in terms of X, Y, and Z operators, show that the Toffoli gate is not in the normalizer of G_3.

3. Let us consider a simple two-qubit circuit consisting of two Hadamard gates and one CNOT gate. The first qubit goes first through one of the Hadamard gates, then through the control input of the CNOT gate, and then through the second Hadamard gate. The second qubit only goes through the target input of the CNOT gate. Both qubits start in the state $|0\rangle$. Describe the state of the system by specifying the generators of its stabilizer group after the first Hadamard gate, after the CNOT gate, and after the second Hadamard gate.

Further Reading

1. D. Gottesman, *Stabilizer Codes and Quantum Error Correction*, Caltech Ph.D. thesis, https://arxiv.org/abs/quant-ph/9705052
2. S. Aaronson, D. Gottesman, Improved simulation of stabilizer circuits. Phys. Rev. A **70**, 052328 (2004)

Chapter 11
Quantum Information Theory

We have already come across the Shannon and von Neumann entropies in the chapter on entanglement. The Shannon entropy assesses the information content of a classical probability distribution while the von Neumann entropy does the same for its quantum counterpart, the quantum state (density matrix) of a quantum system. We have also come across the trace distance and the fidelity in the chapter on quantum states. These quantities proved very useful when we wanted to compare different states of a quantum system. However, these distance measures were introduced ad hoc, without introducing their classical counterpart first and then showing that these are the proper, or at least the obvious, quantum generalizations. Therefore, in the first two sections of this chapter, we will introduce these quantities in a more systematic manner. Then, we want to use the entropic quantities to say something about quantum communication, in particular, sending classical messages using quantum systems. In order to do so, we will need to define Shannon and von Neumann entropies for two random variables, in the classical case, and two systems, in the quantum case, which we do in the third section. It is also in order here to introduce consistent notation: entropic quantities related to classical probability distributions will be denoted by H and then further specified by their arguments, and their quantum counterparts, entropic quantities related to quantum states (density matrices), will be denoted by S and, again, further specified by their arguments. In the remaining sections, we will apply these quantities to studying quantum signatures of communication, establishing some important bounds and discussing an illustrative example.

© Springer Nature Switzerland AG 2021

J. A. Bergou et al., *Quantum Information Processing*, Graduate Texts in Physics,
https://doi.org/10.1007/978-3-030-75436-5_11

11.1   Comparing Classical Probability Distributions

It is a frequent task to compare probability distributions in the classical case and their quantum counterparts, states, in the quantum case. In this section we look at classical quantities that are frequently used for this task and then, in the next section, we look at the corresponding quantum expressions.

If we have two different probability distributions, $\{p_X(x)\}$ and $\{q_X(x)\}$, for a single random variable, X, then one such quantity is the (classical) trace distance between them, which we define as

$$D(p_X, q_X) \equiv \frac{1}{2} \sum_x |p_X(x) - q_X(x)|. \tag{11.1}$$

The trace distance is also called the L_1 distance or Kolmogorov distance. Furthermore, the trace distance is a metric, which follows from the following two observations. First, it is obviously symmetric in its variables, $D(p_X, q_X) = D(q_X, p_X)$. Second, if we have three different probability distributions, p, q and r, over the same random variable, X, the trace distance satisfies the triangle inequality, $D(p_X, r_X) \leq D(p_X, q_X) + D(q_X, r_X)$. Here we state the triangle inequality without proof but, actually, the proof is rather straightforward and we leave it to the reader as an end-of the-chapter problem.

Another quantity that is used for the comparison of two probability distributions is the fidelity. If we have two different probability distributions, $\{p_X(x)\}$ and $\{q_X(x)\}$, for a single random variable, X, then we can define two vectors in the event space, $\vec{u}$ and $\vec{v}$, such that their x component is given by $\vec{u}_x = \sqrt{p_X(x)}$ and $\vec{v}_x = \sqrt{q_X(x)}$. From the definition it is obvious that $\vec{u} \cdot \vec{u} = \vec{v} \cdot \vec{v} = 1$, i.e., both vectors are normalized to 1, $\|u\| = \|v\| = 1$. We now introduce the scalar product of these two vectors and call it the (classical) fidelity,

$$F(p_X, q_X) \equiv \vec{u} \cdot \vec{v} = \sum_x \sqrt{p_X(x)q_X(x)}. \tag{11.2}$$

Clearly, $\vec{u} \cdot \vec{v} = \|u\|\|v\|\cos(\theta) = \cos(\theta)$, since the norms are 1. From here, $0 \leq F(p_X, q_X) \leq 1$ and F is symmetric in its arguments. It is still not a metric since it does not satisfy a triangle inequality. However, it is easy to see that $\cos^{-1}(F) = \theta$ does, so F is very closely related to a proper distance measure. Note also that the fidelity is zero if $q_X(x) = 0$, for those values of x for which $p_X(x) \neq 0$, i.e., when the two distributions are as different as possible. In all other cases, the fidelity is between 0 and 1, so, intuitively, the fidelity is a good quantifier of how the distributions compare.

Finally, we also introduce an entropic quantity which is frequently used for the comparison of probability distributions. First, we recall that for a single random variable X, which takes the value x with probability $p_X(x)$, the Shannon entropy was defined as

$$H(X) = -\sum_x p_X(x) \log p_X(x), \qquad (11.3)$$

where the logarithm is base 2. We already encountered the Shannon entropy in Chap. 3 where we also encountered its quantum version, the von Neumann entropy. There are many entropic quantities that are related to the Shannon entropy.

Here we introduce one such quantity, the relative entropy (also called the Kullback–Leibler divergence, information gain or information divergence), which is the quantity used for the comparison of classical probability distributions. Actually, we have also encountered the quantum version of this quantity in Chap. 3. Classically, the relative entropy of two probability distributions, $p_X(x)$ and $q_X(x)$, is given by

$$H(p\|q) = \sum_x p_X(x) \log \frac{p_X(x)}{q_X(x)}. \qquad (11.4)$$

It can also be written as $H(p\|q) = -H(X) - \sum_x p_X(x) \log q_X(x)$, where $H(X)$ is once again the Shannon entropy. This quantity can act as a kind of distance between two quantum states. Note that if the states are identical, the relative entropy is zero. Note also that this quantity can be infinite if there is an x for which $q_X(x) = 0$, but $p_X(x) \neq 0$. Further, just like its quantum version, the classical relative entropy is non-negative, $H(p\|q) \geq 0$. The proof follows from the considerations after Eq. (3.46), along the same lines as the Klein inequality.

Clearly, the relative entropy is not a metric. For example, it is not symmetric in its arguments. But, because of the above properties, it is a useful quantity for the comparison of probability distributions. Its main usefulness, however, lies in the fact that other entropic quantities can be regarded as special cases of the relative entropy.

11.2 Comparing Quantum States

In this brief section we summarize the quantum counterparts of the classical quantities that were introduced in the preceding section where we also noted that all of the quantum quantities were introduced previously, mostly in Chaps. 2 and 3, so they will not need a detailed introduction.

We start by stating that the quantum counterpart of a classical probability distribution is the state of a quantum system. We studied quantum states in Chap. 2 in detail. So, if we have two different quantum states, ρ and σ, for a quantum system living in the Hilbert space, $\mathcal{H}$, then the analog of the (classical) trace distance is the (quantum) trace distance, which was defined in Chap. 2 as

$$D(\rho, \sigma) \equiv \frac{1}{2} Tr|\rho - \sigma| \equiv \frac{1}{2}\|\rho - \sigma\|. \tag{11.5}$$

It has the usual properties: It is symmetric in its arguments and satisfies the triangle inequality.

A natural candidate for the quantum analog of the classical fidelity would be the quantity $Tr(\rho^{1/2}\sigma^{1/2})$. However, the operator $\rho^{1/2}\sigma^{1/2}$ is not a always positive, therefore we introduce the quantum fidelity between the two states, ρ and σ, as

$$F(\rho, \sigma) \equiv Tr(|\rho^{1/2}\sigma^{1/2}|). \tag{11.6}$$

A word of caution is in order here. The absolute value of an operator, A, is defined as $\sqrt{A^\dagger A}$. Applying this to the above expression, we obtain the more explicit form,

$$F(\rho, \sigma) \equiv Tr(\sqrt{\sigma^{1/2}\rho\sigma^{1/2}}), \tag{11.7}$$

and in this we recognize the quantum fidelity that was introduced in Chap. 2. Its properties are very similar to the classical fidelity: $0 \leq F(\rho, \sigma) \leq 1$ and, from the first of these two definitions, it is clear that the quantum fidelity is symmetric in its arguments. It is still not a metric since it does not satisfy a triangle inequality. However, it is easy to see that $\cos^{-1}(F) = \theta$ does, so F is very closely related to a proper distance measure. Note also that the fidelity is one if $\rho = \sigma$ and zero if the support of ρ (the subspace spanned by the eigenvectors of ρ) is orthogonal to the support of σ, in which case the states are as different as possible. In all other cases, the fidelity is between 0 and 1, so, intuitively, the quantum fidelity is a good quantifier of how the states compare.

Finally, in analogy with the classical relative entropy, the quantum relative entropy can be defined as follows. First, we recall that for a quantum state ρ, the von Neumann entropy (the quantum counterpart of the classical Shannon entropy) was defined as in Chap. 3 as

$$S(\rho) = -Tr(\rho \log \rho), \tag{11.8}$$

where the logarithm is base 2. With its help, the quantum relative entropy for two quantum states, ρ and σ, was defined in Chap. 3 as

$$S(\rho\|\sigma) = Tr[\rho(\log \rho - \log \sigma)]. \tag{11.9}$$

It can also be written as $S(\rho\|\sigma) = -S(\rho) - Tr[\rho(\log \sigma)]$, where $S(\rho)$ is once again the von Neumann entropy. This quantity can act as a kind of distance between two quantum states. Note that if the states are identical, the relative entropy is zero. We recall that the quantum relative entropy is non-negative, $S(\rho\|\sigma) \geq 0$, known as the Klein inequality.

Clearly, just like the classical relative entropy, the quantum relative entropy is also not a metric. For example, it is not symmetric in its arguments. But, because of its properties, it is still a useful quantity for the comparison of quantum states. Its main usefulness, however, lies in the fact that other quantum entropic quantities can be regarded as special cases of the quantum relative entropy.

11.3 Entropies for Two Classical Random Variables

Suppose that we have two random variables, X and Y, with a joint distribution $p_{XY}(x, y)$. It is natural to define the joint entropy, in analogy with the Shannon entropy for a single variable, as

$$H(X, Y) = -\sum_x \sum_y p_{XY}(x, y) \log p_{XY}(x, y), \tag{11.10}$$

but there are other possibilities. We can define a conditional entropy that tells us about the entropy of Y if we know X

$$H(Y|X) = \sum_x p_X(x) H(Y|X = x) = -\sum_x \sum_y p_X(x) p_{XY}(y|x) \log p_{XY}(y|x)$$

$$= -\sum_x \sum_y p_{XY}(x, y) \log p_{XY}(y|x), \tag{11.11}$$

where $p_X(x) = \sum_y p_{XY}(x, y)$ and $p_{XY}(y|x) = p_{XY}(x, y)/p_X(x)$. The joint entropy and the conditional entropy are related;

$$H(X, Y) = -\sum_x \sum_y p_{XY}(x, y) \log[p_{XY}(y|x) p_X(x)]$$

$$= H(X) + H(Y|X). \tag{11.12}$$

For the case of two random variables, we can define an important quantity, the mutual information as

$$I(X : Y) = H(p_{XY}(x, y)\|p_X(x)p_Y(y)) = \sum_{x,y} p_{XY}(x, y) \log \frac{p_{XY}(x, y)}{p_X(x)p_Y(y)}$$

$$= \sum_{x,y} p_{XY}(x, y) \log \frac{p_{XY}(x|y)}{p_Y(x)}. \tag{11.13}$$

Here $p_X(x)$ and $p_Y(y)$ are the marginal distributions for X and Y, respectively.

What we see here is that the mutual information is a special case of the relative entropy: we compare the joint distribution of two random variables to the case when the joint distribution is a product of the marginal distributions, i.e., the variables are assumed independent. Thus, the mutual information is a measure of the correlations between X and Y. If X and Y are independent, then $p_{XY}(x, y) = p_X(x)p_Y(y)$, and $I(X : Y) = 0$. The other extreme is if X and Y are perfectly correlated. Suppose they both take values in the same set, and that $p_{XY}(x|y) = \delta_{x,y}$. We then have that $p_X(x) = p_Y(x)$, and

$$I(X : Y) = \sum_{x,y} \delta_{x,y} p_Y(y)[\log \delta_{x,y} - \log p_X(x)]$$

$$= -\sum_x p_X(x) \log p_X(x) = H(X) = H(Y). \qquad (11.14)$$

In this case, the mutual information is just the information in one of the random variables, since they are perfectly correlated.

Mutual information can be related to conditional entropy. We have that

$$I(X : Y) = \sum_{x,y} p_{XY}(x, y)[\log p_{XY}(x|y) - \log p_X(x)]$$

$$= H(X) - H(X|Y). \qquad (11.15)$$

We also find that

$$I(X : Y) = H(Y) - H(Y|X)$$

$$= H(X) + H(Y) - H(X, Y)$$

$$= I(Y : X). \qquad (11.16)$$

The mutual information is an important quantity that can tell us the rate at which we can send information through a channel. A channel is represented by a sender, X, a receiver, Y, and a conditional probability $p_{XY}(y|x)$, which gives the probability that the message y is received if the message x was sent. An ideal channel would simply transmit the message unaltered, but most channels are not ideal and introduce errors. The sender sends a message x with probability $p_X(x)$, and the receiver receives the message y with probability $p_Y(y) = p_{XY}(y|x)p_X(x)$. The mutual information $I(Y : X)$ tells us about the correlation between the channel input and the channel output. It is used to define the channel capacity, C, which is given by the maximum of $I(Y : X)$ over the input probability distribution, $p_X(x)$. It can be shown, after a considerable amount of work, that C is the maximum rate (bits per channel use) at which bits can be sent through the channel with an arbitrarily small probability of error.

11.4   Quantum Properties

So far, we have only discussed the sending of classical information using classical systems. We would now like to generalize this to the quantum case. We consider a Hilbert space $\mathcal{H} = \mathcal{H}_A \otimes \mathcal{H}_B$ and a state on that space, ρ_{AB}, and we define the conditional entropy and mutual information as

$$S(A|B) = S(\rho_{AB}) - S(\rho_B)$$

$$S(A : B) = S(\rho_A) + S(\rho_B) - S(\rho_{AB}), \tag{11.17}$$

where ρ_A and ρ_B are the reduced density matrices on $\mathcal{H}_A$ and $\mathcal{H}_B$, respectively.

We would like to prove some results about the quantum mutual information, and to do so we need a result called strong subadditivity. The proof of this result is complicated, so we will just state it. Consider a state ρ_{ABC} on $\mathcal{H}_A \otimes \mathcal{H}_B \otimes \mathcal{H}_C$. It is then the case that

$$S(\rho_A) + S(\rho_B) \leq S(\rho_{AC}) + S(\rho_{BC})$$

$$S(\rho_{ABC}) + S(\rho_B) \leq S(\rho_{AB}) + S(\rho_{BC}), \tag{11.18}$$

where the subscripts on the density matrices in the above equations indicate the systems that are not traced out. Using these results, we would like to prove two things. First, that discarding a quantum system does not increase the mutual information, that is, $S(A : B) \leq S(A : B, C)$. To see this we start with the second inequality above, rewrite it as

$$S(\rho_B) - S(\rho_{AB}) \leq S(\rho_{BC}) - S(\rho_{ABC}), \tag{11.19}$$

add $S(\rho_A)$ to both sides, and finally make use of the definition of quantum mutual information. This proves the result.

The next thing we wish to prove, that quantum operations on a single party do not increase quantum mutual information, takes a bit more work. We start with a state, ρ_{AB} on $\mathcal{H}_A \otimes \mathcal{H}_B$. Now apply a trace-preserving quantum operation, $\mathcal{E}$, to system B, so that $\rho_{A'B'} = (I_A \otimes \mathcal{E})(\rho_{AB})$. What we want to show is that $S(A' : B') \leq S(A : B)$. The quantum operation, $\mathcal{E}$, can be implemented by appending an ancilla, C, in a pure state, $|\phi\rangle_C$, to A and B, so that we have $\rho_{AB} \otimes |\phi\rangle_C \langle\phi|$, acting on systems B and C with a unitary U_{BC}, and then tracing out system C. Because C is initially in a pure state, and the entropy of a pure state is zero, we have that $S(\rho_{AB}) = S(\rho_B)$ and $S(\rho_{ABC}) = S(\rho_{AB})$. This implies that $S(A : B) = S(A : B, C)$. After applying $I_A \otimes U_{BC}$ to ρ_{ABC}, the state is $\rho_{A'B'C'}$. Applying this operator does not change $S(\rho_A)$, $S(\rho_{BC})$, or $S(\rho_{ABC})$, so that $S(A : B, C) = S(A' : B', C')$. If we now discard C, we can use the previous result to obtain $S(A' : B') \leq S(A : B)$.

11.5   The Holevo Theorem

Now that we have some properties of quantum mutual information, we would like to use them to prove an important result in quantum communication theory, the Holevo theorem. This theorem places an upper bound on the mutual information between Alice, who prepares and sends quantum systems, and Bob, who receives and measures those quantum systems. Let us be more precise. Alice sends one of the quantum states, ρ_x, where $x \in \{1, 2, \ldots n\}$, with probability p_x and sends it to Bob. The random variable corresponding to Alice is X, which assumes the value x with probability p_x. Bob measures the state Alice sent with a POVM with elements $\{\Pi_0, \Pi_1, \ldots \Pi_{m-1}\}$. He obtains output $y \in \{0, 1, \ldots m - 1\}$ with probability $p_y = p(y|x)p_x = \mathrm{Tr}(\Pi_y \rho_x)p_x$. The random variable Y, which takes the value y with probability p_y, corresponds to Bob's output. The Holevo theorem states that

$$I(X : Y) \leq S(\rho) - \sum_x p_x S(\rho_x), \tag{11.20}$$

where $\rho = \sum_x p_x \rho_x$.

Our proof of this theorem is taken from the book by Nielsen and Chuang. We start by introducing two ancillas, P and M, and we will let Q be the original system. P labels the inputs, and has the orthonormal basis $\{|x\rangle_P \mid x = 1, 2, \ldots n\}$ and M labels the output, and has the orthonormal basis $\{|y\rangle_M \mid y = 0, 1, \ldots m - 1\}$. We now start with the state

$$\rho^{PQM} = \sum_x p_x |x\rangle_P \langle x| \otimes \rho_x^Q \otimes |0\rangle_M \langle 0|. \tag{11.21}$$

In order to describe Bob's measurement, we define the quantum map, $\mathcal{E}$, which acts on QM, by

$$\mathcal{E}(\sigma^Q \otimes |0\rangle_M \langle 0|) = \sum_y \sqrt{\Pi_y}\sigma^Q\sqrt{\Pi_y} \otimes |y\rangle_M \langle y|, \tag{11.22}$$

where σ^Q is an arbitrary density matrix in Q. This corresponds to measuring system Q and storing the result in M. We do need to show that this is a trace-preserving map, and, in particular, that we can find Kraus operators that realize it. Let U_y acting on M have the action $U_y|y'\rangle = |y + y'\rangle$, where the addition is modulo m. Define Kraus operators $K_y = \sqrt{\Pi_y} \otimes U_y$, which do satisfy $\sum_y K^\dagger K_y = I_{QM}$, as they should. This gives us a Kraus representation for $\mathcal{E}$, which insures that it is a valid quantum map.

Denote the system before the application of $\mathcal{E}$ as PQM and afterwards by $P'Q'M'$. We have that $S(P : Q) = S(P : Q, M)$ as M is initially uncorrelated with PQ. We then have that $S(P : Q, M) \geq S(P' : Q', M')$, as applying $\mathcal{E}$ does not increase the mutual information. Finally, $S(P' : Q', M') \geq S(P' : M')$, since discarding Q' does not increase the mutual information. Summarizing, we have that $S(P' : M') \leq S(P : Q)$.

In order to complete the proof of the theorem, we just need to calculate the quantities on both sides of this last inequality. First, note that

$$S(P : Q) = S(P) + S(Q) - S(P, Q), \qquad (11.23)$$

and

$$\rho^{PQ} = \sum_x p_x |x\rangle_P \langle x| \otimes \rho_x^Q. \qquad (11.24)$$

From this we see that

$$S(P) = H(X)$$
$$S(Q) = S(\rho)$$
$$S(P, Q) = H(X) + \sum_x p_x S(\rho_x). \qquad (11.25)$$

Putting this together,

$$S(P : Q) = S(\rho) - \sum_x p_x S(\rho_x). \qquad (11.26)$$

Next, we have

$$S(P' : M') = S(P') + S(M') - S(P', M'), \qquad (11.27)$$

and

$$\rho^{P'M'} = \sum_{x,y} p_x \operatorname{Tr}(\Pi_y \rho_x) |x\rangle_P \langle x| \otimes |y\rangle_M \langle y|. \qquad (11.28)$$

Now $\operatorname{Tr}(\Pi_y \rho_x) = p_{XY}(y|x)$ so we can express $\rho^{P'M'}$ as

$$\rho^{P'M'} = \sum_{x,y} p_{XY}(x, y) |x\rangle_P \langle x| \otimes |y\rangle_M \langle y|. \qquad (11.29)$$

Therefore, $S(P') = H(X)$, $S(M') = H(Y)$, and $S(P', M') = H(X, Y)$. Finally, we have that $S(P' : M') = I(X : Y)$, and the theorem is proved.

11.6   Example: Unambiguous Discrimination

We can use the Holevo bound to get a lower bound for the failure probability for unambiguous state discrimination. In order to do so, we make use of the communication scenario in which Alice sends one of the states $\{|\psi_x\rangle \,|\, x = 1, 2, \ldots, N\}$ to Bob, and the state $|\psi_x\rangle$ is sent with probability p_x. Bob then applies an unambiguous state discrimination measurement to see which state Alice sent. Let X be the random variable corresponding to the state Alice sent, and Y be the random variable corresponding to Bob's measurement result. X takes values in the set $\{1, 2, \ldots, N\}$ and Y takes values in the set $\{1, 2, \ldots, N, f\}$, where f corresponds to the measurement failing. The mutual information between Alice and Bob is

$$I \equiv I(X:Y) = \sum_{x,y} p(x, y) \log\left[\frac{p(y|x)}{p_Y(y)}\right], \tag{11.30}$$

where the logarithms are base 2. The conditional probabilities, $p(y|x)$, for X and Y are $p(y|x) = \delta_{x,y} p_{xx}$ for $y \neq f$, and $p(f|x) = 1 - p_{xx}$, where p_{xx} is the probability that if $|\psi_x\rangle$ is sent, then $|\psi_x\rangle$ is detected. For the joint distribution, we then have that $p(x, y) = \delta_{x,y} p_{xx} p_x$ for $y \neq f$, and $p(x, f) = (1 - p_{xx}) p_x$. The distribution for Y is given by

$$p_Y(y) = \sum_x p(x, y) = \begin{cases} p_{yy} p_y, & y \neq f \\ \sum_x (1 - p_{xx}) p_x, & y = f \end{cases}. \tag{11.31}$$

Note that $p_Y(f)$ is the total failure probability for the measurement.

We can now compute the mutual information. We begin with

$$I = \sum_{x,y \neq f} p(x, y) \log\left[\frac{p(y|x)}{p_Y(y)}\right] + \sum_x p(x, f) \log\left[\frac{p(f|x)}{p_Y(f)}\right]. \tag{11.32}$$

This gives us that

$$I = -\sum_x p_x p_{xx} \log p_x + \sum_x p_x (1 - p_{xx}) \log\left[\frac{1 - p_{xx}}{p_Y(f)}\right]. \tag{11.33}$$

Now note that

$$\log\left[\frac{1 - p_{xx}}{p_Y(f)}\right] = \log\left[\frac{p_x (1 - p_{xx})}{p_Y(f)}\right] - \log p_x, \tag{11.34}$$

and that

$$- p_x p_{xx} \log p_x - p_x (1 - p_{xx}) \log p_x = -p_x \log p_x. \tag{11.35}$$

Making use of these equations and setting $q_x = p_x(1 - p_{xx})/p_Y(f)$, we have that

$$I = H(\{p_x\}) + p_Y(f) \sum_x q_x \log q_x, \tag{11.36}$$

where $H(\{p_x\})$ is the Shannon entropy of the distribution $\{p_x\}$. Note that $\sum_x q_x = 1$, and $q_x \geq 0$, so that we can express the mutual information as

$$I = H(\{p_x\}) - p_Y(f) H(\{q_x\}). \tag{11.37}$$

We can now apply Holevo's theorem. Defining the density matrix,

$$\rho = \sum_x p_x |\psi_x\rangle \langle \psi_x|, \tag{11.38}$$

the theorem implies that $I \leq S(\rho)$, since Alice is sending pure states, or

$$H(\{q_x\}) p_Y(f) \geq H(\{p_x\}) - S(\rho), \tag{11.39}$$

and making use of the fact that $H(\{q_x\}) \leq \log N$, we finally have the bound,

$$\frac{H(\{p_x\}) - S(\rho)}{\log N} \leq p_Y(f) = P_f, \tag{11.40}$$

where, as we noted previously, we took into account that $p_Y(f)$ is just the total failure probability for the measurement, P_f, so this inequality gives us our desired lower bound. Note that in the case that all of the states are orthogonal, the left-hand side is equal to zero. This is as expected, of course, since orthogonal states are perfectly distinguishable, and the failure probability is zero.

11.7 Problems

1. We want to compare two different measurements on the trine states (see Chap. 5),

$$|\psi_0\rangle = -\frac{1}{2}(|0\rangle + \sqrt{3}|1\rangle)$$

$$|\psi_1\rangle = -\frac{1}{2}(|0\rangle - \sqrt{3}|1\rangle)$$

$$|\psi_2\rangle = |0\rangle.$$

(a) Alice sends a qubit to Bob in one of the trine states, and each of the states is equally probable. Bob measures the qubit with the POVM $\{\Pi_j = (2/3)|\psi_j\rangle\langle\psi_j| \mid j = 0, 1, 2\}$. Bob's measurement result is described by the random variable Y, and the qubit that Alice sent is described by the random variable X. We have that $p(Y = j|X = k) = \langle\psi_k|\Pi_j|\psi_k\rangle$. Find the mutual information between X and Y.

(b) Now Bob uses the POVM $\{\Pi_j = (2/3)|\psi_j^\perp\rangle\langle\psi_j^\perp| \mid j = 0, 1, 2\}$, where

$$|\psi_{0\perp}\rangle = -\frac{1}{2}(\sqrt{3}|0\rangle - |1\rangle)$$

$$|\psi_1^\perp\rangle = -\frac{1}{2}(\sqrt{3}|0\rangle + \sqrt{3}|1\rangle)$$

$$|\psi_2^\perp\rangle = |1\rangle.$$

We have that $\langle\psi_j|\psi_j^\perp\rangle = 0$. Note that in this case, if $Y = j$, Bob knows that Alice did not send $|\psi_j\rangle$, but the other two possibilities are equally likely. Find the mutual information between X and Y for this measurement.

2. Show that, in complete analogy with the classical case, the quantum mutual information can be expressed as $S(\rho_{AB}\|\rho_A \otimes \rho_B)$, where ρ_A and ρ_B are the reduced density matrices of ρ_{AB}.

3. We start with the Bell state $(1/\sqrt{2})(|00\rangle_{AB} + |11\rangle_{AB})$ and send the second qubit through a depolarizing channel (see Chap. 4). Find the quantum mutual information between the two qubits after we have done so.

4. Show that the trace distance for classical probability distributions, introduced in Eq. (11.5), satisfies the triangle inequality, $D(p_X, r_X) \leq D(p_X, q_X) + D(q_X, r_X)$.

Further Reading

1. E. Bagan, J. Bergou, M. Hillery, Phys. Rev. A **102**, 022224 (2020)
2. C.E. Shannon, W. Weaver, *The Mathematical Theory of Communication* (University of Illinois Press, Urbana, 1963)
3. T.M. Cover, J.A. Thomas, *Elements of Information Theory* (Wiley, New York, 1991)
4. G. Jaeger, *Quantum Information: An Overview* (Springer, New York, 2007)
5. S.M. Barnett, *Quantum Information* (Oxford University Press, Oxford, 2009)

Chapter 12
Implementations

In the remaining chapters of the book we will survey the leading approaches for implementing quantum computation. This is a vast, and very active field of research. The presentation will not be comprehensive, but will strive to explain the basic physical principles underlying the most promising approaches, as well as providing a snapshot of the current status. Before proceeding to descriptions of the different physical approaches we introduce some general tools and concepts including the Lindblad equation, T_1 and T_2 coherence times, Rabi oscillations, and Ramsey interference.

12.1 DiVincenzo Criteria

There are several capabilities that are essential for implementation of quantum computation. As enumerated by DiVincenzo [1] these are

- Scalability with well-characterized qubits
- Initializing qubits to a simple fiducial state
- A qubit-specific measurement capability
- Long relevant decoherence times
- A "universal" set of quantum gates
- The ability to interconvert stationary and flying qubits
- The ability to transmit flying qubits between specified locations

The first five items are necessary for quantum computation. The last two items are required for distributed quantum computation, as well as other quantum information tasks including quantum communication and quantum enhanced sensing. In the material that follows we will describe how these criteria are realized in the leading approaches. Atomic qubits including neutral atoms and trapped ions will be treated in Chap. 13, optical qubits in Chap. 14, and solid state approaches using supercon-

© Springer Nature Switzerland AG 2021

J. A. Bergou et al., *Quantum Information Processing*, Graduate Texts in Physics,
https://doi.org/10.1007/978-3-030-75436-5_12

ducting circuits and quantum dots in semiconductor heterostructures in Chap. 15. There are also other qubit platforms that have been, or are under development. Spins in solids, including nitrogen-vacancy centers in diamond are important for quantum sensing, and have been used for quantum communication tasks. Small scale quantum registers have been demonstrated with this platform [2] although it does not appear they are particularly well suited for scalable computation. Other approaches including electrons floating on liquid helium [3], carbon nanotubes, and topologically protected qubits [4] are all under development, but are not yet at a stage comparable to the platforms we treat in detail.

12.2   Quantum Dynamics

One of the central challenges inherent in realizing quantum computation is the need to isolate qubits from the environment, while also providing the capability for control and measurement of qubits by the classical systems that are needed to operate the quantum computer. Environmental isolation is never perfect which leads to a generalized, non-unitary dynamics. In Chap. 4 non-unitary dynamics were described using Kraus operators. Another, related approach is to establish a differential equation for the continuous time evolution of the reduced density operator of the quantum system. This is generally referred to as a master equation which often appears in Lindblad form [5, 6] involving quantum jump (Lindblad) operators.

Consider a quantum system S and an environment E as shown in Fig. 12.1. The density matrix describing system and environment is $\hat{\rho}_{SE}$ and the Hamiltonian is $\hat{\mathcal{H}} = \hat{\mathcal{H}}_S + \hat{\mathcal{H}}_E + \hat{\mathcal{H}}_{SE}$. Here $\hat{\mathcal{H}}_S$ is the system Hamiltonian, $\hat{\mathcal{H}}_E$ is the Hamiltonian of the environment, and $\hat{\mathcal{H}}_{SE}$ describes the coupling between the system and the environment. Let's assume that the system and the environment are initially in a separable state, $\hat{\rho}_{SE}(t_0) = \hat{\rho}_S(t_0) \otimes \hat{\rho}_E(t_0)$. At a later time t

Fig. 12.1 Quantum system and environment. The total Hamiltonian $\hat{\mathcal{H}}$ and the combined density operator $\hat{\rho}_{SE}$ consist of parts describing the system, the environment, and their interaction

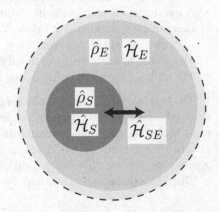

we have $\hat{\rho}_{SE}(t) = U\hat{\rho}_{SE}(t_0)U^\dagger$ where the time evolution operator U is defined by the total Hamiltonian $\hat{\mathcal{H}}$. The state of the·system alone at a later time is $\hat{\rho}_S(t) = \text{Tr}_E[\hat{\rho}_{SE}(t)] = \text{Tr}_E[U\hat{\rho}_S(t_0) \otimes \hat{\rho}_E(t_0)U^\dagger]$. As has been derived in Chap. 4 the time evolution can be cast in an operator sum representation

$$\hat{\rho}_S(t) = \sum_j \hat{A}_j\hat{\rho}_S(t_0)\hat{A}_j^\dagger$$

where $\hat{A}_j$ are non-unitary Kraus operators.

We can now use the operator sum representation to find a general form for the time evolution of the system density operator when coupled to an environment. Consider an infinitesimal time evolution

$$\hat{\rho}(t + \delta t) = \hat{\rho}(t) + \delta\hat{\rho} = \sum_j \hat{A}_j\hat{\rho}(t)\hat{A}_j^\dagger \tag{12.1}$$

with $\delta\hat{\rho} \sim \delta t$ and we have dropped the subscript S for brevity. To proceed assume one of the Kraus operators $\hat{A}_0$ is dominant and is given by $\hat{A}_0 = \hat{I} + \hat{L}_0\delta t$ and the other operators are $\hat{A}_j = \hat{L}_j\sqrt{\delta t}, \quad j \neq 0$. Then $\hat{A}_0\hat{\rho}\hat{A}_0^\dagger = \hat{\rho} + (\hat{L}_0\hat{\rho} + \hat{\rho}\hat{L}_0^\dagger)\delta t + \mathcal{O}(\delta t^2)$ and $\hat{A}_j\hat{\rho}\hat{A}_j^\dagger = \hat{L}_j\hat{\rho}\hat{L}_j^\dagger\delta t$. To order δt (12.1) becomes

$$\hat{\rho} + \delta\hat{\rho} = \hat{\rho} + \left(\hat{L}_0\hat{\rho} + \hat{\rho}\hat{L}_0^\dagger + \sum_{j\neq 0}\hat{L}_j\hat{\rho}\hat{L}_j^\dagger\right)\delta t.$$

This implies the differential equation

$$\frac{d\hat{\rho}}{dt} = \hat{L}_0\hat{\rho} + \hat{\rho}\hat{L}_0^\dagger + \sum_{j\neq 0}\hat{L}_j\hat{\rho}\hat{L}_j^\dagger.$$

Unitary evolution of the density operator is described by $\frac{d\hat{\rho}}{dt} = \frac{i}{\hbar}[\hat{\rho}, \hat{\mathcal{H}}]$, so to recover the Hamiltonian part of the dynamics we put $\hat{L}_0 = -i\hat{\mathcal{H}}/\hbar + L_0'$ giving

$$\frac{d\hat{\rho}}{dt} = \frac{i}{\hbar}[\hat{\rho}, \hat{\mathcal{H}}] + \hat{L}_0'\hat{\rho} + \hat{\rho}\hat{L}_0'^\dagger + \sum_{j\neq 0}\hat{L}_j\hat{\rho}\hat{L}_j^\dagger.$$

Since $\text{Tr}[\hat{\rho}] = 1$ at all times we require $\text{Tr}[d\hat{\rho}/dt] = 0$ which results in the condition $\hat{L}_0' = -\frac{1}{2}\sum_j \hat{L}_j^\dagger\hat{L}_j$. Plugging in we get

$$\frac{d\hat{\rho}}{dt} = \frac{i}{\hbar}[\hat{\rho}, \hat{\mathcal{H}}] + \sum_j \hat{L}_j\hat{\rho}\hat{L}_j^\dagger - \frac{1}{2}\hat{L}_j^\dagger\hat{L}_j\hat{\rho} - \frac{1}{2}\hat{\rho}\hat{L}_j^\dagger\hat{L}_j.$$

This is known as the Lindblad equation [5] and is widely used to study open quantum system dynamics. The specific form of the Lindblad jump operators $\hat{L}_j$ depends on the problem being treated.

12.3   Bloch Equations and Qubit Coherence

In almost all cases the quantum state of a qubit is controlled by coupling it to an electromagnetic field.[1] An important example of the Lindblad formalism is therefore that of a two-level quantum system (a qubit) coupled to an external field, and with the upper level having a finite lifetime due to spontaneous relaxation to the lower level. The physical mechanism causing relaxation does not need to be specified to write down the general form of the equations, but could be due, for example, to spontaneous emission or interactions between several two-level systems.

Denote the levels by $|g\rangle, |e\rangle$ and assume the upper level decays exponentially with rate γ so $\rho_{ee}(t) = \rho_{ee}(0)e^{-\gamma t}$. At short times we have

$$\rho_{ee}(\delta t) = \rho_{ee}(0) - \rho_{ee}(0)\gamma \delta t + \mathcal{O}(\delta t^2).$$

In a closed system any population leaving the excited state must enter the ground state to preserve the trace of ρ. Therefore we can write

$$\rho_{gg}(\delta t) = \rho_{gg}(0) + \rho_{ee}(0)\gamma \delta t + \mathcal{O}(\delta t^2).$$

For a pure state $|\psi\rangle = c_g|g\rangle + c_e|e\rangle$ the populations are $\rho_{gg} = |c_g|^2, \rho_{ee} = |c_e|^2$. Relaxation of the excited state corresponds to the transition operator $|g\rangle\langle e|$ and since the change in population at time δt is proportional to $\gamma \delta t$ which is proportional to $\hat{L}_j \rho \hat{L}_j^\dagger \delta t$ the jump operator must be $\hat{L} = \sqrt{\gamma}|g\rangle\langle e|$. The master equation is then

$$\frac{d\hat{\rho}}{dt} = \frac{i}{\hbar}[\hat{\rho}, \hat{\mathcal{H}}] + \hat{L}\hat{\rho}\hat{L}^\dagger - \frac{1}{2}\hat{L}^\dagger\hat{L}\hat{\rho} - \frac{1}{2}\hat{\rho}\hat{L}^\dagger\hat{L}$$

$$= \frac{i}{\hbar}[\hat{\rho}, \hat{\mathcal{H}}] + \gamma \begin{pmatrix} \rho_{ee} & -\rho_{ge}/2 \\ -\rho_{ge}^*/2 & -\rho_{ee} \end{pmatrix}. \tag{12.2}$$

In the last line we have introduced $\hat{\rho} = \begin{pmatrix} \rho_{gg} & \rho_{ge} \\ \rho_{ge}^* & \rho_{ee} \end{pmatrix}$ and used $\rho_{gg} + \rho_{ee} = 1$.

We see that the coherence decays at half the rate of the population, or $\gamma/2$. This is expected since $\rho_{ge} = c_g c_e^*$ and $c_e(t) \sim c_e(0)e^{-\gamma t/2}$. The $1/e$ lifetime of the

---

[1]The most important counterexample is that of qubits encoded in single photons to be discussed in Chap. 14.

population difference between ground and excited states is referred to as the T_1 or longitudinal relaxation time. The $1/e$ lifetime of the ground and excited state coherence is referred to as the T_2 or transverse relaxation time. In this example with exponential population decay rate γ we find $T_1 = \frac{1}{\gamma}$, $T_2 = \frac{2}{\gamma} = 2T_1$. In a real situation there may be additional dephasing mechanisms that cause the coherence to decay faster without affecting the populations so that, more generally, $T_2 \leq 2T_1$.

Let us now derive the equations of motion for the density matrix elements including coherent coupling to a driving field. The results are referred to as Bloch equations since they were first introduced by F. Bloch in his analysis of nuclear magnetic resonance [7]. The Hamiltonian for the interaction of a two-level quantum system with a radiation field can be written as $\hat{\mathcal{H}} = \hat{\mathcal{H}}_q + \hat{\mathcal{H}}_1$. Here $(\hat{\mathcal{H}}_q)_{ij} = \delta_{ij} U_i$ with U_i the unperturbed energy of level $|i\rangle$ ($i = e, g$) and $\hat{\mathcal{H}}_1$ is the field-qubit interaction.

Specializing to the case of electric dipole coupling between the field and qubit $\hat{\mathcal{H}}_1 = -\hat{\mathbf{d}} \cdot \mathbf{E}$, with $\hat{\mathbf{d}}$ the dipole moment operator and $\mathbf{E}$ the classical electric field. In a real physical system the interaction may additionally cause shifts of the unperturbed levels due to the ac Stark effect. When the energy eigenstates have definite parity the diagonal elements of $\hat{\mathbf{d}}$ vanish at first order in the field. Neglecting second-order AC Stark shifts and vectorial aspects of the interaction we find $\hat{\mathcal{H}}_1 = -E \begin{bmatrix} 0 & d^* \\ d & 0 \end{bmatrix}$ where $d = |\langle e|\hat{\mathbf{d}}|g\rangle|$ and $E = |\mathbf{E}|$. Inserting into Eq. (12.2) we arrive at the Bloch equations

$$\frac{d\rho_{gg}}{dt} = \frac{\rho_{ee}}{T_1} - \frac{i}{\hbar} E \left(d\rho_{eg}^* - d^* \rho_{eg} \right) \tag{12.3a}$$

$$\frac{d\rho_{ee}}{dt} = -\frac{\rho_{ee}}{T_1} - \frac{i}{\hbar} E \left(d^* \rho_{eg} - d\rho_{eg}^* \right) \tag{12.3b}$$

$$\frac{d\rho_{eg}}{dt} = -\left(\frac{1}{T_2} + i\omega_q \right) \rho_{eg} - \frac{i}{\hbar} Ed \left(\rho_{ee} - \rho_{gg} \right). \tag{12.3c}$$

where $\hbar\omega_q = U_e - U_g$ is the separation of the qubit levels and $\rho_{eg} = \rho_{ge}^*$. We see that $\frac{d\rho_{gg}}{dt} + \frac{d\rho_{ee}}{dt} = 0$ which follows from $\rho_{gg} + \rho_{ee} = 1$.

We wish to solve for the qubit dynamics in the presence of a monochromatic driving field $E = |\mathcal{E}| \cos(\omega t + \theta) = \frac{\mathcal{E}}{2} e^{-i\omega t} + c.c.$, where $\mathcal{E} = |\mathcal{E}| e^{-i\theta}$. Exact solutions of the resulting equations are difficult to come by. When the driving field is near resonant with the qubit transition frequency the detuning $\Delta = \omega - \omega_q$ satisfies $|\Delta| \ll \omega, \omega_q$, and it is a good approximation to neglect rapidly oscillating terms in the Bloch equations. This is known as the rotating wave approximation (RWA) which leads to the set

$$\frac{dw}{dt} = -\frac{1+w}{T_1} + i \left(\Omega \tilde{\rho}_{eg}^* - \Omega^* \tilde{\rho}_{eg} \right) \tag{12.4a}$$

$$\frac{d\tilde{\rho}_{eg}}{dt} = \left(i\Delta - \frac{1}{T_2}\right)\tilde{\rho}_{eg} - i\frac{\Omega}{2}w. \tag{12.4b}$$

Here we have introduced the complex Rabi frequency $\Omega = d\mathcal{E}/\hbar$, the inversion $w = \rho_{ee} - \rho_{gg}$, and $\tilde{\rho}_{eg} = \rho_{eg}e^{i\omega t}$. The population inversion w acts as a source term for the coherence ρ_{eg} and the coherence acts as a source term for changes in the population inversion. The coherent dynamics evolves on the frequency scale set by the Rabi frequency Ω. Steady state solutions of the Bloch equations are a balance between coherent driving and decay processes governed by the time constants T_1, T_2. Setting the time derivatives to zero we find the stationary solutions

$$w = -\frac{1 + \Delta^2 T_2^2}{1 + \Delta^2 T_2^2 + T_1 T_2 |\Omega|^2} \tag{12.5a}$$

$$\tilde{\rho}_{eg} = -iw\frac{\Omega T_2}{2}\frac{1 - i\Delta T_2}{1 + \Delta^2 T_2^2}. \tag{12.5b}$$

12.4   Rabi Oscillations

When the qubit is well isolated from the environment T_1, T_2 tend to infinity. The dynamics are then completely coherent and unitary. In this limit the density matrix formalism is not needed and we can use the simpler Schrödinger equation approach.

Writing $|\psi\rangle = c_g(t)e^{-i\omega_g t}|g\rangle + c_e(t)e^{-i\omega_e t}|e\rangle$ and making the rotating wave approximation, the Schrödinger equation can be expressed as a pair of ordinary differential equations for the qubit amplitudes in the form

$$\frac{dc_g}{dt} = i\frac{\Omega^*}{2}c_e e^{i\Delta t}, \tag{12.6a}$$

$$\frac{dc_e}{dt} = i\frac{\Omega}{2}c_g e^{-i\Delta t}. \tag{12.6b}$$

These are readily solved for constant amplitude Rabi frequency as $\mathbf{c}(t) = \mathbf{U}(t)\mathbf{c}(0)$ with $\mathbf{c}(t) = \begin{pmatrix} c_g(t) \\ c_e(t) \end{pmatrix}$ and

$$\mathbf{U}(t) = \begin{pmatrix} e^{i\frac{\Delta t}{2}}\left[\cos\left(\frac{\Omega' t}{2}\right) - i\frac{\Delta}{\Omega'}\sin\left(\frac{\Omega' t}{2}\right)\right] & ie^{i\frac{\Delta t}{2}}\frac{\Omega^*}{\Omega'}\sin\left(\frac{\Omega' t}{2}\right) \\ ie^{-i\frac{\Delta t}{2}}\frac{\Omega}{\Omega'}\sin\left(\frac{\Omega' t}{2}\right) & e^{-i\frac{\Delta t}{2}}\left[\cos\left(\frac{\Omega' t}{2}\right) + i\frac{\Delta}{\Omega'}\sin\left(\frac{\Omega' t}{2}\right)\right] \end{pmatrix}. \tag{12.7}$$

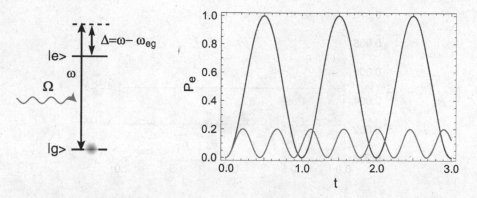

Fig. 12.2 Rabi oscillations of a two-level system with $\Omega = 2\pi$, $\Delta = 0$ (blue curve with full amplitude) and $\Omega = 2\pi$, $\Delta = 4\pi$ (red curve with small amplitude)

with $\Omega' = \sqrt{|\Omega|^2 + \Delta^2}$ the generalized Rabi frequency. If the qubit is initially in the ground state the time dependent probabilities to be in the ground and excited states are

$$P_{\mathrm{g}} = |c_{\mathrm{g}}(t)|^2 = \cos^2\left(\frac{\Omega' t}{2}\right) + \frac{\Delta^2}{|\Omega|^2 + \Delta^2} \sin^2\left(\frac{\Omega' t}{2}\right),$$

$$P_{\mathrm{e}} = |c_{\mathrm{e}}(t)|^2 = \frac{|\Omega|^2}{|\Omega|^2 + \Delta^2} \sin^2\left(\frac{\Omega' t}{2}\right). \tag{12.8}$$

The population oscillations are shown in Fig. 12.2. If the driving field is resonant, $\Delta = 0$, $\Omega' = \Omega$ and the amplitude to be in the excited state will reach unity at time $\Omega t/2 = \pi/2$ or $t = \pi/\Omega$. This defines a so-called π pulse of duration $t_\pi = \pi/\Omega$ which inverts the population between the two states.

It is also interesting to examine the phase evolution of a quantum state under Rabi oscillations. A resonant 2π pulse, $\Omega t = 2\pi$, has the evolution matrix $\mathbf{U} = \begin{pmatrix} -1 & 0 \\ 0 & -1 \end{pmatrix}$ which results in the transformation

$$|\psi\rangle \rightarrow -|\psi\rangle = e^{\imath\pi}|\psi\rangle.$$

Recall that rotation of a spin $1/2$ object about the z axis by 2π results in the same transformation

$$|\psi\rangle \rightarrow e^{-\imath \hat{J}_z \theta/\hbar}|\psi\rangle = e^{-\imath(\pm 1/2)(2\pi)}|\psi\rangle = e^{\imath\pm\pi}|\psi\rangle = e^{\imath\pi}|\psi\rangle.$$

A 2π pulse applied to a two-level system is analogous to rotation of a spin $1/2$ through an angle of 2π. In both cases we are rotating between the two quantum states of the object. Classically a 2π rotation has no observable effect. In quantum

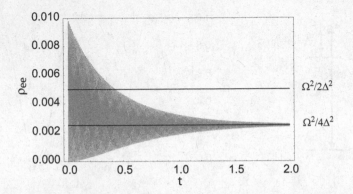

Fig. 12.3 Excited state population from solution of Bloch equations for a two-level system with $\Omega = 2\pi$, $\Delta = 2\pi \times 10$, $T_1 = 2.5$, $T_2 = 5$ so $\gamma = 2/T_2 = 0.4$

mechanics the phase of the wave function changes, an effect which is widely used as an essential ingredient in protocols for preparing entangled states.

To conclude this section let's compare the dynamics found from the Rabi oscillation solutions of the Schrödinger equation with solutions of the Bloch equations that account for decoherence. When $T_1, T_2 \rightarrow \infty$ the two descriptions must give the same results. With finite decay times significant differences are seen as shown in Fig. 12.3. The Schrödinger equation solution for the time averaged excited state population at large detuning is

$$\langle P_e \rangle = \frac{|\Omega|^2}{|\Omega|^2 + \Delta^2}\frac{1}{2} \simeq \frac{|\Omega|^2}{2\Delta^2}.$$

The factor of $1/2$ comes from taking the time average of $\sin^2(\Omega' t/2)$. The stationary solution of the density matrix equations at large detuning is

$$\rho_{ee} \simeq \frac{|\Omega|^2}{4\Delta^2},$$

which is twice smaller. We see that at short times the average of ρ_{ee} is indeed close to $\langle P_e \rangle$. However, at long times, there is a loss of coherence, and the population oscillations decay towards the stationary solution $\rho_{ee} = \langle P_e \rangle/2$. Accurate prediction of the excited state population requires solving the density matrix equations, even at time scales several times shorter than T_1.

12.4.1 One-Qubit Gates

The importance of Rabi oscillations for quantum computing is that they can be used to implement gates. With judicious choice of parameters we can use the solution (12.7) to implement $X = \begin{pmatrix} 0 & 1 \\ 1 & 0 \end{pmatrix}$, $Y = \begin{pmatrix} 0 & -i \\ i & 0 \end{pmatrix}$ gates and other rotations about axes in the equatorial plane of the Bloch sphere.

To see this recall that rotations about the x and y axes through an angle θ are described by the operators

$$R_x(\theta) = \begin{pmatrix} \cos(\theta/2) & -i\sin(\theta/2) \\ -i\sin(\theta/2) & \cos(\theta/2) \end{pmatrix}$$

$$R_y(\theta) = \begin{pmatrix} \cos(\theta/2) & -\sin(\theta/2) \\ \sin(\theta/2) & \cos(\theta/2) \end{pmatrix}.$$

If we choose $\Delta = 0$ we get

$$\mathbf{U}(t) = \begin{pmatrix} \cos\left(\frac{|\Omega|t}{2}\right) & i\frac{\Omega^*}{|\Omega|}\sin\left(\frac{|\Omega|t}{2}\right) \\ i\frac{\Omega}{|\Omega|}\sin\left(\frac{|\Omega|t}{2}\right) & \cos\left(\frac{|\Omega|t}{2}\right) \end{pmatrix}.$$

Setting Ω to be real and positive and $t = \theta/|\Omega|$ we get

$$\mathbf{U}(t) = \begin{pmatrix} \cos\left(\frac{\theta}{2}\right) & -i\sin\left(\frac{\theta}{2}\right) \\ -i\sin\left(\frac{\theta}{2}\right) & \cos\left(\frac{\theta}{2}\right) \end{pmatrix} = R_x(\theta).$$

We then use $R_x(\pi) = -iX$ to generate the X gate up to a global phase. Setting Ω to be imaginary with $i\Omega > 0$ and $t = \theta/|\Omega|$ we get

$$\mathbf{U}(t) = \begin{pmatrix} \cos\left(\frac{\theta}{2}\right) & -\sin\left(\frac{\theta}{2}\right) \\ \sin\left(\frac{\theta}{2}\right) & \cos\left(\frac{\theta}{2}\right) \end{pmatrix} = R_y(\theta).$$

We then use $R_y(\pi) = -iY$ to generate the Y gate up to a global phase. When Ω is complex we can implement rotations about an arbitrary equatorial axis in an analogous fashion.

Z gates cannot be directly implemented with Rabi rotations. One possibility is to rely on a physical mechanism that imparts a differential phase to the qubit states. This can be achieved using strongly off-resonant interactions that impart a differential energy shift to the qubit levels, but do not transfer populations. Alternatively we may combine X and Y gates to get $Z = iYX$.

12.4.2   Composite Pulses

The fidelity of a gate operation depends on the correct application of a desired Rabi frequency amplitude and phase, as well as pulse time and detuning. Unavoidable variations in all of these parameters lead to gate errors. A widely used approach for reducing gate errors is to divide the pulses into several pulses that combine to reduce the sensitivity to imperfections. This is known as a composite pulse.

As a simple example suppose we wish to perform a single qubit π rotation about the x axis using $U = R_x(\pi)$. This can be implemented with a Rabi pulse of amplitude Ω and time t such that $\theta = \Omega t = \pi$. If the control system is noisy the Rabi pulses will have fluctuating areas $\theta' \neq \pi$ resulting in gate errors. The sensitivity to pulse area errors can be reduced using a composite sequence $U' = R_z(-\pi)R_x(\pi/2)R_y(\pi)R_x(\pi/2)$. It can be shown that if the pulse error is $\epsilon = \theta' - \theta$ the error in $\mathbf{U}$ is $\mathcal{O}(\epsilon)$ whereas the error in $\mathbf{U}'$ is $\mathcal{O}(\epsilon^2)$, provided ϵ is the same for each operation in the composite sequence.

Using a composite pulse we reduce the sensitivity to errors in the pulse area. Many different composite pulses have been designed. One of the most famous is the so-called BB1 for Broadband 1 [8]. The theory of composite pulses shows that this idea can be extended to reducing the sensitivity to arbitrary powers of ϵ at the cost of a longer pulse sequence [9]. In practice the advantage of the reduced error sensitivity has to be weighed against the longer time it takes to implement the gate which will expose the qubit to more decoherence.

12.5   Ramsey Interferometry and a Qubit in a Noisy Environment

A qubit can be represented as a point on the surface of the Bloch sphere parameterized by two angles as

$$|\psi\rangle = \cos(\theta/2)|0\rangle + \sin(\theta/2)e^{i\phi}|1\rangle.$$

Noise acting on the qubit will change both θ and ϕ. If we prepare the qubit in state $|0\rangle$ and wait a time t the probability to be in the initial state will decay. The $1/e$ time constant for the population decay is called the T_1 time, or longitudinal decay time. This can be determined by measuring the population at different delay times after the qubit is prepared.

Similarly if we prepare the qubit with a particular value of ϕ and wait a time t the state will change due to the presence of noise. The value of ϕ cannot be determined directly from a measurement in the computational or Z basis which is only sensitive to θ. To determine ϕ we need to either measure in the X or Y basis or use interference to make a Z basis measurement sensitive to ϕ. The latter approach was introduced by Ramsey [10, 11] and is commonly referred to

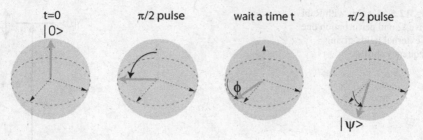

Fig. 12.4 Pulse sequence for Ramsey interferometry

as Ramsey interferometry based on the pulse sequence shown in Fig. 12.4. The characteristic time for the decay of the phase is called the T_2 time or transverse decay time. The notation of T_1 and T_2 times stems from the field of nuclear magnetic resonance. When talking about phase decay there is an additional subtlety in that two different times T_2 and T_2^* are often reported. The inhomogeneous coherence time T_2^* is dependent on the noise characteristics of the perturbing fields and can often be lengthened using pulse sequences that average out the effects of noise. Doing so reveals a longer homogeneous decay time T_2 that is intrinsic to the qubit itself. In addition to its' importance for qubit characterization Ramsey interferometry forms the basis for operation of atomic clocks [12].

Let's calculate the output state after the sequence of Fig. 12.4. We start with the qubit in state $|\psi\rangle = |0\rangle$. We then perform a rotation $R_y(\pi/2)$ giving $|\psi\rangle = \frac{|0\rangle+|1\rangle}{\sqrt{2}}$. After waiting a time t the state becomes $|\psi\rangle = \frac{|0\rangle+e^{i\phi}|1\rangle}{\sqrt{2}}$. The final state after the second $\pi/2$ pulse is

$$|\psi\rangle = R_y(\pi/2)\frac{1}{\sqrt{2}}\begin{pmatrix} 1 \\ e^{i\phi} \end{pmatrix} = e^{i\phi/2}\begin{pmatrix} -i\sin(\phi/2) \\ \cos(\phi/2) \end{pmatrix}.$$

The probability of observing $|1\rangle$ as the final state is $\cos^2(\phi/2)$ which allows the phase ϕ to be determined. If $\phi(t)$ is a linear function of t we will observe sinusoidal oscillation of the output probability as a function of the delay time t. If ϕ has a stochastic noise component the population oscillations will decay. The time after which the amplitude of the oscillations has decayed to $1/e$ of the initial amplitude defines T_2^*.

To make these notions more concrete let's calculate qubit decoherence in the presence of noise. Consider a qubit with levels $|0\rangle$, $|1\rangle$ subject to a "Rabi" field that couples the levels with frequency Ω as shown in Fig. 12.5. The Rabi field is turned on a for a time t which results in a $R_x(\theta)$ rotation of the qubit with $\theta = \Omega t$. We have the equality $R_x(\theta) = e^{-i\theta\sigma_x/2} = e^{-i\hat{\mathcal{H}}_x t/\hbar}$ where $\hat{\mathcal{H}}_x = \hbar\Omega\sigma_x/2$ is the effective Hamiltonian due to the Rabi drive. The actual value of Ω depends on the strength of the driving field and the matrix element connecting the qubit levels.

Fig. 12.5 A qubit with Rabi
drive Ω and perturbation due
to external fields giving
frequency shift Δ

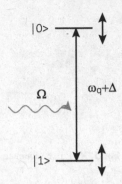

In addition we will assume there are some background noise fields that change
the energy separation of the qubit levels. Let's define the zero of energy as midway
between the levels so

$$\hat{\mathcal{H}}_z |0\rangle = \left(U_{\text{zero}} + \frac{\hbar\omega_q}{2} \right) |0\rangle,$$

$$\hat{\mathcal{H}}_z |1\rangle = \left(U_{\text{zero}} - \frac{\hbar\omega_q}{2} \right) |1\rangle,$$

with ω_q the unperturbed qubit frequency. We can write $\hat{\mathcal{H}}_z$ as $\hat{\mathcal{H}}_z = U_{\text{zero}}\hat{I} + \frac{\hbar\omega_q}{2}\sigma_z$.
Due to additional electric or magnetic background fields $\omega_q \rightarrow \omega_q + \Delta(E, B)$.
Constant terms in the Hamiltonian can be ignored and working in a rotating frame
at frequency ω_q we can write $\hat{\mathcal{H}}_z = \frac{\hbar\Delta}{2}\sigma_z$.

Combining the Rabi drive and the background fields we get the total Hamiltonian
acting on the qubit

$$\hat{\mathcal{H}} = \frac{\hbar}{2} \left(\Omega\sigma_x + \Delta\sigma_z \right) = \frac{\hbar}{2} \begin{pmatrix} \Delta & \Omega \\ \Omega & -\Delta \end{pmatrix}. \tag{12.9}$$

Equation (12.9) has eigenvalues and eigenvectors

$$\lambda_\pm = \pm\frac{\hbar\Omega'}{2}$$

$$v_+ = \frac{1}{\left[\Omega^2 + (\Omega' + \Delta)^2 \right]^{1/2}} \begin{pmatrix} \Omega' + \Delta \\ \Omega \end{pmatrix}$$

$$v_- = \frac{1}{\left[\Omega^2 + (\Omega' - \Delta)^2 \right]^{1/2}} \begin{pmatrix} \Omega \\ -\Omega' + \Delta \end{pmatrix}.$$

One way to solve for the time evolution is to express the initial state in terms of the
eigenstates and then evolve in time with eigenvalue dependent phase factors.

A more convenient approach is based on returning to the Hamiltonian and developing a geometrical picture of the evolution. We can write

$$\hat{\mathcal{H}} = \frac{\hbar\Omega'}{2}\left[\cos(\xi)\sigma_x + \sin(\xi)\sigma_z\right]$$

with $\xi = \tan^{-1}(\Delta/\Omega)$, $\cos(\xi) = \Omega/\Omega'$, $\sin(\xi) = \Delta/\Omega'$. The Hamiltonian is now in the same form as that describing a spin $1/2$ particle in an effective magnetic field $\mathbf{B}$

$$\hat{\mathcal{H}} = \frac{\hbar}{2}\mathbf{B} \cdot \boldsymbol{\sigma} = \frac{\hbar}{2}\left(B_x\sigma_x + B_z\sigma_z\right)$$

with $B_x = \Omega'\cos(\xi) = \Omega$, $B_z = \Omega'\sin(\xi) = \Delta$. The effective driving field is in the $x - z$ plane and makes an angle ξ with respect to the x axis.

We are now ready to calculate the qubit dynamics due to fields $\Omega(t)$ and $\Delta(t)$. Let the qubit be in an initial state $|\psi\rangle = a_0|0\rangle + a_1|1\rangle$. The density operator can be written as

$$\rho = \frac{I + \mathbf{M} \cdot \boldsymbol{\sigma}}{2} = \frac{I + M_x\sigma_x + M_y\sigma_y + M_z\sigma_z}{2}$$

with

$$M_x = 2\text{Re}(a_0 a_1^*), \quad M_y = -2\text{Im}(a_0 a_1^*), \quad M_z = |a_0|^2 - |a_1|^2.$$

The equation of motion $\frac{d\rho}{dt} = \frac{i}{\hbar}[\rho, \hat{\mathcal{H}}]$ then takes the form

$$\frac{d\mathbf{M}}{dt} = \mathbf{B} \times \mathbf{M}. \tag{12.10}$$

This is just the equation for the precession of a magnetic moment $\mathbf{M}$ about a magnetic field $\mathbf{B}$. We see that the quantum dynamics of a qubit (any two-level system) can be recast in the form of an equation for the classical motion of a magnet subjected to a torquing field. This geometrical correspondence is useful for solving problems in magnetism and in laser dynamics [13].

Let's use Eq. (12.10) to study qubit decoherence. Assume the effective field $\mathbf{B}$ is composed of a control field $\mathbf{B}_0(t)$ that may be time dependent, plus a noise field $\mathbf{b}(t)$. We will further assume that $\mathbf{b}$ is parallel to $\mathbf{B}_0$. This "longitudinal" noise causes dephasing of the qubit as we proceed to show. With these definitions $\mathbf{B} = \mathbf{B}_0 + \mathbf{b}$ has a constant direction in space but changes in magnitude. The magnetization vector $\mathbf{M}$ precesses about $\mathbf{B}$ at a rate that is proportional to $|\mathbf{B}|$. In addition the magnitude of the magnetization, averaged over $\mathbf{b}(t)$ will decay. To see this put $\mathbf{M} = \mathbf{M}_\parallel + \mathbf{M}_\perp$

where $\mathbf{M}_\parallel, \mathbf{M}_\perp$ are the components parallel and perpendicular to $\mathbf{B}$ respectively. The equation of motion (12.10) then takes the form

$$\frac{d\mathbf{M}}{dt} = \frac{d\mathbf{M}_\parallel}{dt} + \frac{d\mathbf{M}_\perp}{dt}$$

$$= \mathbf{B} \times (\mathbf{M}_\parallel + \mathbf{M}_\perp)$$

$$= \mathbf{B} \times \mathbf{M}_\perp. \tag{12.11}$$

An immediate consequence is $\frac{d\mathbf{M}_\parallel}{dt} = 0$ so $\mathbf{M}_\parallel$ is a constant and $\frac{d\mathbf{M}_\perp}{dt} = \mathbf{B} \times \mathbf{M}_\perp$. To proceed we define a coordinate system with $\hat{z}$ along $\mathbf{B}$, transverse coordinates x, y and complex magnetization amplitude $M_+ = M_x + iM_y$. The equations of motion for the components are

$$\frac{dM_x}{dt} = (\mathbf{B} \times \mathbf{M})_x = -BM_y$$

$$\frac{dM_y}{dt} = (\mathbf{B} \times \mathbf{M})_y = BM_x$$

so $\frac{dM_+}{dt} = iBM_+$ which is solved by $M_+(t) = e^{i\phi(t)}M_+(0)$ with $\phi(t) = \int_0^t dt' B(t')$. We then divide ϕ into a deterministic and random part as $\phi = B_0 t + \chi(t)$ with $\chi(t) = \int_0^t dt' b(t')$. The time averaged value of M_+ is then

$$\langle M_+(t) \rangle = \left\langle e^{i\phi(t)} M_+(0) \right\rangle$$

$$= \left\langle e^{iB_0 t} e^{i\chi(t)} M_+(0) \right\rangle$$

$$= e^{iB_0 t} M_+(0) \left\langle e^{i\chi(t)} \right\rangle.$$

The decay of the magnetization vector can be found from the average value $\left\langle e^{i\chi(t)} \right\rangle$. One way to proceed is to assume a probability distribution $P(\chi)$ that is Gaussian, $P(\chi) = \frac{1}{\sqrt{2\pi}\sigma} e^{-\chi^2/2\sigma^2}$ with $\sigma = \sqrt{\langle \chi^2 \rangle}$ the RMS fluctuation level. Using $P(\chi)$ we find $\left\langle e^{i\chi(t)} \right\rangle = \int_{-\infty}^{\infty} d\chi e^{i\chi} P(\chi) = e^{-\sigma^2/2}$ and

$$\langle M_+(t) \rangle = e^{iB_0 t} M_+(0) e^{-\sigma^2/2}.$$

The justification for the assumption of Gaussian statistics is that if the time t at which we are interested in knowing $\langle M_+(t) \rangle$ is long compared to the noise correlation time then $\chi(t)$ will be the sum of many random parts and the central limit theorem says $\chi(t)$ has Gaussian statistics

To finish the calculation we need to relate σ to the noise $b(t)$. We have $\sigma^2(t) = \langle \chi^2 \rangle = \int_0^t dt_1 \int_0^t dt_2 \, \langle b(t_1)b(t_2) \rangle$. Using the Wiener–Khinchin theorem it can be shown that

$$\sigma^2(t) \approx 2\pi t |\tilde{b}(0)|^2.$$

The magnetization therefore decays as $|\langle M_+(t) \rangle| = M_+(0) e^{-\pi t |\tilde{b}(0)|^2}$ and defining T_2^* by $|M_+(t)/M_+(0)| = e^{-t/T_2^*}$ we find

$$T_2^* = \frac{1}{\pi |\tilde{b}(0)|^2}. \tag{12.12}$$

The stronger the noise is, the larger $|b(0)|^2$, and the shorter the coherence time. Note that for some types of noise the result (12.12) is ambiguous. For example for $1/f$ noise $|\tilde{b}(0)|^2$ diverges and $T_2^* \to 0$. For a careful treatment of this case see [14]. For the case considered we found a finite T_2^* but no longitudinal decay. If we had made other assumptions about the direction of $\mathbf{b}(t)$ relative to $\mathbf{B}_0$ finite T_1 and T_2 would have resulted.

12.5.1   Dynamical Decoupling

As we have seen environmental influences such as electromagnetic fields cause the qubit frequency to be time dependent. Generically the qubit frequency can be written as $\omega_q = \omega_q(E, B)$ with E, B electric and magnetic field amplitudes. If $E = E(t)$ and $B = B(t)$ then effectively $\omega_q = \omega_q(t)$. In such a situation a qubit prepared in the state

$$|\psi(t = 0)\rangle = c_0|0\rangle + c_1|1\rangle$$

evolves (suppressing a global phase) to

$$|\psi(t)\rangle = c_0|0\rangle + c_1 e^{-i \int_0^t dt' \, \Delta(t')}|1\rangle$$

where

$$\Delta(t) = \omega_q(t) - \omega_{q0}$$

with ω_{q0} the unperturbed qubit frequency. As we showed in the previous section the observed coherence time scales as $T_2^* \sim 1/|\tilde{\Delta}(\omega = 0)|^2$ with $|\tilde{\Delta}(\omega)|^2$ the power spectrum of the fluctuations of the qubit frequency. The T_2^* time is referred to as the inhomogeneous coherence time due to the time dependent, or inhomogeneous, noise.

We can mitigate the effects of noise using pulse sequences that serve to average out the field induced energy shifts. This approach goes under the general name of dynamical decoupling [15]. With decoupling the qubit acquires a longer coherence time which is the T_2 or homogeneous coherence which is limited by intrinsic properties of the qubit, not perturbations from the environment. Dynamical decoupling seeks to reduce decoherence due to temporal variation of the environment by modulation in time. A related physical approach to mitigation of decoherence based on decoherence free subspace (DFS) encoding in multiple physical qubits was introduced in Sect. 9.3. Dynamical decoupling can be interpreted as a form of temporal DFS encoding.

The simplest possible approach is based on applying a sequence of π pulses about the x axis at time intervals

$$\tau \cdot X \cdot \tau \cdot X \cdot \tau \cdot X \quad \ldots$$

where τ stands for wait a time τ. This is often referred to as applying echo pulses.

The qubit state at time $t = \tau$ after the first X pulse will be

$$|\psi(\tau)\rangle = X \left(c_0|0\rangle + c_1 e^{-\imath \int_0^\tau dt' \, \Delta(t')}|1\rangle \right) = c_1 e^{-\imath \int_0^\tau dt' \, \Delta(t')}|0\rangle + c_0|1\rangle.$$

The qubit state at time $t = 2\tau$ after the second X pulse will be

$$|\psi(2\tau)\rangle = X \left(c_1 e^{-\imath \int_0^\tau dt' \, \Delta(t')}|0\rangle + c_0 e^{-\imath \int_\tau^{2\tau} dt' \, \Delta(t')}|1\rangle \right)$$

$$= c_0 e^{-\imath \int_\tau^{2\tau} dt' \, \Delta(t')}|0\rangle + c_1 e^{-\imath \int_0^\tau dt' \, \Delta(t')}|1\rangle$$

$$= e^{-\imath \int_\tau^{2\tau} dt' \, \Delta(t')} \left(c_0|0\rangle + c_1 e^{-\imath \int_0^\tau dt' \, \Delta(t')} e^{\imath \int_\tau^{2\tau} dt' \, \Delta(t')}|1\rangle \right).$$

Neglecting the unimportant global phase the qubit phase is

$$\phi(2\tau) = \int_\tau^{2\tau} dt' \, \Delta(t') - \int_0^\tau dt' \, \Delta(t').$$

Provided the fluctuations $\Delta(t)$ are statistically stationary at timescale τ the perturbed qubit phase will tend to zero. The bandwidth of the pulse sequence is $\omega_\tau \sim 1/\tau$. If the noise is bandwidth limited such that $\tilde{\Delta}(\omega) \to 0$ for $\omega > \omega_n$ and $\omega_\tau > \omega_n$ the noise will be effectively suppressed. Detailed analysis of the degree of noise suppression can be found in [15].

One challenge associated with this method is that the X pulses may not be perfect. A long sequence of pulses will then lead to uncorrected errors in the qubit state. If the error is that the X pulse is slightly too long or too short this can be corrected for by applying the sequence

$$\tau \cdot X \cdot \tau \cdot (-X) \cdot \tau \cdot X \cdot \tau \cdot (-X) \quad \ldots$$

Many other compensating pulse sequences have been proposed and demonstrated. For details we refer to the review [16].

12.6 Problems

1. Rabi oscillations are described by the solution of the Schrödinger equation for a two-level system interacting with a driving field as in Eqs. (12.6). In those equations the state vector was defined in a frame relative to the unperturbed atomic states. For some problems it is more convenient to work in a frame that rotates with the applied field at frequency ω. Find the equations for the slowly varying state amplitudes in this frame and solve them for the time evolution operator analogous to Eq. (12.7). Compare your result with (12.7) and show that the two operators are equivalent up to a frame transformation.

2. Qubit decoherence may be modeled as an exponential decay with time constants T_1 for the population and T_2 for the coherence. It is often the case that

$$\frac{1}{T_2} = \frac{1}{2T_1} + \frac{1}{T_\phi}$$

where $1/T_\phi$ is a dephasing rate that leads to loss of coherence without changing the population. The evolution of the density matrix can be written as

$$\rho = \begin{pmatrix} 1 - \rho_{11} & \rho_{01} \\ \rho_{01}^* & \rho_{11} \end{pmatrix} \rightarrow \rho' = \begin{pmatrix} 1 - \rho_{11}e^{-t/T_1} & \rho_{01}e^{-t\left(\frac{1}{2T_1}+\frac{1}{T_\phi}\right)} \\ \rho_{01}^* e^{-t\left(\frac{1}{2T_1}+\frac{1}{T_\phi}\right)} & \rho_{11}e^{-t/T_1} \end{pmatrix}.$$

Find the Kraus operators corresponding to this decoherence channel. Write your answers in terms of I, X, Y, Z and T_1, T_ϕ.

3. Consider a qubit with states $|g\rangle, |e\rangle$ having energies U_g, U_e and energy separation $U_e - U_g = \hbar\omega_q$. Assume $U_e > U_g$. The qubit is illuminated with radiation at frequency ω where $\omega - \omega_q = \Delta$, and assume $\Delta > 0$. The radiation couples the two levels through the electric dipole Hamiltonian $H_{E1} = -\hat{d}E = -\hat{d}\left(\frac{\mathcal{E}^*}{2}e^{-\imath\omega t} + \frac{\mathcal{E}}{2}e^{\imath\omega t}\right)$. For this problem assume $\mathcal{E} = \mathcal{E}^*$, and use the Rabi frequency $\Omega = \mathcal{E}d/\hbar$ with $d = \langle e|\hat{d}|g\rangle$ to characterize the strength of the qubit-radiation coupling.

   (a) Assume at $t = 0$ the qubit state is $|\psi\rangle = |g\rangle$. Give a formula for the probability to be in state $|e\rangle$ as a function of time t using the Schrödinger equation solution (Rabi oscillations).

(b) Assume at $t = 0$ the qubit state is $|\psi\rangle = |g\rangle$. Give a formula for the probability to be in state $|e\rangle$ as a function of time t using first order time dependent perturbation theory and the rotating wave approximation.

(c) Same as (b) but do not make the rotating wave approximation.

(d) Compare the results found in (a), (b), and (c) for short times such that $\Omega t \ll 1$, $\Delta t \ll 1$, and $\omega_q t \ll 1$. If the results are not all the same explain why.

4. A qubit is continuously driven by a control field with Rabi frequency Ω that is resonant with the qubit transition ($\Delta = \omega - \omega_q = 0$). Due to noise in the control system the Rabi frequency is not constant. We model the intensity of the noisy control field as

$$I = I_0 + \delta I$$

with $\Omega = \sqrt{I}$ the Rabi frequency, $\Omega_0 = \sqrt{I_0}$ is the average Rabi frequency, and the noise distribution

$$P(\delta I) = \frac{1}{\sqrt{2\pi\sigma^2}} e^{-(\delta I)^2/(2\sigma^2)}.$$

The root mean square intensity noise is σ.

Due to the intensity noise the amplitude of Rabi oscillation cycles will decay with time. Calculate the number of Rabi cycles $N = N(\sigma)$ that can be observed before the oscillation amplitude has decayed to $1/e$. If the qubit starts in state $|0\rangle$ how small does σ need to be such that the qubit population in $|0\rangle$ after a single 2π Rabi cycle is >0.999.

5. Composite pulses. We wish to perform single qubit π rotations on a qubit about the x axis.

(a) Write down the operator $U(\pi) = R_x(\pi)$ corresponding to a π pulse in the basis $\{|0\rangle, |1\rangle\}$.

(b) The control system is not entirely stable so the pulses have fluctuating pulse areas. Write down the error operator $U_\epsilon = U(\pi(1 + \epsilon)) - U(\pi)$ to leading order in ϵ assuming a pulse area of $\pi(1 + \epsilon)$, where ϵ is a small fractional error.

(c) Let's try to reduce the sensitivity to possible pulse area errors by using a composite sequence $U' = R_z(\phi)R_x(\pi/2)R_y(\pi)R_x(\pi/2)$. Find ϕ such that $U' = U(\pi)$.

(d) In the presence of a pulse area error the effective operator is $U'(\epsilon) = R_z(\phi(1 + \epsilon))R_x(\pi/2(1 + \epsilon))R_y(\pi(1 + \epsilon))R_x(\pi/2(1 + \epsilon))$. Write down the error operator $U'_\epsilon = U'(\epsilon) - U'(0)$ to leading order in ϵ .

(e) Is U'_ϵ less sensitive to pulse area errors than U_ϵ?

6. In Sect. 12.5 we solved for the time dependence of $\langle M_+(t)\rangle$. For a qubit in the initial state $|\psi\rangle = \cos(\theta/2)|0\rangle + \sin(\theta/2)e^{i\phi}|1\rangle$ find the magnitude of the coherence $|\rho_{01}|$ as a function of time, θ, and ϕ.

7. Consider two qubits interacting with the Hamiltonian

$$H = g\sigma_z^{(1)} \otimes \sigma_z^{(2)}.$$

This is referred to as an Ising interaction and is a typical interaction between physical spins, e.g., between nuclear spins in liquids. Show that

$$X^{(2)} U(t) X^{(2)} = U^{-1}(t),$$

where

$$U(t) = e^{-\iota H t}$$

and $X^{(2)} = \hat{I} \otimes X$ is an X gate acting on the second qubit. This result implies that $X^{(2)} U(t) X^{(2)} U(t) = 1$. The single qubit operations have effectively removed the interaction between the two qubits. This is referred to as 'refocusing' and can be used to remove undesired time evolution of interacting qubits when the interaction cannot be switched off.

References

1. D. DiVincenzo, The physical implementation of quantum computers. Fort. Phys. **48**, 771 (2000)
2. C.E. Bradley, J. Randall, M.H. Abobeih, R.C. Berrevoets, M.J. Degen, M.A. Bakker, M. Markham, D.J. Twitchen, T.H. Taminiau, A ten-qubit solid-state spin register with quantum memory up to one minute. Phys. Rev. X **9**, 031045 (2019)
3. P.M. Platzman, M.I. Dykman, Quantum computing with electrons floating on liquid helium. Science **284**, 1967 (1999)
4. S.D. Sarma, M. Freedman, C. Nayak, Majorana zero modes and topological quantum computation. npj Qu. Inf. **1**, 15001 (2015)
5. G. Lindblad, On the generators of quantum dynamical semigroups. Commun. Math. Phys. **48**, 119 (1976)
6. B. Schumacher, M. Westmoreland, *Quantum Processes Systems, & Information* (Cambridge University Press, Cambridge, 2010)
7. F. Bloch, Nuclear induction. Phys. Rev. **70**, 460 (1946)
8. S. Wimperis, Broadband, narrowband, and passband composite pulses for use in advanced NMR experiments. J. Mag. Res. **109**, 221 (1994)
9. G.H. Low, T.J. Yoder, I.L. Chuang, Optimal arbitrarily accurate composite pulse sequences. Phys. Rev. A **89**, 022341 (2014)
10. N.F. Ramsey, A new molecular beam resonance method. Phys. Rev. **76**, 996 (1949)
11. N.F. Ramsey, A molecular beam resonance method with separated oscillating fields. Phys. Rev. **78**, 695 (1950)
12. M.A. Lombardi, T.P. Heavner, S.R. Jefferts, NIST primary frequency standards and the realization of the SI second. Measure **2**, 74 (2007)
13. R.P. Feynman, F.L. Vernon, R.W. Hellwarth, Geometrical representation of the Schrödinger equation for solving maser problems. J. Appl. Phys. **28**, 49 (1957)

14. E. Paladino, Y.M. Galperin, G. Falci, B.L. Altshuler, $1/f$ noise: implications for solid-state quantum information. Rev. Mod. Phys. **86**, 361 (2014)
15. L. Viola, S. Lloyd, Dynamical suppression of decoherence in two-state quantum systems. Phys. Rev. A **58**, 2733–2744 (1998)
16. A.M. Souza, G.A. Álvarez, D. Suter, Robust dynamical decoupling. Philos. Trans. R. Soc. A. **370**, 4748 (2012)

Chapter 13
Atomic Qubits

In this chapter we treat quantum computing with atomic qubits. Atoms are a very natural setting for encoding qubits. Although they intrinsically have more than two energy levels it is possible to prepare, measure, and control a two-dimensional subspace of atomic states using optical techniques that have been developed over the last 50 years. The most precise instrument developed by humankind, the optical atomic clock, exploits the remarkable coherence properties of atoms. The availability of well developed techniques for cooling and trapping both neutral atoms and ions using electromagnetic fields, together with the availability of long coherence times, motivates their use as qubits for quantum computation and simulation. Images of arrays of neutral atom and trapped ion qubits are shown in Fig. 13.1.

There is much commonality in the atomic and optical physics underlying neutral atom and trapped ion approaches, as well as significant differences in the methods for trapping and implementing two-qubit gates. We will present neutral atoms starting in Sect. 13.1 and positively charged trapped ions starting in Sect. 13.6. The section on neutral atoms starts with background material on the techniques of laser cooling and trapping, as well as state preparation and measurement. We then proceed to discussions of one- and two-qubit gate protocols using microwave and optical frequency fields. For trapped ions we will summarize the electromagnetic trapping methods followed by techniques for two-qubit gates and remote entanglement.

13.1 Neutral Atoms

Long lived states of individual neutral atoms can be used to encode qubits. The most widely used approach selects two hyperfine-Zeeman states of the ground electronic configuration to represent $|0\rangle$ and $|1\rangle$. The energy separation of the states, depending

© Springer Nature Switzerland AG 2021
J. A. Bergou et al., *Quantum Information Processing*, Graduate Texts in Physics,
https://doi.org/10.1007/978-3-030-75436-5_13

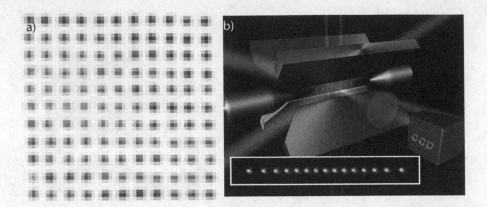

Fig. 13.1 Arrays of atomic qubits. (**a**) Fluorescence image of two-dimensional array of optically trapped Cs atom qubits. The atomic spacing is 3 μm. Image courtesy of Minho Kwon and Trent Graham. (**b**) Linear array of trapped Ca ions in a Paul trap. The spacing of the central ions is about 5 μm. Image courtesy of Rainer Blatt.

on the atomic species, is a few GHz. When the qubit states are associated with the same electronic orbital the transitions between them are magnetic dipole to leading order, and hence relatively weak. For this reason it is possible to achieve long coherence times with GHz frequency atomic qubits in a room temperature environment. As we will see in Chap. 15 the situation is different in sold state qubits that require mK cryostats to prevent transitions induced by background blackbody radiation.

Atoms are laser cooled to μK temperatures and then trapped in optical or magnetic potential wells. Atoms are prepared in a desired substate using optical pumping techniques. State measurements are performed using resonant light scattering that distinguishes the quantum states. Single qubit gate operations can be performed with microwave transitions that directly couple the qubit states or with a two-photon stimulated Raman optical transition. Many different ideas have been proposed for implementation of two-qubit entangling gates. The most widely used approach relies on optical excitation to highly-excited Rydberg states which have strong interaction energies [1].

Many different atomic elements have been laser cooled, see Fig. 13.2, and are candidates for qubit encoding. Until very recently experimental demonstrations have been limited to the heavy alkali atoms Rb and Cs which have large ground and excited state hyperfine splittings. These atoms have a simple electronic structure due to the presence of only a single valence electron. In the last several years there has been rapid progress with the alkaline earth and alkaline earth like atoms Sr and Yb which have two valence electrons [2, 3]. While their electronic structure is more complex than that of the alkali atoms, the complexity provides some desirable features including narrow optical transitions that enable stronger laser cooling and metastable excited states that are essential for atomic clock applications.

1 H																	2 He
3 Li	4 Be											5 B	6 C	7 N	8 O	9 F	10 Ne
11 Na	12 Mg											13 Al	14 Si	15 P	16 S	17 Cl	18 Ar
19 K	20 Ca	21 Sc	22 Ti	23 V	24 Cr	25 Mn	26 Fe	27 Co	28 Ni	29· Cu	30 Zn	31 Ga	32 Ge	33 As	34 Se	35 Br	36 Kr
37 Rb	38 Sr	39 Y	40 Zr	41 Nb	42 Mo	43 Tc	44 Ru	45 Rh	46 Pd	47 Ag	48 Cd	49 In	50 Sn	51 Sb	52 Te	53 I	54 Xe
55 Cs	56 Ba	57 La	72 Hf	73 Ta	74 W	75 Re	76 Os	77 Ir	78 Pt	79 Au	80 Hg	81 Tl	82 Pb	83 Bi	84 Po	85 At	86 Rn
87 Fr	88 Ra	89 Ac	104 Rf	105 Db	106 Sg	107 Bh	108 Hs	109 Mt									

58 Ce	59 Pr	60 Nd	61 Pm	62 Sm	63 Eu	64 Gd	65 Tb	66 Dy	67 Ho	68 Er	69 Tm	70 Yb	71 Lu
90 Th	91 Pa	92 U	93 Np	94 Pu	95 Am	96 Cm	97 Bk	98 Cf	99 Es	100 Fm	101 Md	102 No	103 Lr

Fig. 13.2 Shaded elements have been laser cooled and in many cases trapped in either neutral or singly ionized form

The great potential of neutral atoms for quantum computing is due to several favorable characteristics. Atoms of the same element are all identical so there is no need to calibrate the qubits or to keep track of individual phase evolutions. This statement must be modified when atomic qubits are trapped, which does modify the qubit frequency, but the modifications are small. Neutral atoms interact very weakly when in their electronic ground state, and it is therefore possible to have a dense array of qubits that provide a stable quantum memory. By exciting atoms to Rydberg states the interaction strength is increased by up to 12 orders of magnitude. Thus atomic qubits combine the features of a non-interacting memory and strong interactions for logic gates in a single platform.

A detailed understanding of the operation of atomic qubits involves atomic structure theory and the theory of coherent interactions between atomic states and electromagnetic fields. We do not have time for a complete discussion of the underlying atomic physics. References to relevant background and current literature are included in the following sections.

13.2   Laser Cooling and Trapping

Single atoms have very little mass so an atom in equilibrium with a room temperature bath has a thermal speed $\frac{1}{2}mv^2 = \frac{3}{2}k_BT$ where v is the velocity, m is the atomic mass, T is the temperature, and k_B is the Boltzmann constant. Plugging in numbers for a Cs atom in a 300K environment we find $v = 240$ m/s. Even in a dilution refrigerator at $T = 10$ mK the velocity is 1.4 m/s. Clearly to be useful as part of a computer we want the qubits to be stationary. This is achieved by first laser cooling the atoms to μK temperatures and then localizing them in potential wells defined by optical or magnetic fields. Laser cooling uses photon momentum to reduce the kinetic energy of atoms. The most basic form of laser cooling relies on the Doppler effect.

Interaction of an atom with a photon must conserve energy and momentum. It is instructive to evaluate the consequences of energy and momentum conservation using a particle like picture of the interaction. Initially the atom is in the ground state with energy $U_{ai} = U_i + \frac{1}{2}mv_i^2$ and momentum $\mathbf{p}_i = m\mathbf{v}_i$. The photon has energy $U_v = \hbar\omega$ and momentum $\mathbf{p}_v = \hbar\mathbf{k}$. After absorption there is no photon but the atom has $U_a = \hbar\omega_a + \frac{1}{2}mv_a^2$ and $\mathbf{p}_a = m\mathbf{v}_a$ where $\hbar\omega_a$ is the energy separation of the atomic levels, and $\mathbf{v}_a$ is the atomic center of mass velocity. A photon is then spontaneously emitted in a random direction with energy $U_{sp} = \hbar\omega_{sp}$ and momentum $\mathbf{p}_{sp} = \hbar\mathbf{k}_{sp}$. After spontaneous emission the atom is in a ground state and has energy $U_{af} = U_f + \frac{1}{2}mv_f^2$ and momentum $\mathbf{p}_f = m\mathbf{v}_f$. We allow for a change of atomic state with U_i and U_f possibly different.

Energy and momentum conservation between the initial and final states implies

$$U_i + \frac{1}{2}mv_i^2 + \hbar\omega = U_f + \frac{1}{2}mv_f^2 + \hbar\omega_{sp} \tag{13.1}$$

$$m\mathbf{v}_i + \hbar\mathbf{k} = m\mathbf{v}_f + \hbar\mathbf{k}_{sp}. \tag{13.2}$$

The parameters of the atomic motion after the interaction are

$$\frac{1}{2}mv_f^2 = (U_i - U_f) + \frac{1}{2}mv_i^2 + \hbar(\omega - \omega_{sp}) \tag{13.3}$$

$$m\mathbf{v}_f = m\mathbf{v}_i + \hbar(\mathbf{k} - \mathbf{k}_{sp}). \tag{13.4}$$

In the presence of a resonant field the atom will cycle between the ground and excited states at a maximum rate of $r_{max} = 1/(2\tau)$ with τ the spontaneous lifetime of the excited state. This result follows from the optical Bloch equations of Sect. 12.3. In each absorption and emission cycle momenta $\mathbf{k}$ and $-\mathbf{k}_{sp}$ are transferred to the atom. Since the directions of $\mathbf{k}_{sp}$ are randomly distributed the net effect is to accelerate the atom in a direction parallel to $\mathbf{k}$ and simultaneously to heat the atom due to the random momentum kicks from spontaneous emission.

After a time t the atom will have a mean kinetic energy along $\mathbf{k}$ of $\bar{K} = U_r(rt)^2$ and a kinetic temperature along each direction of motion of

$$T_j \equiv \frac{\text{var}(p_j)}{mk_B} \simeq \frac{2U_r}{3k_B}rt, \quad j = x, y, z \tag{13.5}$$

where $U_r = (\hbar k)^2/(2m)$ is the single photon recoil energy and r is the scattering rate which depends on the intensity and detuning of the light. Even though the momentum carried by a photon is tiny, due to the very small mass of a single atom, and the fact that the scattering rate can be large, the force on an atom due to light can be as much as 10^4 times larger than that due to gravity.

A single optical beam heats an atom at a rate proportional to the interaction time t. In order to cool an atom we apply a pair of counterpropagating beams, both detuned from the atomic transition by an amount $\Delta = \omega - \omega_a < 0$ (ω is the optical frequency and ω_a is the atomic transition frequency). The scattering rate, and therefore the optical force, decreases with the magnitude of the detuning. Due to the Doppler effect $|\Delta|$ becomes smaller for an atom moving towards the light source and increases for an atom moving away from the light source. The net effect is that the atom always scatters more photons from the beam opposing its' motion which leads to an effective frictional force that is referred to as optical molasses.

The momentum damping effect from optical molasses is countered by the heating from the random direction of scattered photons. The equilibrium atomic temperature depends on the detuning and can be shown to be [4] $k_B T_D = \hbar\gamma/2$ for a detuning of $\Delta = -\gamma/2$, with $\gamma = 1/\tau$. Using three pairs of molasses beams aligned along the x, y, z axes, cooling to the Doppler temperature T_D is achieved in three dimensions. Some characteristic parameters for different atoms are given in Table 13.1. In practice subDoppler temperatures by factors of up to about ten are routinely achieved. The explanation for this lies in the presence of additional atomic levels together with light induced optical pumping that redistributes the atomic population between the levels. A lower limit for cooling would appear to be set by the recoil temperature corresponding to the momentum kick of a single scattered photon. However, even recoil can be evaded and their are methods for reaching sub-recoil temperatures using light scattering. Details may be found in [4].

Cooling the motion is not sufficient for atomic particles to be used as qubits. It is also necessary to trap them at a well defined location. The workhorse for cooling and trapping is the magneto-optical trap (MOT) which uses a combination of optical molasses, to cool the atomic momentum, and a quadrupole magnetic field which provides position dependent forces, leading to simultaneous cooling and trapping of atoms. The MOT was first demonstrated in 1987 at Bell Laboratories [5] using Na atoms and has since become a standard tool of atomic physics due to its

Table 13.1 Light scattering and cooling parameters for several atoms. N_s is the number of scattered photons and the last column is for the narrow intercombination cooling transition of Sr

Parameter	Units	^{87}Rb	^{133}Cs	Sr	Sr
Transition		$5s_{1/2} - 5p_{3/2}$	$6s_{1/2} - 6p_{3/2}$	$5s^2\,^1S_0$ $-5s5p\,^1P_1$	$5s^2\,^1S_0$ $-5s5p\,^3P_1$
$\lambda_{cooling}$	(μm)	0.7802	0.8523	0.461	0.689
$\gamma/2\pi$	(MHz)	6.07	5.23	30.5	0.0074
$r_{s,\,max} = \gamma/2$	$(10^6\,\text{s}^{-1})$	19.1	16.4	95.8	0.023
t_{min} ($N_s = 100$)	(μs)	5.2	6.1	1.0	4350
$T_{Doppler} = \hbar\gamma/2k_B$	(μK)	146	125	720	1.02
$T_r = 2U_r/k_B$	(μK)	0.362	0.198	1.02	0.46

simplicity and robustness. Even in the face of moderate misalignments and intensity imbalances cold atoms can be trapped.

The basic idea can be summarized briefly. We arrange for each pair of molasses beams to have orthogonal polarization states that preferentially couple to atomic transitions that experience Zeeman shifts of opposite sign. The combination of negative detuning, a quadrupole magnetic field, and different polarizations results in the atom scattering more photons from the beam pushing it towards the center of the quadrupole field. This leads to a cloud of cold atoms forming in a stationary location at the zero of the quadrupole field. With cm sized optical beams and magnetic gradients of a few G/cm, samples of a few million atoms at temperatures of tens of μK can readily be prepared.

Although the MOT easily prepares a sample of cold atoms it is not well suited for qubit experiments. The light used to cool the atoms in a MOT is tuned relatively close to resonance so the atomic state is subject to decoherence from photon scattering. A trap suitable for holding qubits should provide the same potential for both qubit states and have a small rate of photon scattering. This can be accomplished using magnetic or far-detuned optical traps. While arrays of magnetic traps are possible [6], a substantial gradient is needed to hold an atom against gravity, and defining an array of traps with a micron scale period to accommodate the typical range of atomic interactions, implies that the trapping potential will be close to a material surface. This tends to be problematic due to light scattering from the beams that will be used for qubit control.

The bulk of work with neutral atom qubits has used optical traps. The basic idea is that the rate of photon scattering scales with detuning as $r \sim 1/|\Delta|^2$ but the optical potential, or light shift, induced by far detuned light scales as $U \sim 1/|\Delta|$. It is therefore possible to operate at detunings that are large enough to reach scattering rates at the level of $1\,\mathrm{s}^{-1}$ or lower, and simultaneously have a sufficient optical potential to trap an atom that has been laser cooled. Furthermore in the limit of large detuning the trapping potential can be almost perfectly identical for both qubit states which leads to long coherence times. The simplest type of this kind of optical trap is formed by taking a single red detuned ($\Delta < 0$) laser beam and focusing it to a spot of width $w \sim 1\,\mu$m. Atoms are attracted to the region of highest intensity at the center of the beam. To load an atom the trap is overlapped with the atoms in a MOT, and after turning off the quadrupole field one or a few atoms are left in the optical trap. Parameters can be tuned to probabilistically prepare single trapped atoms. This type of trap can be extended to an array of dipole traps in one- or two-dimensions leading to large arrays of single atom qubits. Alternatively, light that is blue detuned ($\Delta > 0$) which provides a repulsive potential for the atoms, can be patterned to form dark spots surrounded by light that trap atoms. This has been realized for arrays of atomic qubits in two- and three-dimensions [7, 8]. The advantage of the blue detuned traps is that the atom is localized in a region of low intensity and is thus less perturbed by the trapping light than in red detuned traps for which the atom is located at the high intensity point. For both red and blue arrays typical atom spacings are in the range of 2–10 μm which implies that many thousands of qubits can be held in mm sized regions. An example of a two-dimensional array of trapped atoms is shown in Fig. 13.1.

The loading into optical traps follows a Poissonian, or in the case of very small traps, sub-Poissonian distribution [9], but is intrinsically probabilistic. Although more than one atom can be loaded with certainty the reduction to a single atom, for use as a qubit, proceeds via collisions which tend to remove atoms in pairs. Thus, if the initial atom number is even, the final state after collisions will be no atoms in the trap. The problem of probabilistic loading of atom arrays has been solved by using atom rearrangement techniques. The idea is to partially load an array of traps, take an image of which sites hold an atom, and then rearrange the sites [10], or rearrange the atoms [11, 12] to create a fully occupied region. With these methods it has been possible to prepare fully occupied arrays of more than a hundred atoms [13] which serve as a starting point for quantum computation.

A challenge for the neutral atom optical trap array approach is the fact that the trap depths are less than the energy of untrapped background atoms and molecules. Collisions with hot, untrapped particles remove trapped atoms. Even with excellent vacuum conditions it is difficult to reach trap lifetimes longer than about 10 minutes. In order to scale up the array and allow for long computations it is necessary to track atom loss, which can be done as part of an error correction protocol, and replace atoms as needed. Analysis of this requirement suggests that arrays with thousands of atoms will be feasible with available technology [14].

13.3 Qubit State Preparation, Measurement, and Coherence

After loading single atoms into traps optical pumping is used to prepare a fiducial qubit state. A widely used choice of atom is one of the heavy alkalis, either Rb or Cs. The lowest energy levels of Cs are shown in Fig. 13.3. The clock states $|f_+, m = 0\rangle$, $|f_-, m = 0\rangle$ provide a convenient qubit basis. These states have no linear Zeeman shift at small magnetic fields and only a weak quadratic shift. A combination of π polarized light coupling f_+ to an excited state with the same f value, together with a repumper to remove population from the f_- hyperfine level is used to optically pump the atoms. The matrix element $|f, m = 0\rangle \rightarrow |f' = f, m = 0\rangle$ vanishes due to electric dipole selection rules so the atoms accumulate in $|f_+, m = 0\rangle$ which is a dark state for the pumping light. With Raman techniques via one of the excited states cooling to the motional ground state of the optical trap can be achieved [15].

An alternative which has attracted growing attention is encoding in alkaline earth like atoms such as Sr [2] and Yb [3]. The low lying levels of Sr are shown in Fig. 13.3. The alkaline earths have two valence electrons and a $J = 0$ ground state with no electronic angular momentum. The spectrum splits into singlet and triplet manifolds. Doppler cooling is performed on the singlet transition to the first excited state. Second stage cooling on a narrow triplet transition conveniently enables preparation in the motional ground state. For isotopes with no nuclear spin

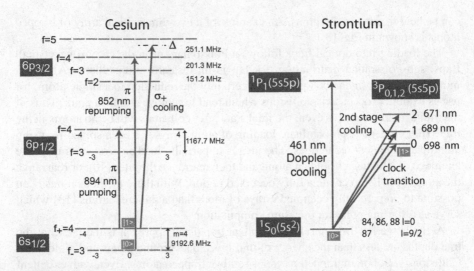

Fig. 13.3 Structure of low lying energy levels of Cs (left) and Sr (right). Cs hyperfine qubits are encoded in Zeeman sublevels of $f_+ = 4$ and $f_- = 3$. Optical transitions to the lowest lying excited states are used for cooling and state preparation. Sr is Doppler cooled on the $^1P_1$ transition followed by second stage cooling via the transition to $^3P_1$. The bosonic isotopes have no nuclear spin so optical qubits can be encoded in the single ground state and the metastable $^3P_0$ state. The fermionic isotope has nuclear spin $I = 9/2$ so qubits can be encoded in nuclear spin states

the qubit transition is at optical frequencies using a ground and metastable excited state. For isotopes with nuclear spin, qubits can be encoded in the nuclear spin projection [16]. Due to the small value of the nuclear magneton these qubits have only small energy separations in modest magnetic fields, so selective addressing relies on optical polarization selection rules.

The presence of a single atom can be detected by imaging scattered light onto a sensitive detector and counting photons. This was first done for atoms in a high field gradient MOT [17] and has become routine for atoms in an optical dipole trap. For a single trapped atom and no background noise the recorded distribution of photoelectron counts is described by a Poisson distribution. The probability of counting n photoelectrons is $f_s(n) = P(n, \bar{n})$ where $P(n, \bar{n}) = \frac{e^{-\bar{n}} \bar{n}^n}{n!}$ is a Poisson distribution with $\bar{n} = r_s t_s$ the expected mean number of counts, where t_s is the signal counting time and r_s is the count rate produced by a single atom.

This idealized Poisson distribution is modified in an actual experiment by uncertainty about the number of atoms and a nonzero rate of noise or background counts. Assuming the counts are additive when there is more than one atom present, and the atom number is also Poissonian distributed with mean $\bar{n}_a$ we have $f_s(n) = \sum_{n_a=0}^{\infty} P(n, r_s t_s n_a) P(n_a, \bar{n}_a)$. For the purposes of calculation the sum over n_a can be truncated at a realistic upper limit $n_{a,\max}$ that will depend on the trap parameters. In the presence of Poissonian background noise processes with rate r_b the probability for recording n background counts is $f_b(n) = P(n, r_b t_s)$. The

expected photoelectron count distribution is then the convolution of the signal and
the background

$$f(n) = \sum_{m=0}^{n} f_s(m) f_b(n-m)$$

$$= \sum_{n_a=0}^{n_{a,\max}} \sum_{m=0}^{n} P(m, r_s t_s n_a) P(n_a, \bar{n}_a) P(n-m, r_b t_s). \qquad (13.6)$$

We see that in general the photocount distribution is described by a triple Poisson
process: one for the scattering statistics, one for the atom distribution, and one for
the background statistics.

To detect an atom a cutoff n_c is introduced. An observed count number $<n_c$
corresponds to no atom while a count number $\geq n_c$ corresponds to one or more atoms
present. Detection of the presence or absence of an atom can be made to correspond
to the detection of a qubit state by using a cycling transition that is resonant with
one of the states, while the other state is dark [18]. Using a fixed cutoff and fixing
$n_a = 0$ or 1 in (13.6) we define a measurement error for the qubit state $c_0|0\rangle + c_1|1\rangle$
as

$$E(n_c) = |c_0|^2 \sum_{n=n_c}^{\infty} f(n)|_{n_a=0} + |c_1|^2 \sum_{n=0}^{n_c-1} f(n)|_{n_a=1}$$

$$= |c_0|^2 \left[1 - \frac{\Gamma(n_c, r_b t)}{(n_c-1)!} \right] + |c_1|^2 \frac{\Gamma(n_c, (r_b+r_s)t)}{(n_c-1)!}.$$

Here $\Gamma(n,x) = \int_x^{\infty} dy\, y^{n-1} e^{-y}$ is the incomplete Gamma function. Measurement
errors below 2% in an integration time of 1.5 ms have been reported [18].

Atomic qubits are subject to several primary sources of decoherence. Scattering
of trap light leads to Raman transitions which change the qubit state. This implies
a limit on the T_1 coherence time. Using atoms cooled close to the motional ground
state in blue detuned traps T_1 times at the 10 s scale have been observed [8]. The
T_2 time is primarily limited by magnetic noise, as in Sect. 12.5, and fluctuating
differential light shifts due to noise of the trapping light and atomic motion at
finite temperature which causes the atom to see a time dependent optical intensity.
Also elastic Rayleigh scattering, which does not change the atomic state can lead to
decoherence in some cases [19].

Motional decoherence provides an instructive example of undesired entanglement between the qubit and other dynamical variables. The qubit may be encoded in
a spin degree of freedom, but the atom has additional quantum numbers associated
with the center of mass motion. If the total energy is correlated to both the spin state
and other degrees of freedom decoherence can arise. Consider an atom in a harmonic
trap with vibrational frequency ω. The energies of the harmonic oscillator levels are
$U(n) = \hbar\omega(1/2 + n)$. Assuming a Maxwell Boltzmann thermal distribution the

probability of occupation of level n is $P_n = ae^{-U(n)/k_B T}$, with a a normalization constant. The expected value of n is $\langle n \rangle = \frac{1}{e^{\hbar\omega/k_B T}-1}$. The temperature can therefore be expressed as

$$k_B T = \frac{\hbar\omega}{\ln(1 + 1/\langle n \rangle)}. \tag{13.7}$$

In three dimensions this generalizes to $\langle \mathbf{n} \rangle = (\langle n_x \rangle, \langle n_y \rangle, \langle n_z \rangle)$ with the Cartesian components n_j dependent on ω_j.

Consider a qubit encoded in hyperfine ground states $|0\rangle, |1\rangle$ with a thermal state for the center of mass atomic motion. The density operator describing this state is

$$\rho(0) = |0\rangle\langle 0| \otimes \sum_{\mathbf{n}} P_{\mathbf{n}} |\mathbf{n}\rangle\langle \mathbf{n}|.$$

Here $\mathbf{n} = (n_x, n_y, n_z)$ is a set of vibrational quantum numbers specifying the motional state. The T_2 time can be measured with Ramsey spectroscopy which involves the sequence $U = R_{\pi/2} U_{\mathbf{n}}(t) R_{\pi/2}$. We will assume R only acts on the electronic state and does not change $|\mathbf{n}\rangle$. The free evolution operator for a time t is $U_{\mathbf{n}}(t) = e^{-\imath(\Delta_1(\mathbf{n})-\Delta)\sigma_z t/2}$. Here $\Delta = \omega - \omega_{10}$ is the detuning of the field driving R from the qubit frequency ω_{10}, $\Delta_1(\mathbf{n}) = \frac{1}{\hbar}[U_1(\mathbf{n}) - U_0(\mathbf{n})] = (1/2+\mathbf{n})\cdot(\omega_1-\omega_0)$, and ω_j are the trap frequencies when the atom is in electronic state $|j\rangle$. The trap frequencies depend on the electronic state due to the fractional differential light shift $\bar{\Delta}_{\mathrm{LS}} = (U_1 - U_0)/\frac{1}{2}(U_0+U_1)$. The situation is shown in Fig. 13.4. The energy difference $\Delta U = U_1 - U_0$ scales as

$$\Delta U \sim \bar{U}\left(\frac{\Delta}{\Delta - \omega_q/2} - \frac{\Delta}{\Delta + \omega_q/2}\right) \simeq \bar{U}\frac{\omega_q}{\Delta}$$

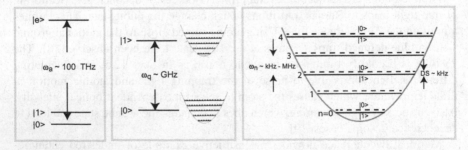

Fig. 13.4 A spin qubit encoded in hyperfine ground states $|0\rangle, |1\rangle$ of an atom with an electronically excited state $|e\rangle$. The qubit has a GHz scale frequency ω_q and is confined in a harmonic trap with vibrational levels n separated by ω_n. Coupling between the spin and center of mass degrees of freedom results in a small spin dependent differential shift (DS) of the vibrational energies

with $\bar{U}$ the average trap potential and Δ the detuning of the trap light from the excited state. Using $\bar{U} \sim k_B \times 0.1\,\mathrm{mK}$, $\omega_q \sim 2\pi \times 10\,\mathrm{GHz}$ and $\Delta \sim 2\pi \times 10\,\mathrm{THz}$ gives $\Delta U \sim k_B \times 0.1\,\mu\mathrm{K} \sim h \times 2\,\mathrm{kHz}$. Accurate calculation of the differential shift for alkali atoms requires a more detailed analysis that accounts for the induced tensor polarizability of the ground state [20].

With the assumption of $\bar{\Delta}_{LS} \ll 1$, which is generally the case, we can write $\omega_1 \simeq \omega_0(1 + \bar{\Delta}_{LS}/2)$ so $\Delta_1(\mathbf{n}) = (1/2 + \mathbf{n}) \cdot \omega_0\bar{\Delta}_{LS}/2$. Renormalizing ω_{10} by $\sum_{j=x,y,z} \omega_{j0}\bar{\Delta}_{LS}/4$ to account for the trap shift of the transition we obtain $\Delta_1(\mathbf{n}) = \frac{\bar{\Delta}_{LS}}{2}\mathbf{n} \cdot \omega_0$.

The density matrix after a Ramsey sequence is therefore

$$\rho(t) = U\rho(0)U^\dagger = R_{\pi/2}U_\mathbf{n}(t)R_{\pi/2}\left(|0\rangle\langle 0| \otimes \sum_\mathbf{n} P_\mathbf{n}|\mathbf{n}\rangle\langle\mathbf{n}|\right) R_{\pi/2}^\dagger U_\mathbf{n}(t)^\dagger R_{\pi/2}^\dagger$$

$$= \sum_\mathbf{n} P_\mathbf{n}|\mathbf{n}\rangle\langle\mathbf{n}|R_{\pi/2}U_\mathbf{n}(t)R_{\pi/2}|0\rangle\langle 0|R_{\pi/2}^\dagger U_\mathbf{n}^\dagger(t)R_{\pi/2}^\dagger. \tag{13.8}$$

The probability of measuring the atom in $|1\rangle$ after time t is $P_{|1\rangle}(t) = \mathrm{Tr}_{vib}[\langle 1|\rho(t)|1\rangle] = \sum_\mathbf{n} P_\mathbf{n}|\langle 1|R_{\pi/2}U_\mathbf{n}(t)R_{\pi/2}|0\rangle|^2$. Evaluating the operator product we find

$$P_{|1\rangle}(t) = \frac{1}{2} + \frac{1}{2}\sum_\mathbf{n} P_\mathbf{n}\cos(\theta_\mathbf{n}t) \tag{13.9}$$

with $\theta_\mathbf{n} = \frac{\bar{\Delta}_{LS}}{2}\mathbf{n} \cdot \omega_0 - \Delta$. When the atom is in the motional ground state $P_0 = 1$ and $P_{|1\rangle}(t) = \frac{1}{2}[1 + \cos(\Delta t)]$.

For thermal states with many occupied vibrational modes evaluation of the sum in (13.9) is inefficient. In this limit we can let $\hbar/T \to 0$, use $P_\mathbf{n} \simeq \omega_x\omega_y\omega_z\beta^3 e^{-\beta\mathbf{n}\cdot\omega_0}$, and approximate the sum by an integral. This semiclassical approximation leads to

$$P_{|1\rangle}(t) = \frac{1}{2} + \frac{1}{2\left[1 + 0.948(t/T_2^*)^2\right]^{3/2}}\cos(\Delta t - \kappa) \tag{13.10}$$

with $T_2^* = \sqrt{e^{2/3} - 1}\frac{2\hbar}{k_B T\bar{\Delta}_{LS}} = 1.947\frac{\hbar}{\bar{\Delta}_{LS}k_B T}$ and $\kappa = \tan^{-1}\left(\frac{12\beta^2\bar{\Delta}_{LS}t - \bar{\Delta}_{LS}^3 t^3}{8\beta^3 - 6\beta\bar{\Delta}_{LS}^2 t^2}\right)$. The factor κ gives a small correction to the Ramsey frequency. The expression for T_2^* was first derived in [21] starting from a semiclassical model. A comparison of the quantum and semiclassical calculations is given in Fig. 13.5. This dephasing is inhomogeneous, and can be reversed using sequences of echo pulses, as described in Sect. 12.5.1.

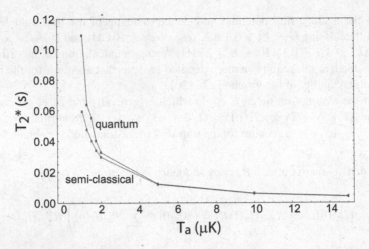

Fig. 13.5 T_2^* calculated with quantum and semiclassical models for $\omega_x = \omega_y = 2\pi \times 50\,\text{kHz}$, $\omega_z = 2\pi \times 10\,\text{kHz}$, and $\bar{\Delta}_{\text{LS}} = 2.5 \times 10^{-4}$ which corresponds to Cs atoms with trap light at 780 nm

13.4 One-Qubit Gates

Gates acting on qubits encoded in hyperfine states can be implemented with microwave or optical frequency fields. As presented in Sect. 12.4.1 resonant Rabi rotations provide $R_x(\theta)$ or $R_y(\theta)$ gates for real and imaginary Rabi frequencies. The rotation angle is $\theta = t|\Omega|$ with t the gate time. Rotations about other axes in the equatorial plane can be obtained with an appropriate choice of the phase of Ω. Z gates can be implemented using the operator identity $Z = iYX$.

Gates acting on qubits encoded in hyperfine states can be simply implemented using microwave fields that are resonant with the qubit frequency. When the hyperfine states are connected with the same electronic orbital the transition is of magnetic dipole character and the Rabi frequency is $\Omega = \boldsymbol{\mu} \cdot \mathcal{B}/\hbar$ where μ is the matrix element of the magnetic moment operator and $\mathcal{B}$ is the amplitude of the magnetic field. For radiation of intensity I the field amplitude is $\mathcal{B} = \mathcal{E}/c = \sqrt{2I/(\epsilon_0 c^3)}$. Typical Rabi frequencies for a few W of microwave power driving a hyperfine transition in an alkali atom are $\sim 10\,\text{kHz}$.

All qubits in an array can be rotated in parallel by turning on a resonant microwave field. Since the microwave wavelength is much longer than a typical atomic array size all qubits are addressed. Site selective addressing can be achieved with a combination of microwaves and a focused optical beam that Stark shifts selected sites. In one approach the microwaves are detuned and the Stark beam shifts selected sites into resonance [22]. In another approach the microwaves are kept on resonance and a Stark shifting beam is used to implement a site specific z

rotation [8]. Site specific x or y rotations are then obtained using the identities

$$R_x(\theta) = \bar{R}_y(\pi/2)R_z(\theta)\bar{R}_y(-\pi/2),$$
$$R_y(\theta) = \bar{R}_x(-\pi/2)R_z(\theta)\bar{R}_x(\pi/2).$$

Here $\bar{R}_x, \bar{R}_y$ are global rotations which cancel at those sites that do not receive a local R_z rotation. Single qubit gate operations with microwaves acting globally on neutral atom qubits have been demonstrated with fidelity $\mathcal{F} > 0.99995$, as determined by randomized benchmarking [23]. Microwave gates with optical Stark shifting for addressing single qubits have been demonstrated with $\mathcal{F} > 0.99$ [8, 22].

Faster Rabi frequencies can be achieved using optical transitions. Either a one-photon transition for optical frequency qubits or a two-photon stimulated Raman transition via an opposite parity excited state for hyperfine encoded qubits. This involves a light field consisting of two optical frequencies with the difference equal to the qubit frequency. If the one-photon Rabi frequencies are Ω_1, Ω_2 with respect to an intermediate state that is detuned by Δ then the two-photon Rabi frequency is $\Omega = \Omega_1\Omega_2^*/(2\Delta)$. With the optical Raman approach Rabi rates of several MHz can be readily achieved [24].

There is a decoherence cost associated with this approach since photons are scattered from the intermediate level. It can be shown that the probability of photon scattering during a π pulse is $P_{\text{scat}} \sim \gamma/|\Delta|$ where γ is the radiative decay rate of the intermediate state. To see this we recall that the time for a π pulse is $t_\pi = \pi/|\Omega| = 2\pi|\Delta|/|\Omega_1\Omega_2^*|$. For simplicity assume $|\Omega_1| = |\Omega_2|$ so $t_\pi = 2\pi|\Delta|/|\Omega_1|^2$. The average excited state population during the pulse is $\rho_{22} \sim |\Omega_1|^2/\Delta^2$ and the integrated scattering probability is $P_{\text{scat}} \sim \gamma\rho_{22}t_\pi \sim \gamma/|\Delta|$.

13.5 Two-Qubit Gates

Interactions between qubits are needed to implement two-qubit gates. Neutral atoms are a promising platform for qubit encoding because they are neutral and interact only weakly. This makes them a good choice for a quantum memory with long coherence time. In order to also use neutral atoms for two-qubit gates interactions are needed. There have been many proposals for entangling gates between neutral atom qubits. The earliest ideas date back to the 1990s and involve atom-photon coupling in optical resonators [25], atomic collisions [26], and dipole mediated interactions [27].

The approach that has been most successful and is currently under active development is based on Rydberg blockade [28, 29]. In this approach atoms are excited with laser beams to high lying states, called Rydberg states, which possess large electronic wavefunctions that have a spatial extent $r \sim a_0n^2$ where a_0 is a Bohr radius and n is the principal quantum number. Although atomic energy eigenstates have definite parity, and therefore no permanent dipole moment, two

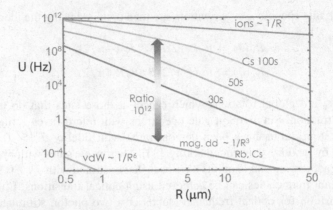

Fig. 13.6 Atomic interaction strength as a function of interatomic distance. Ground state alkali atoms interact through a $1/R^6$ van der Waals potential and $1/R^3$ magnetostatic dipole term. Increasing the Rydberg principal quantum number gives stronger interactions that are still less than the $1/R$ Coulomb interaction of ions

closely spaced Rydberg excited atoms interact strongly through an effective dipole interaction that scales as $U_{\mathrm{dd}} \sim d^2/R^3$ where R is the interatomic spacing and d is a transition matrix element between neighboring Rydberg states of opposite parity. The transition matrix element scales as n^2 so the interaction energy scales as n^4. When the atoms are further apart the interaction is no longer resonant and has a van der Waals form [1] $U_{\mathrm{vdW}} \sim U_{\mathrm{dd}}^2/\Delta \sim d^4/\Delta R^6$. Here $\Delta \sim 1/n^3 - 1/n^4$ is an energy mismatch between neighboring Rydberg levels so $U_{\mathrm{vdW}} \sim n^{11} - n^{12}$.

The rapid scaling of interaction strength with n provides a switchable interatomic coupling that can be turned on and off with laser pulses. The variation with atomic separation is shown in Fig. 13.6. As we see the interaction strength of Rydberg excited atoms can be 12 orders of magnitude larger than that of atoms in the ground state. The ability to turn the interaction on and off with this extremely large contrast is enabling for quantum information applications.

Strong Rydberg interactions result in what is called Rydberg blockade which is analogous to the Coulomb blockade of charges in a narrow conductor. An atom can be excited to a Rydberg state using a laser that is resonant with the ground-Rydberg transition. If we then attempt to excite a second atom the transition will be shifted due to the Rydberg interaction and the excitation will be blocked. The blockade condition is simply $U(R_B) = \hbar\Omega$ where R_B is the blockade distance and Ω is the excitation Rabi frequency. For $R > R_B$ we can excite both atoms, but for $R < R_B$ only one atom at a time can be excited to the Rydberg state. When the qubit states have a large energy difference only one of the states is resonantly coupled to the Rydberg state. The result is that we can engineer a sequence of laser pulses that give a dynamical evolution which is dependent on the quantum states of both qubits. In this way an entangling phase gate can be designed.

Fig. 13.7 The Rydberg blockade gate uses a three pulse sequence: π pulse on the control qubit, 2π pulse on the target qubit, π pulse on the control qubit. The pulses have Rabi frequency Ω, the interaction gives a blockade shift B, the qubit frequency is ω_q, and the Rydberg state decays at rate γ

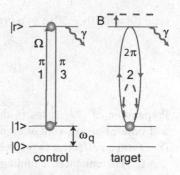

The original Rydberg gate protocol [28] illustrated in Fig. 13.7 relies on a three pulse sequence acting on the control and target qubits. The action of the gate can understood by considering the transformation of all two-qubit states in the computational basis. The state $|00\rangle$ is not Rydberg coupled and experiences no change. The states $|01\rangle$ and $|10\rangle$ acquire a π phase shift due to the Rydberg excitation and deexcitation. The Rydberg interaction comes into play for $|11\rangle$. The first pulse on the control atom puts population in the Rydberg state which blocks the excitation of the target atom. Therefore this state also acquires a π shift. The gate matrix in this ideal limit is $U = \text{diag}\,(1, -1, -1, -1)$ which is an entangling phase gate.

In practice the fidelity is less than perfect due to decay from the Rydberg state and finite blockade strength. Analysis shows [1, 30] that optimal fidelity is obtained by choosing a Rabi frequency $\Omega \sim B^{2/3}/\tau^{1/3}$ in order to balance the contributions of spontaneous decay and blockade errors. The resulting gate error scales as $1/(B\tau)^{2/3}$. Both the blockade interaction and the Rydberg lifetime τ increase as the principal quantum number increases. However, this does not imply that the gate error can be arbitrarily small since the blockade shift is limited by the distance to the neighboring Rydberg level which scales as $1/n^3$. Numerical studies [30] suggest that optimal states lie in the range $n = 80$–120 giving gate errors as low as ~ 0.001.

The gate protocol described here is the original proposal from the year 2000 [28]. Since then many variations have been proposed [1, 14, 31] including ideas based on shaped pulses that predict errors as low as 10^{-5} [32]. The best results achieved to date are preparation of Bell states with fidelity 0.89 in a 2D array of Cs atoms [33] and 0.97 in a 1D array of Rb atoms [31]. Quantum algorithms using multiple qubits have not yet been demonstrated with neutral atoms, although this will likely occur soon. On the other hand there have been numerous quantum simulation experiments that use the long range Rydberg interaction to implement Ising type models. These experiments have been scaled to systems with more than 50 atoms [34].

There are many interesting consequences of Rydberg blockade which can be used not only for engineering two-qubit gates, but also multi-qubit interactions, many particle entanglement, and non-classical quantum states. If N atoms are all within a blockaded region and are all simultaneously coupled to the Rydberg state only one of the atoms will be excited. This results in the coupling of a many particle ground state $|\bar{0}\rangle = |0_1 0_2 \ldots 0_N\rangle$ to a symmetric entangled state

$$|W\rangle = \frac{1}{\sqrt{N}} \sum_{j=1}^{N} |00..r_j..0\rangle.$$

Preparation of $|W\rangle$ states in this way has been demonstrated in both disordered and ordered arrays with atom numbers exceeding 100 [35, 36]. The states $|\bar{0}\rangle$ and $|W\rangle$ form an effective two-level system with an enhanced coupling strength $\Omega_N = \sqrt{N}\Omega$. The enhanced coupling strength allows for new capabilities for tailoring atom-light interactions that transfer entanglement between atoms and photons.

13.6 Trapped Ions

Qubits encoded in trapped ions are another approach that is being actively developed for quantum information applications [37, 38]. Analogously to the case of neutral atoms hyperfine-Zeeman ground states or a combination of ground and metastable excited states can be used to encode qubits. In the first approach the energy separation of the states, depending on the atomic species, is a few GHz. When the qubit states are associated with the same electronic orbital the transitions between them are magnetic dipole to leading order, and hence relatively weak. For this reason it is possible to achieve long coherence times with GHz frequency qubits in a room temperature environment. As we will see in Chap. 15 the situation is different in solid state qubits that require mK cryostats to prevent transitions induced by background blackbody radiation. In the second approach qubits are encoded in a ground state and in a metastable electronically excited state. The qubit frequency is then several hundred THz. Both approaches have been successfully demonstrated and trapped ions hold the record for the highest fidelity quantum gates and longest coherence times demonstrated on any platform [39, 40].

As with neutral atoms there are certain elements that have primarily been used. Alkaline earth atoms that occupy the second column of the periodic table have two valence electrons. After removing one electron to give a positive ion with unit charge there is a single valence electron remaining. The energy level spectrum of the positively charged alkaline earth ion is therefore relatively simple and convenient for laser cooling, optical pumping, and state measurement, using techniques that are similar to those we have discussed for neutral alkali atoms. A primary difference in the alkaline earth ions is the existence of metastable D states that can be used for laser cooling, and for encoding of qubits. These are analogous to the metastable triplet P states of neutral alkaline earth atoms.

13.7 Ion Traps

Although optical traps are possible for ions, by far the most widely used approach is an electromagnetic trap based on radio and microwave frequency fields with Paul and Penning traps the standard techniques. Consider a positively charged ion. We would like to design a trapping potential that stably confines the ion at a fixed position in space. Let $\phi(\mathbf{r})$ be a static trapping potential so the energy of the ion is $U(\mathbf{r}) = q\phi(\mathbf{r})$ with q the charge. The local electric field is $\mathbf{E} = -\nabla\phi$, and in free space $\nabla \cdot \mathbf{E} = -\nabla^2\phi = 0$. In order to have a stable trap we require that ϕ has a local maximum or minimum which implies that $\mathbf{E} = -\nabla\phi$ must be either negative or positive along all lines originating at the extremum. This is not possible since $\nabla \cdot \mathbf{E} = 0$ so there can be no local maximum or minimum of ϕ, only a saddle point. This result is known as Earnshaw's theorem. A saddle point can be generated in a quadrupole configuration as in Fig. 13.8. Choosing, for example, $r_c = \sqrt{2}z_c$ the potential is $U = \frac{qV}{4z_c^2}\left(x^2 + y^2 - 2z^2\right)$. For $V > 0$ the potential has a quadratic maximum at $z = 0$ and minimum at $x = y = 0$.

To achieve stability it is necessary to add additional fields. In the so-called Penning trap, demonstrated first by Dehmelt [41], we take $V < 0$ and add a strong axial magnetic field. Negative V localizes the ions in the $z = 0$ plane while the gyromagnetic orbits radially confine the ions. This type of configuration has been used for quantum simulation [42], but presents a challenge for quantum gate operations since the ions rotate rapidly about the z axis at the gyro frequency.

The Paul trap [43, 44] provides a solution that confines the ions to well defined spatial positions with only minimal additional motion. To do so we apply a potential $V(t) = V_0 \cos(\omega_{\mathrm{rf}} t)$ with typical values $\omega_{\mathrm{rf}}/2\pi \sim 10$–$100\,\mathrm{MHz}$. The potential changes sign at frequency ω_{rf} which stabilizes the ion, apart from a small micromotion which we proceed to analyze.

Fig. 13.8 Quadrupole trap geometry

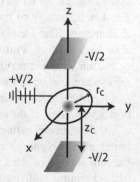

In the geometry of Fig. 13.8 $U(\mathbf{r}) = U_x(x) + U_y(y) + U_z(z)$ so the motion is decoupled along each axis. For the z motion we find

$$\frac{d^2 z}{d\tau^2} = 2\tilde{q}_z \cos(2\tau) z$$

with $\tau = \omega_{rf} t/2$ a dimensionless time and $\tilde{q}_z = 2q V_0/(m z_c^2 \omega_{rf}^2)$ and m is the mass. A similar equation describes the x and y motion with $\tilde{q}_x = \tilde{q}_y = -\tilde{q}_z/2$. This is a Mathieu equation and it can be shown that the motion is bounded for particular value ranges of $\tilde{q}_z$. The primary stability region is $0 \le \tilde{q}_z \le 0.908$ which corresponds to $\omega_{rf} \ge 1.48 \left[q V_0/(m z_c^2)\right]^{1/2}$. This stability condition can be understood qualitatively from the requirement that the potential change sign before the accelerated ion reaches an electrode.

The motion of the ion can be decomposed into an averaged "macromotion" Z and a "micromotion" ζ. We then put $z = Z + \zeta$ and assume $|\zeta| \ll |Z|$ but $|d^2\zeta/dt^2| \gg |d^2 Z/dt^2|$ to arrive at the micromotion equations

$$\frac{d^2\zeta}{dt^2} \simeq \frac{q V_0}{m z_c^2} \cos(\omega_{rf} t) Z,$$

$$\zeta = -\frac{\tilde{q}_z}{2} \cos(\omega_{rf} t) Z.$$

The kinetic energy of the micromotion along z is $U_\zeta = \frac{1}{2} m \dot{\zeta}^2 = \frac{1}{2} m \omega_z^2 Z^2$ with $\omega_z = \frac{1}{2^{3/2}} \tilde{q}_z \omega_{rf}$. A similar result holds along x and y with $\omega_x = \omega_y = \omega_z/2$. This result shows that the energy associated with the micromotion increases quadratically with the distance from the origin. This provides an effective harmonic trapping potential for the macromotion.

Putting in numbers for a Be^+ ion in a Paul trap ($m = 1.5 \times 10^{-26}$ kg, $V_0 = 100$ V, $\omega_{rf}/2\pi = 10$ MHz, $z_c = 1$ mm) we find $\tilde{q}_z = 0.54$ and $\omega_z/2\pi = 1.9$ MHz. The kinetic energy of the micromotion is then $U_\zeta/k_B = 7.8 \times 10^{10} Z^2$ (K), with Z in meters. The localization of the ion depends on the residual motional energy after laser cooling. Putting $T = 10\,\mu$K and $U_\zeta = \frac{1}{2} k_B T$ we find $Z \simeq 8$ nm and $\zeta \simeq 2$ nm which is negligible compared to the wavelength of light used to control the electronic state of the ion.

Of particular importance is the fact that the effective potential for the macromotion is very deep. If the ion were to move a tenth of the distance to the z electrode or $z_c/10$ the corresponding potential depth would be ~ 800 K which is significantly larger than the energy of untrapped atoms or molecules in a room temperature vacuum apparatus. This implies that ions in a Paul trap are stable in the presence of background collisions. Indeed single ions have been trapped for as long as 1 month. Although chemical reactions with untrapped elements also lead to trap loss, lifetimes can still be many hours, which is significantly longer than for optically trapped neutral atoms. The actual electrode geometry that is used in practice for ion traps is quite different than that illustrated in Fig. 13.8, including chip scale geometries that may include integrated electronics and optics for qubit control and measurement [38], but the principles remain the same.

13.7.1 Multiple Ions

One dimensional arrays of trapped ion qubits can be established in Paul traps by loading more than one ion. The positions of the ions are then determined by a balance between the Coulomb repulsion and the confining potential. For the simplest case of two ions the total potential along the trap axis at the equilibrium spacing Z_{eq} is

$$U_{\text{total}}(Z_{eq}) = 2 \times \frac{1}{2} m \omega_z^2 (Z_{eq}/2)^2 + \frac{q^2}{4\pi \epsilon_0 Z_{eq}}.$$

Minimizing the total energy we find $Z_{eq} = \left(\frac{q^2}{2\pi \epsilon_0 m \omega_z^2} \right)^{1/3} \simeq 6 \, \mu\text{m}$. This is a convenient spacing that is large enough for optical addressing of individual ions.

When the ions are cold so $|\zeta| \ll Z_{eq}$ collective motional modes are observed. For two ions these are the center of mass mode with both ions moving parallel along the trap axis, the stretch mode where the ions move in opposite directions along the trap axis, and transverse modes. Paul trap electrodes are typically designed so the trapping potential is anisotropic with transverse trap frequencies much higher than the axial frequency. The transverse motion is then frozen out and a one-dimensional string of ions forms along the "soft" axis with the lowest vibrational frequency.

In this one-dimensional limit we can decompose the motion into the two lowest frequency normal modes: the center of mass motion Z_1 and the inter-ion separation Z_2. The total energy can then be written as

$$U(Z_1, Z_2, \dot{Z}_1, \dot{Z}_2) = m \left(\dot{Z}_1^2 + \omega_z^2 Z_1^2 \right) + \frac{m}{4} \left[\dot{Z}_2^2 + 3\omega_z^2 (Z_2 - Z_{eq})^2 \right].$$

This expression can be generalized to multiple ions. However, as the number of ions is increased the normal mode spectrum becomes increasingly dense [45]. In practice this has so far limited Paul trap experiments to less than about 100 ions.

13.8 Coupling of Electronic and Motional Degrees of Freedom

Electromagnetic fields at either microwave or optical frequencies are used to implement gates. One-qubit gates are implemented using methods analogous to those already discussed for neutral atom qubits. Two-qubit gates for trapped ion

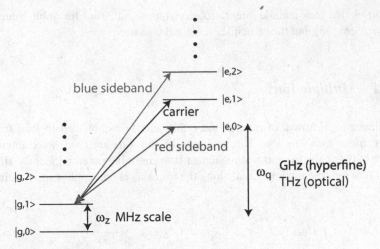

Fig. 13.9 Electronic and motional energy levels of a trapped ion

qubits rely on coupling between the internal electronic degrees of freedom and the center of mass motional state. A π pulse between the qubit states can leave the motional state unchanged, remove one or more quanta of motional excitation, or add one or more quanta of motional excitation. These are referred to as carrier, red-sideband, and blue-sideband transitions respectively.

Consider the energy levels shown in Fig. 13.9. The quantum state of an ion $|g/e, n\rangle$ is specified with two variables, the spin state g or e and the number of motional quanta n. Here we are assuming the transverse motion is frozen out and n refers to the motion along the trap axis. A field at frequency ω_q which is resonant with the qubit level spacing drives a carrier transition $|g, n\rangle \leftrightarrow |e, n\rangle$. A field at frequency $\omega_r = \omega_q - \omega_z$ drives a red-sideband transition $|g, n\rangle \leftrightarrow |e, n-1\rangle$ and a field at frequency $\omega_b = \omega_q + \omega_z$ drives a blue-sideband transition $|g, n\rangle \leftrightarrow |e, n+1\rangle$. Note that if the ion is in state $|g, 0\rangle$ there is no red sideband transition since the state $|e, -1\rangle$ does not exist. Also there is no blue sideband transition starting in $|e, 0\rangle$ since $|g, -1\rangle$ does not exist. These non-allowed transitions will turn out to be crucial for creating entanglement.

In what follows we will describe the coupling between states in terms of a Rabi frequency Ω that may be due to microwave fields driving a magnetic dipole (M1) transition or optical fields driving electric dipole (E1) or electric quadrupole (E2) transitions. A detailed treatment of multipole couplings is outside the scope of this discussion (a good reference is [46]). Nevertheless the type of coupling, electric or magnetic, has some important consequences for our ability to change the motional quantum number with a Rabi pulse.

For an E1 or E2 transition the Rabi frequency is proportional to a matrix element of the form $\langle e, n'|\hat{\mathcal{H}}_{\text{int}}|g, n\rangle$ with $\hat{\mathcal{H}}_{\text{int}}$ the interaction Hamiltonian, and n, n' the initial and final motional quantum numbers of the harmonic oscillator eigenstates. Denoting the center of mass coordinate of the ion with Z and the position of the valence electron relative to the ion core with $\mathbf{r}$ the matrix element can be separated as [47]

$$\langle e, n' | \hat{\mathcal{H}}_{\text{int}} | g, n \rangle \sim \langle e | r^q | g \rangle \langle n' | \hat{\mathcal{H}}_{\text{cm}}(Z) | n \rangle$$

where $q = 1$ or 2 for dipole or quadrupole transitions. The matrix element between center of mass motional states is of the form $\langle n' | \hat{\mathcal{H}}_{\text{cm}}(Z) | n \rangle \sim \langle n' | e^{i\mathbf{k} \cdot Ze_z} | n \rangle$ where $\mathbf{k}$ is the optical wavevector. These matrix elements are generally nonzero. However, for a M1 transition with a microwave field at GHz frequencies the relevant matrix element between position eigenstates is negligible since the photon momentum is much less than the center of mass momentum. Therefore microwaves cannot be used to directly change the motional state. As a side remark we note that this limitation can be overcome by spatially displacing the harmonic oscillator states of different n [48].

Let us now extend the previous description of Rabi oscillations to include the motional states. The Hamiltonian can be written as $\hat{\mathcal{H}} = \hat{\mathcal{H}}_0 + \hat{\mathcal{H}}_{\text{int}}$ with $\hat{\mathcal{H}}_0$ accounting for the motional and qubit terms

$$\hat{\mathcal{H}}_0 = \hat{\mathcal{H}}_z + \hat{\mathcal{H}}_a = \hbar\omega_z(1/2 + \hat{a}^\dagger \hat{a}) + \frac{\hbar\omega_q}{2}\hat{\sigma}_z.$$

The phonon field is described by $\hat{\mathcal{H}}_z$ with $\hat{a}, \hat{a}^\dagger$ bosonic annihilation and creation operators. In the following we will drop the constant offset of $\hbar\omega_z/2$. The qubit energies are described by $\hat{\mathcal{H}}_a$ with $\hat{\sigma}_z = |0\rangle\langle 0| - |1\rangle\langle 1|$ acting on the qubit degrees of freedom and the zero of energy set to the midpoint between the levels.

The interaction Hamiltonian can be written as

$$\hat{\mathcal{H}}_{\text{int}} = -i\frac{\hbar\Omega}{2}e^{-i\omega t}e^{ik\cos(\theta)\hat{z}}e^{-i\phi}\hat{\sigma}_+ + \text{H.c.}$$

In this expression Ω is the Rabi frequency, the optical field propagates along $\mathbf{k} = k\cos(\theta)\mathbf{e}_z$ with k the wavenumber and θ the angle with respect to the $\mathbf{e}_z$ unit vector, $\omega = kc$ is the field frequency, ϕ is the phase of the field at $z = 0$ where the ion is located, $\hat{z}$ is the position operator and $\hat{\sigma}_+ = |e\rangle\langle g|$ is the atomic raising operator. This Hamiltonian is a semi-classical approximation where we have quantized the qubit levels and the motional state, but treat the electromagnetic field amplitude as a classical quantity. This is a good approximation since the number of photons interacting with the ion during a pulse is very large. Concerns about the emergence of undesired ion-field entanglement during a pulse have been considered by several authors [49, 50] who have shown that the effect is negligible for realistic experimental parameters.

The part of the interaction Hamiltonian that couples to the ion center of mass is $e^{ik\cos(\theta)\hat{z}}$. As has been explained above this term is present for optical frequency transitions. A similar term also appears for qubits encoded in hyperfine states of the same electronic level when the coupling is mediated by a two-photon Raman process. The position operator is given by $\hat{z} = z_0(\hat{a} + \hat{a}^\dagger)$ with $z_0 = \sqrt{\hbar/(2m\omega_z)}$ the harmonic oscillator length scale. The position dependent term in the interaction Hamiltonian can then be written as $e^{ik\cos(\theta)\hat{z}} = e^{i\eta(\hat{a}+\hat{a}^\dagger)}$ with $\eta = kz_0\cos(\theta) = \eta_0\cos(\theta)$. The parameter $\eta_0 = kz_0 = 2\pi z_0/\lambda$ with λ the transition wavelength is known as the Lamb–Dicke parameter. It quantifies the extent of the atomic wavefunction relative to the wavelength of the transition. As we will see the simplest form of trapped ion entangling gates assumes a small Lamb–Dicke parameter.

With these definitions we can rewrite the exponential in the form $e^{ik\cos(\theta)\hat{z}} = e^{-\eta^2/2}e^{i\eta\hat{a}^\dagger}e^{i\eta\hat{a}}$ where we have used the operator identity $e^{\hat{A}+\hat{B}} = e^{\hat{A}}e^{\hat{B}}e^{-[\hat{A},\hat{B}]/2}$ which is true provided $[\hat{A}, [\hat{A}, \hat{B}]] = [\hat{B}, [\hat{A}, \hat{B}]] = 0$. Expanding the exponentials and rearranging we find

$$
e^{ik\cos(\theta)\hat{z}} = e^{-\eta^2/2} \sum_{m,n=0}^{\infty} (i\eta)^{m+n} \frac{(\hat{a}^\dagger)^m}{m!} \frac{(\hat{a})^n}{n!}
$$

$$
= e^{-\eta^2/2} \sum_{m=0}^{\infty} (i\eta)^{2m} \frac{(\hat{a}^\dagger)^m}{m!} \frac{(\hat{a})^m}{m!} \qquad \text{(carrier)}
$$

$$
+ e^{-\eta^2/2} \sum_{n=0}^{\infty} \sum_{m>n} (i\eta)^{m+n} \frac{(\hat{a}^\dagger)^m}{m!} \frac{(\hat{a})^n}{n!} \qquad \text{(blue sidebands)}
$$

$$
+ e^{-\eta^2/2} \sum_{n=1}^{\infty} \sum_{m<n} (i\eta)^{m+n} \frac{(\hat{a}^\dagger)^m}{m!} \frac{(\hat{a})^n}{n!}. \qquad \text{(red sidebands)}.
$$

The carrier interaction does not change the vibrational quantum number, the blue sidebands add $m-n$ phonons and the red sidebands remove $n-m$ phonons.

When $\eta \ll 1$ only first order motional sidebands are nonnegligible and we can write the interaction Hamiltonian as

$$
\hat{\mathcal{H}}_{\text{int}} = -i\frac{\hbar\Omega}{2}e^{-i\omega t}e^{-i\phi}\hat{\sigma}_+ \left(1 + i\eta\hat{a} + i\eta\hat{a}^\dagger\right) + i\frac{\hbar\Omega^*}{2}e^{i\omega t}e^{i\phi}\hat{\sigma}_- \left(1 - i\eta\hat{a} - i\eta\hat{a}^\dagger\right).
$$

The Hamiltonian can be decomposed into parts corresponding to different motional state changes as $\hat{\mathcal{H}}_{\text{int}} = \hat{\mathcal{H}}_{\text{int},0} + \hat{\mathcal{H}}_{\text{int},1} + \hat{\mathcal{H}}_{\text{int},-1}$ with

$$
\hat{\mathcal{H}}_{\text{int},0} = -i\frac{\hbar}{2}(\Omega e^{-i\omega t}e^{-i\phi}\hat{\sigma}_+ - \Omega^*e^{i\omega t}e^{i\phi}\hat{\sigma}_-),
$$

$$
\hat{\mathcal{H}}_{\text{int},1} = \frac{\hbar\eta}{2}(\Omega e^{-i\omega t}e^{-i\phi}\hat{a}^\dagger\hat{\sigma}_+ - \Omega^*e^{i\omega t}e^{i\phi}\hat{a}\hat{\sigma}_-),
$$

$$
\hat{\mathcal{H}}_{\text{int},-1} = \frac{\hbar\eta}{2}(\Omega e^{-i\omega t}e^{-i\phi}\hat{a}\hat{\sigma}_+ - \Omega^*e^{i\omega t}e^{i\phi}\hat{a}^\dagger\hat{\sigma}_-).
$$

The three terms correspond to carrier, blue-sideband, and red-sideband transitions. Note that the carrier transition has Rabi frequency Ω whereas the sideband transitions have Rabi frequency $\eta\Omega$ which for small η is much slower.

A small Lamb–Dicke parameter not only facilitates separation of the interaction into resolved carrier and sideband transitions which is important for gate operations, but also enables cooling of the ion to its motional ground state. To see this we note that the Lamb–Dicke parameter can also be expressed as

$$\eta_0 = \sqrt{E_R/E_{h.o.}}$$

where $E_R = \hbar^2 k^2/2m$ is the photon recoil energy and $E_{h.o.} = \hbar\omega_z$ is the harmonic oscillator energy level spacing. When $\eta_0 \ll 1$, which corresponds to $E_R \ll E_{h.o.}$, absorption or emission of a photon does not provide sufficient momentum exchange to change the motional state of the trapped ion. This implies that photons can be scattered without heating. Furthermore, in this limit, ions can be cooled to their motional ground state using a method known as sideband cooling [47].

Suppose we start with an ion in $|g, n\rangle$ and we drive a π pulse on the red sideband to $|e, n - 1\rangle$. The ion will then spontaneously decay back to g while emitting a photon. Provided $\eta_0 \ll 1$ the motional state cannot change so the ion decays to $|g, n - 1\rangle$. In ions with more than one ground state it may be necessary to repump back to $|g\rangle$ before the next π pulse. Repeating this process n times the ion will end up in $|g, 0\rangle$ from which there is no red-sideband transition. In this way the ion can be prepared in the ground state [51] which is a prerequisite for the Cirac–Zoller CNOT gate protocol.

13.9 Two-Qubit Gates

One-qubit gates can be performed using microwave or optical fields analogously to what has been discussed for neutral atom qubits. Two-qubit gates are in contrast very different for trapped ions compared to neutral atom qubits. We will discuss two gate protocols. The Cirac–Zoller (CZ) gate [52], which was the original proposal for a two-qubit entangling gate, requires that the ion is prepared in the motional ground state in order to achieve high fidelity. The Mølmer–Sørensen (MS) gate [53] relaxes this requirement, and can be used for high fidelity entanglement with ions in thermal motion. The MS gate is also of interest for preparing multi-qubit entangled states [54].

13.9.1 Cirac–Zoller Gate

With the preliminaries of coupling between electronic and motional degrees of freedom out of the way we are ready to show how two ions can be entangled following the seminal proposal of Cirac and Zoller [52]. The first step is to laser cool the ions to the motional ground state and optically pump them into the state $|g\rangle$. The joint state of the ions is then

$$|\psi\rangle = |gg; 0\rangle.$$

For simplicity we assume only one of the axial motional modes plays a role. We then shine a laser on the first ion and perform a $\pi/2$ pulse on the blue sideband which results in the state[1]

$$|\psi\rangle = \frac{|gg; 0\rangle + |eg; 1\rangle}{\sqrt{2}}.$$

Then apply a π pulse on the carrier transition of ion 2 to get

$$|\psi\rangle = \frac{|ge; 0\rangle + |ee; 1\rangle}{\sqrt{2}}.$$

Then apply a π pulse on the blue-sideband transition of ion 2 to get

$$|\psi\rangle = \frac{|ge; 0\rangle + |eg; 0\rangle}{\sqrt{2}} = \frac{|ge\rangle + |eg\rangle}{\sqrt{2}}|0\rangle.$$

Crucially the blue-sideband pulse does not affect $|ge; 0\rangle$ since there is no $n = -1$ state. The final state is a maximally entangled Bell state in the spin basis and is disentangled from the motional state, although the coupling that creates entanglement is mediated by the collective motional state. A full CNOT gate can be implemented with the same ideas but a more complicated pulse sequence. A CNOT gate between spin and motional degrees of freedom was first demonstrated in Boulder in 1995 [55] and the full CZ protocol for a CNOT gate was demonstrated in Innsbruck in 2003 [56]. Note that the gate is intrinsically relatively slow since the Rabi frequency of the sideband transition scales as $\eta\Omega$ and we need η to be small for preparation of the ground state and to effectively suppress higher order carrier and sideband transitions. Typical experimental gate times are $\sim$100 μs.

13.9.2 Mølmer–Sørensen Gate

The CZ gate is of historical importance, but has some drawbacks. In order for the protocol to succeed in creating entanglement the ions must be prepared in the motional ground state. An ion in motional state n has matrix elements

---

[1] In this and subsequent expressions we have suppressed phase factors associated with $\pi/2$ and π rotations.

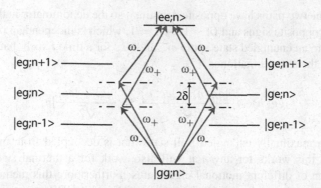

Fig. 13.10 The MS gate uses a bichromatic laser field with frequencies ω_+, ω_- to couple ions from the $|gg\rangle$ to the $|ee\rangle$ state with one-photon carrier detuning δ

$\langle n+1|\hat{a}^\dagger|n\rangle = \sqrt{n+1}$, $\langle n-1|\hat{a}|n\rangle = \sqrt{n}$, which are n dependent and therefore not well defined for ions in a thermal state. It is difficult to prepare the motional ground state with 100% efficiency and any heating mechanisms that are present, for example due to sensitivity of the ions to patch potentials on the confining electrodes, will put the ions in a thermal state of motion.

The MS gate solves this issue through an ingenious use of multi-photon transitions. Consider the level diagram of Fig. 13.10. Two ions are simultaneously illuminated by a bichromatic field with frequencies ω_+, ω_-. We choose $\omega_\pm = \omega_c \pm \delta$, ω_c is the carrier frequency, and $|\eta\Omega| \ll \omega_z - \delta$. As is seen in the figure there are four ways in which the two ions can absorb two photons and conserve energy to make the transition $|gg; n\rangle \rightarrow |ee; n\rangle$. The effective Rabi rate allowing for all energy conserving paths is found from second-order perturbation theory to be

$$\tilde{\Omega} = 2\sum_{|j\rangle} \frac{\langle ee; n|\hat{\mathcal{H}}_{\text{int}}|j\rangle\langle j|\hat{\mathcal{H}}_{\text{int}}|gg; n\rangle}{U_{|j\rangle} - (U_{|gg;n\rangle} + \hbar\omega_\pm)}$$

with $|j\rangle = |eg; n \pm 1\rangle, |ge; n \pm 1\rangle$. Summing over the intermediate states we find the remarkable result

$$\tilde{\Omega} = -\frac{(\eta\Omega)^2}{\omega_z - \delta}.$$

The coupling rate is independent of the motional excitation n so Rabi pulses can be accurately applied without knowing n.

We can see the origin of this result by noting that for the transition $|gg; n\rangle \rightarrow |eg; n-1\rangle \rightarrow |ee; n\rangle$ we pick up a factor of $\sqrt{n} \times \sqrt{n} = n$ from the harmonic oscillator matrix elements. For the transition $|gg; n\rangle \rightarrow |eg; n+1\rangle \rightarrow |ee; n\rangle$ we pick up a factor of $\sqrt{n+1} \times \sqrt{n+1} = n+1$ from the harmonic oscillator matrix elements. The two paths have opposite detunings so the denominator in the sum that gives $\tilde{\Omega}$ has opposite signs and $(n + 1) - n = 1$, which is independent of n.

To prepare an entangled state apply ω_+ and ω_- for a time t such that $\tilde{\Omega}t = \pi/2$ which gives the transformation

$$|gg; n\rangle \rightarrow \frac{|gg; n\rangle + |ee; n\rangle}{\sqrt{2}} = \frac{|gg\rangle + |ee\rangle}{\sqrt{2}} |n\rangle.$$

We obtain a maximally entangled Bell state that is decoupled from the motional state. Since this works for any n it will also work for a thermal state that is a superposition of different motional eigenstates. Furthermore this method generalizes in a straightforward way to preparation of N qubit GHZ states $(|gg \ldots g\rangle + |ee \ldots e\rangle)/\sqrt{N}$ which has been demonstrated for $N = 14$ [57].

13.10   Status and Outlook

Trapped ion quantum computing has established records for high fidelity logic operations and continues to be actively developed. Gate model quantum algorithms have been run with up to 11 individually controlled qubits in Paul traps [58] and analog simulators with hundreds of qubits in Penning traps [42]. The biggest challenge facing the trapped ion approach is scaling to large qubit numbers. As mentioned in Sect. 13.7.1 when many ions are confined in a single Paul trap the normal mode spectrum becomes very dense. In order to couple to a desired subset of modes the Rabi frequency should be reduced, which implies slow gates, and more errors due to finite coherence time. Although there is no absolute limit, it is generally assumed that scaling beyond a few hundred ions in a single Paul trap will not be possible, and even that number is well beyond demonstrated capabilities.

There are two primary architectures that are being developed to surpass the limit of a single Paul trap. One is based on mechanical shuttling of ions between different regions of a larger system containing multiple segmented traps [59]. By applying slowly varying voltages to judiciously placed electrodes ions are moved in and out of regions where gates are performed. Only a small number of ions share the motional bus for each gate operation which bypasses the multiple-ion scaling limit. The challenge is in the complexity of the trap system including electrodes for motion, and the fact that mechanical motion tends to be relatively slow. Nevertheless designs have been proposed for very large scale systems based on this architecture [60].

Another approach to scaling is based on a modular architecture where moderately sized qubit registers, each in a single Paul trap, are connected using photonic links that establish entanglement between modules [61]. This is achieved by exciting an ion to a state which can decay along different paths emitting a photon with a polarization that is correlated with the decay path. In this way ion-photon entanglement is prepared of the form $|\psi\rangle \sim |g_1, \epsilon_1\rangle + |g_2, \epsilon_2\rangle$ where $g_{1,2}$ are atomic states and $\epsilon_{1,2}$ are optical polarization states. This is done independently on ions in separate modules which prepares the joint state

$$|\psi\rangle_{ab} \sim \left(|g_1, \epsilon_1\rangle_a + |g_2, \epsilon_2\rangle_a\right) \left(|g_1, \epsilon_1\rangle_b + |g_2, \epsilon_2\rangle_b\right)$$

where a,b refer to the two modules. The photons are then interfered on a beamsplitter followed by photodetection which serves as a probabilistic Bell state analyzer via the Hong–Ou–Mandel effect (see Sect. 14.1). Detection of the photons projects the remote ions into a Bell state via entanglement swapping (see Sect. 3.4.3) which establishes a shared resource that can be used for state teleportation between modules.

Optical crossbars that facilitate switching photons between many different optical fibers have been developed for optical fiber networks and could be used to connect a large number of modules. A primary challenge of this approach, somewhat akin to the challenge of the segmented trap architecture, is that the photonic entanglement links are slow. Recent results [62] have demonstrated intermodule entanglement rates of $180\,\text{s}^{-1}$, which is promising. As with other quantum technologies further development of trapped-ion based architectures is an active area of research and engineering that will undoubtedly lead to further progress in the coming years.

13.11 Problems

1. Using the expressions given in the text for the error of qubit measurements using light scattering calculate the following quantities.

   (a) Assuming $r_1 = 5000.\,\text{s}^{-1}$, $r_b = 200.\,\text{s}^{-1}$, and $t = 1\,\text{ms}$ find the optimal value of n_c which minimizes the error when $a = b = 1/\sqrt{2}$.
   (b) Using the optimal value of n_c from part (a) what is the probability of making an error for the three cases: $(a = 1, b = 0)$, $(a = 0, b = 1)$, and $(a = b = 1/\sqrt{2})$.
   (c) Using $r_1 = 5000.\,\text{s}^{-1}$, $r_b = 200.\,\text{s}^{-1}$, how large must t be, and what is the corresponding value of n_c to make the measurement error $< 10^{-4}$ for $a = b = 1/\sqrt{2}$.

2. The Cirac–Zoller sequence of three pulses used to create a two-ion Bell state is:

   • ion 1, $\pi/2$ pulse on blue sideband,
   • ion 2, π pulse on carrier,
   • ion 2, π pulse on blue sideband.

   (a) Assume that each pulse is implemented as a two-photon transition on a hyperfine state encoded qubit using two-frequency Raman light detuned by Δ from an excited state with lifetime τ. The two-photon Rabi frequency for transitions on the red or blue sidebands is $\Omega_{sb} = \eta^2 \Omega_1^2/(2\Delta)$ where Ω_1 is the one-photon Rabi frequency and η is the Lamb–Dicke parameter. For carrier transitions use $\Omega_c = \Omega_1^2/(2\Delta)$. Calculate P_{se} the probability of spontaneous emission of a photon from the excited state at the end of the three pulse sequence. Express your answer as a function of Δ, τ assuming $P_{se} \ll 1$.

(b) Due to a variety of effects each hyperfine qubit has a coherence that decays. We can model the decay as $C(t) = C(0)e^{-t/T_2}$ where $C(t)$ is the qubit coherence and T_2 is the coherence time. We can approximate the fidelity with which a Bell state can be prepared using a pulse sequence of total duration $T_{Bell} = 3\pi/(2\Omega_{sb}) + \pi/\Omega_c$ as

$$F = C^2(T_{Bells})(1 - P_{se}).$$

Find an expression for Δ that maximizes F. Your answer should depend on Ω_1, τ, T_2.

(c) Assume $\eta = 0.1$, the one-photon Rabi frequency $\Omega_1 = d\mathcal{E}/\hbar$ where $d = ea_0$, e is the electronic charge, a_0 is the Bohr radius, $\mathcal{E}$ is the electric field amplitude and the optical beam is a Gaussian beam focused to a waist (radius to $1/e^2$ intensity point) of $w = 5\,\mu m$. The poor quantum scientist can only afford a laser with a power output of $P = 1\,mW$. Using the optimal value of Δ from part (b) and $T_2 = 1\,s$, $\tau = 10\,ns$, find the value of F and the corresponding gate time.

3. A two-qubit $CNOT$ gate between two neutral atom qubits can be implemented using a C_X operation instead of HC_ZH.

(a) Devise a pulse sequence that does this.
(b) Estimate the success probability of the Rydberg blockade C_X gate, averaged over the four computational basis states, assuming $\Omega \ll \omega_q$, $B \gg \Omega$ and $1/\Omega \ll \tau$. Here Ω is the ground—Rydberg excitation Rabi frequency, ω_q is the frequency separation of the qubit states, B is the blockade shift betwen Rydberg states, and τ is the lifetime of the Rydberg state. You may assume that spontaneous decay from a Rydberg state results in an error.

4. In some quantum computing implementations, such as neutral atoms or ions or photons, qubits can be lost from the circuit. Qubit loss can be diagnosed nondestructively, and without disturbing the qubit state, using one additional ancilla (which we assume has been verified to be present in advance), Pauli and CNOT gates and a single measurement of a qubit state.

(a) Can you devise a circuit which does this? Hint: You may assume that if a qubit is missing a CNOT gate between the qubit and a target acts as though the control qubit were in state $|0\rangle$.
(b) Assuming that all the gates in your circuit and the measurement operation have fidelity 0.99 estimate with what success probability the loss of a qubit can be detected.

References

1. M. Saffman, T.G. Walker, K. Mølmer, Quantum information with Rydberg atoms. Rev. Mod. Phys. **82**, 2313 (2010)
2. A. Cooper, J.P. Covey, I.S. Madjarov, S.G. Porsev, M.S. Safronova, M. Endres, Alkaline-earth atoms in optical tweezers. Phys. Rev. X **8**, 041055 (2018)
3. S. Saskin, J.T. Wilson, B. Grinkemeyer, J.D. Thompson, Narrow-line cooling and imaging of Ytterbium atoms in an optical tweezer array. Phys. Rev. Lett. **122**, 143002 (2019)
4. P. van der Straten, H.J. Metcalf, *Atoms and Molecules Interacting with Light: Atomic Physics for the Laser Era* (Cambridge University Press, Cambridge, 2016)
5. E.L. Raab, M. Prentiss, A. Cable, S. Chu, D.E. Pritchard, Trapping of neutral sodium atoms with radiation pressure. Phys. Rev. Lett. **59**, 2631 (1987)
6. A.L. La Rooij, H.B. V.L. van den Heuvell, R.J.C. Spreeuw, Designs of magnetic atom-trap lattices for quantum simulation experiments. Phys. Rev. A **99**, 022303 (2019)
7. M.J. Piotrowicz, M. Lichtman, K. Maller, G. Li, S. Zhang, L. Isenhower, M. Saffman, Two-dimensional lattice of blue-detuned atom traps using a projected Gaussian beam array. Phys. Rev. A **88**, 013420 (2013)
8. Y. Wang, A. Kumar, T.-Y. Wu, D.S. Weiss, Single-qubit gates based on targeted phase shifts in a 3D neutral atom array. Science **352**, 1562 (2016)
9. N. Schlosser, G. Reymond, I. Protsenko, P. Grangier, Sub-Poissonian loading of single atoms in a microscopic dipole trap. Nature (London) **411**, 1024 (2001)
10. M. Endres, H. Bernien, A. Keesling, H. Levine, E.R. Anschuetz, A. Krajenbrink, C. Senko, V. Vuletic, M. Greiner, M.D. Lukin, Atom-by-atom assembly of defect-free one-dimensional cold atom arrays. Science **354**, 1024 (2016)
11. H. Kim, W. Lee, H.G.Lee, H. Jo, Y. Song, J. Ahn, In situ single-atom array synthesis using dynamic holographic optical tweezers. Nat. Commun. **7**, 13317 (2016)
12. D. Barredo, S. de Leséléuc, V. Lienhard, T. Lahaye, A. Browaeys, An atom-by-atom assembler of defect-free arbitrary two-dimensional atomic arrays. Science **354**, 1021 (2016)
13. D. Ohl de Mello, D. Schäffner, J. Werkmann, T. Preuschoff, L. Kohfahl, M. Schlosser, G. Birkl, Defect-free assembly of 2D clusters of more than 100 single-atom quantum systems. Phys. Rev. Lett. **122**, 203601 (2019)
14. M. Saffman, Quantum computing with atomic qubits and Rydberg interactions: Progress and challenges. J. Phys. B **49**, 202001 (2016)
15. S.E. Hamann, D.L. Haycock, G. Klose, P.H. Pax, I.H. Deutsch, P.S. Jessen, Resolved-sideband Raman cooling to the ground state of an optical lattice. Phys. Rev. Lett. **80**, 4149–4152 (1998)
16. A.J. Daley, M.M. Boyd, J. Ye, P. Zoller, Quantum computing with alkaline-earth-metal atoms. Phys. Rev. Lett. **101**, 170504 (2008)
17. Z. Hu, H.J. Kimble, Observation of a single atom in a magneto-optical trap. Opt. Lett. **19**, 1888 (1994)
18. A. Fuhrmanek, R. Bourgain, Y.R.P. Sortais, A. Browaeys, Free-space lossless state detection of a single trapped atom. Phys. Rev. Lett. **106**, 133003 (2011)
19. H. Uys, M.J. Biercuk, A.P. VanDevender, C. Ospelkaus, D. Meiser, R. Ozeri, J.J. Bollinger, Decoherence due to elastic Rayleigh scattering. Phys. Rev. Lett. **105**, 200401 (2010)
20. A.W. Carr, M. Saffman, Doubly magic optical trapping for Cs atom hyperfine clock transitions. Phys. Rev. Lett. **117**, 150801 (2016)
21. S. Kuhr, W. Alt, D. Schrader, I. Dotsenko, Y. Miroshnychenko, A. Rauschenbeutel, D. Meschede, Analysis of dephasing mechanisms in a standing-wave dipole trap. Phys. Rev. A **72**, 023406 (2005)
22. T. Xia, M. Lichtman, K. Maller, A.W. Carr, M.J. Piotrowicz, L. Isenhower, M. Saffman, Randomized benchmarking of single-qubit gates in a 2D array of neutral-atom qubits. Phys. Rev. Lett. **114**, 100503 (2015)

23. C. Sheng, X. He, P. Xu, R. Guo, K. Wang, Z. Xiong, M. Liu, J. Wang, M. Zhan, High-fidelity single-qubit gates on neutral atoms in a two-dimensional magic-intensity optical dipole trap array. Phys. Rev. Lett. **121**, 240501 (2018)

24. C. Knoernschild, X.L. Zhang, L. Isenhower, A.T. Gill, F.P. Lu, M. Saffman, J. Kim, Independent individual addressing of multiple neutral atom qubits with a MEMS beam steering system. Appl. Phys. Lett. **97**, 134101 (2010)

25. T. Pellizzari, S.A. Gardiner, J.I. Cirac, P. Zoller, Decoherence, continuous observation, and quantum computing: A cavity QED model. Phys. Rev. Lett. **75**, 3788 (1995)

26. D. Jaksch, H.-J. Briegel, J.I. Cirac, C.W. Gardiner, P. Zoller, Entanglement of atoms via cold controlled collisions. Phys. Rev. Lett. **82**, 1975 (1999)

27. G.K. Brennen, C.M. Caves, P.S. Jessen, I.H. Deutsch, Quantum logic gates in optical lattices. Phys. Rev. Lett. **82**, 1060 (1999)

28. D. Jaksch, J.I. Cirac, P. Zoller, S.L. Rolston, R. Côté, M.D. Lukin, Fast quantum gates for neutral atoms. Phys. Rev. Lett. **85**, 2208–2211 (2000)

29. M.D. Lukin, M. Fleischhauer, R. Cote, L.M. Duan, D. Jaksch, J.I. Cirac, P. Zoller, Dipole blockade and quantum information processing in mesoscopic atomic ensembles. Phys. Rev. Lett. **87**, 037901 (2001)

30. X.L. Zhang, A.T. Gill, L. Isenhower, T.G. Walker, M. Saffman, Fidelity of a Rydberg blockade quantum gate from simulated quantum process tomography. Phys. Rev. A **85**, 042310 (2012)

31. H. Levine, A. Keesling, G. Semeghini, A. Omran, T.T. Wang, S. Ebadi, H. Bernien, M. Greiner, V. Vuletić, H. Pichler, M.D. Lukin, Parallel implementation of high-fidelity multiqubit gates with neutral atoms. Phys. Rev. Lett. **123**, 170503 (2019)

32. L.S. Theis, F. Motzoi, F.K. Wilhelm, M. Saffman, A high fidelity Rydberg blockade entangling gate using shaped, analytic pulses. Phys. Rev. A **94**, 032306 (2016)

33. T. Graham, M. Kwon, B. Grinkemeyer, A. Marra, X. Jiang, M. Lichtman, Y. Sun, M. Ebert, M. Saffman, Rydberg mediated entanglement in a two-dimensional neutral atom qubit array. Phys. Rev. Lett. **123**, 230501 (2019)

34. C. Gross, I. Bloch, Quantum simulations with ultracold atoms in optical lattices. Science **357**, 995 (2017)

35. M. Ebert, M. Kwon, T.G. Walker, M. Saffman, Coherence and Rydberg blockade of atomic ensemble qubits. Phys. Rev. Lett. **115**, 093601 (2015)

36. J. Zeiher, P. Schauß, S. Hild, T. Macrí, I. Bloch, C. Gross, Microscopic characterization of scalable coherent Rydberg superatoms. Phys. Rev. X **5**, 031015 (2015)

37. D. Leibfried, R. Blatt, C. Monroe, D. Wineland, Quantum dynamics of single trapped ions. Rev. Mod. Phys. **75**, 281 (2003)

38. C.D. Bruzewicz, J. Chiaverini, R. McConnell, J.M. Sage, Trapped-ion quantum computing: progress and challenges. Appl. Phys. Rev. **6**, 021314 (2019)

39. J.P. Gaebler, T.R. Tan, Y. Lin, Y. Wan, R. Bowler, A.C. Keith, S. Glancy, K. Coakley, E. Knill, D. Leibfried, D.J. Wineland, High-fidelity universal gate set for $^9\text{Be}^+$ ion qubits. Phys. Rev. Lett. **117**, 060505 (2016)

40. C.J. Ballance, T.P. Harty, N.M. Linke, M.A. Sepiol, D.M. Lucas, High-fidelity quantum logic gates using trapped-ion hyperfine qubits. Phys. Rev. Lett. **117**, 060504 (2016)

41. H.G. Dehmelt, Radiofrequency spectroscopy of stored ions I: storage. Adv. Atom. Mol. Phys. **3**, 53 (1968)

42. J.W. Britton, B.C. Sawyer, A.C. Keith, C.-C.J. Wang, J.K. Freericks, H. Uys, M.J. Biercuk, J.J. Bollinger, Engineered two-dimensional Ising interactions in a trapped-ion quantum simulator with hundreds of spins. Nature **484**, 489 (2012)

43. W. Paul, H. Steinwedel, Ein neues massenspektrometer ohne magnetfeld. Z. Naturforsch. A **8**, 448 (1953)

44. W. Paul, Electromagnetic traps for charged and neutral particles. Rev. Mod. Phys. **62**, 531 (1990)

45. D.F.V. James, Quantum dynamics of cold trapped ions with application to quantum computation. Appl. Phys. B **66**, 181 (1998)

46. I.I. Sobelman, *Theory of Atomic Spectra* (Alpha Science International, Oxford, 2006)
47. D.J. Wineland, W.M. Itano, Laser cooling of atoms. Phys. Rev. A **20**, 1521–1540 (1979)
48. L. Förster, M. Karski, J.-M. Choi, A. Steffen, W. Alt, D. Meschede, A. Widera, E. Montano, J.H. Lee, W. Rakreungdet, P.S. Jessen, Microwave control of atomic motion in optical lattices. Phys. Rev. Lett. **103**, 233001 (2009)
49. S.J. van Enk, H.J. Kimble, On the classical character of control fields in quantum information processing. Quant. Inf. Comput. **2**, 1 (2002)
50. A. Silberfarb, I.H. Deutsch, Entanglement generated between a single atom and a laser pulse. Phys. Rev. A **69**, 042308 (2004)
51. F. Diedrich, J.C. Bergquist, W.M. Itano, D.J. Wineland, Laser cooling to the zero-point energy of motion. Phys. Rev. Lett. **62**, 403–406 (1989)
52. J.I. Cirac, P. Zoller, Quantum computations with cold trapped ions. Phys. Rev. Lett. **74**, 4091–4094 (1995)
53. A. Sørensen, K. Mølmer, Quantum computation with ions in thermal motion. Phys. Rev. Lett. **82**, 1971–1974 (1999)
54. K. Mølmer, A. Sørensen, Multiparticle entanglement of hot trapped ions. Phys. Rev. Lett. **82**, 1835–1838 (1999)
55. C. Monroe, D.M. Meekhof, B.E. King, W.M. Itano, D.J. Wineland, Demonstration of a fundamental quantum logic gate. Phys. Rev. Lett. **75**, 4714 (1995)
56. F. Schmidt-Kaler, H. Häffner, M. Riebe, S. Gulde, G.P.T. Lancaster, T. Deuschle, C. Becher, C.F. Roos, J. Eschner, R. Blatt, Realization of the Cirac–Zoller controlled—NOT quantum gate. Nature (London) **422**, 408 (2003)
57. T. Monz, P. Schindler, J.T. Barreiro, M. Chwalla, D. Nigg, W.A. Coish, M. Harlander, W. Hänsel, M. Hennrich, R. Blatt, 14-qubit entanglement: Creation and coherence. Phys. Rev. Lett. **106**, 130506 (2011)
58. K. Wright, K.M. Beck, S. Debnath, J.M. Amini, Y. Nam, N. Grzesiak, J.-S. Chen, N.C. Pisenti, M. Chmielewski, C. Collins, K.M. Hudek, J. Mizrahi, J.D. Wong-Campos, S. Allen, J. Apisdorf, P. Solomon, M. Williams, A.M. Ducore, A. Blinov, S.M. Kreikemeier, V. Chaplin, M. Keesan, C. Monroe, J. Kim, Benchmarking an 11-qubit quantum computer. Nat. Commun. **10**, 5464 (2019)
59. D. Kielpinski, C. Monroe, D.J. Wineland, Architecture for a large-scale ion-trap quantum computer. Nature **417**, 709–711 (2002)
60. B. Lekitsch, S. Weidt, A.G. Fowler, K. Mølmer, S.J. Devitt, C. Wunderlich, W.K. Hensinger, Blueprint for a microwave trapped ion quantum computer. Sci. Adv. **3**, 1601540 (2017)
61. C. Monroe, R. Raussendorf, A. Ruthven, K.R. Brown, P. Maunz, L.-M. Duan, J. Kim, Large-scale modular quantum-computer architecture with atomic memory and photonic interconnects. Phys. Rev. A **89**, 022317 (2014)
62. L.J. Stephenson, D.P. Nadlinger, B.C. Nichol, S. An, P. Drmota, T.G. Ballance, K. Thirumalai, J.F. Goodwin, D.M. Lucas, C.J. Ballance, High-rate, high-fidelity entanglement of qubits across an elementary quantum network. Phys. Rev. Lett. **124**, 110501 (2020)

Chapter 14
Optical Qubits

Photons are the only viable approach for long distance transmission of quantum information. A capability that has been impressively demonstrated with the distribution of quantum keys and entanglement via satellite [1, 2]. It is therefore natural to ask if qubits encoded in photons could also be used for quantum computation [3].

Optical qubits have some attractive features. Photons are Bosonic spin 1 particles that provide degrees of freedom suitable for encoding qubits. The transverse polarization states of a photon, for example horizontal and vertical, or right and left circular, provide a perfectly useful qubit basis. We may use a polarization state encoding $|0\rangle = |H\rangle$, $|1\rangle = |V\rangle$. Other degrees of freedom such as the spatial path can also be used. Since the polarization state can be controlled with standard optical components one-qubit gates are very simple to implement with high fidelity on polarization encoded qubits. In isotropic media photons propagate without changes to their polarization so the T_1 coherence time can be long. Similarly the propagation phase is independent of the polarization state so also the T_2 coherence time can be long. Free space propagation is not well suited for practical implementations. Instead, highly integrated photonic circuits have been developed as shown in Fig. 14.1.

Other aspects of photonic qubits are much less convenient for quantum computing. To start with we do not yet have reliable and deterministic sources of identical single photons. Experiments incorporating a single quantum dot source combined with delay lines have been scaled to demonstrations with 20 simultaneously available photons [4]. Alternatively heralded sources are available that guarantee the presence of a photon based on detection of a second accompanying photon. In order to prepare a large register of photons at one time using heralded sources a memory is needed [5]. Due to the speed of light and the difficulty of building photon boxes with vanishingly low loss it is challenging to store photons in a small space for a long time. Photons can be measured, albeit most typically destructively, using specialized photon counting detectors, but the detection efficiency is generally less

© Springer Nature Switzerland AG 2021

J. A. Bergou et al., *Quantum Information Processing*, Graduate Texts in Physics,
https://doi.org/10.1007/978-3-030-75436-5_14

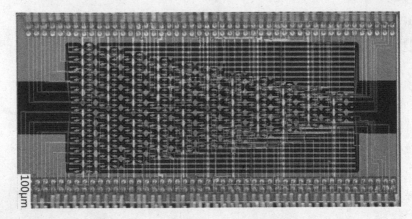

100μm

Fig. 14.1 Photograph of programmable multi-mode interferometer for photonic computation. Image courtesy of Nicholas Harris, Dirk Englund group, MIT

than 100% [6]. This is in contrast to atomic or some solid-state qubits for which quantum states can be measured with close to 100% fidelity.

Finally single photons do not interact with each other, so it is difficult to implement two-qubit gates. It may in fact seem to be impossible to implement a two-qubit gate between photonic qubits. There are two approaches to solving this challenge. Photons can effectively be made to interact via a measurement process. This leads to probabilistic two-photon gates. Surprisingly probabilistic gates can provide the basis for a scalable architecture based on gate teleportation known as KLM for Knill, Laflamme and Milburn [7]. Alternatively, although photons do not interact with each other in vacuum, they do interact in nonlinear optical media. If a nonlinearity can be developed that provides a π phase shift due to interaction of photons deterministic photonic gates will be possible. We will discuss both of these alternatives in this chapter.

14.1   Photons and Beamsplitters

Although a two-qubit gate between photons is difficult to achieve it is very simple to prepare entangled states of single photon qubits. All that is needed is a beamsplitter. A generic beam splitter is shown in Fig. 14.2. There are two input ports (0 and 1) and two output ports (2 and 3). Without making any assumptions about the internal structure of the beamsplitter, apart from the requirement that it is lossless, the transmission and reflection coefficients must satisfy reciprocity relations due to Stokes [8].

The reciprocity relations may be derived from the formal requirement that a lossless beamsplitter results in a unitary transformation of the fields. The scattering

Fig. 14.2 The beamsplitter with reflection and transmission coefficients r', t' and r, t for light incident in ports 0 and 1 respectively

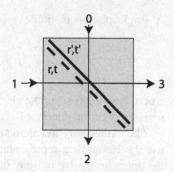

matrix of the beamsplitter is

$$S = \begin{pmatrix} t' & r \\ r' & t \end{pmatrix}$$

and unitarity means $SS^\dagger = S^\dagger S = I$. These conditions result in

$$|r|^2 + |t|^2 = 1, \; |r'|^2 + |t'|^2 = 1, \; |r|^2 + |t'|^2 = 1, \; |r'|^2 + |t|^2 = 1,$$

(14.1a)

$$rt'^* + r'^*t = 0, \quad rt^* + r'^*t' = 0.$$

(14.1b)

It follows that $|r| = |r'|, |t| = |t'|$ and defining $r = |r|e^{i\phi_r}, r' = |r'|e^{i\phi'_r}, t = |t|e^{i\phi_t}, t' = |t'|e^{i\phi'_t}$ leads to

$$\phi_r + \phi'_r - \phi_t - \phi'_t = (2m + 1)\pi$$

with m an integer. Different phase choices are possible. For example we can set $\phi_t = \phi'_t = 0$ so $t = t'$ and both quantities are positive. With this choice it follows that $\phi_r + \phi'_r = (2m + 1)\pi$, which gives $r = |r|e^{i\phi_r}$ and $r' = -|r|e^{-i\phi_r}$. A common choice for calculations is to take $\phi_r = 0$ so $r > 0$, $r' = -r$, and $t = t' > 0$.

The Fresnel coefficients r, t, r', t' refer to the effective behavior of the beamsplitter for a given state of polarization. Relations (14.1) are thus separately valid for both s and p polarizations. It is important to keep in mind that Eqs. (14.1) describe a situation where the index of refraction is the same at all external ports of the beamsplitter. If this were not the case, then we would not find e.g. $|t| = |t'|$.

The relations discussed so far are classical. How do we describe the propagation of single photons through a beamsplitter? To answer this question we need to first have a way of describing photons. The equations of classical electromagnetism can be quantized which reveals that the electromagnetic field can be thought of as an infinite set of uncoupled harmonic oscillator modes with frequencies ω_j.

A single photon in mode j is a quantized excitation of the field that we write as

$$|1\rangle_j = \hat{a}_j^\dagger |\text{vac}\rangle$$

where $|\text{vac}\rangle$ represents the electromagnetic vacuum with no excitations present and $\hat{a}_j^\dagger$ is the creation operator for mode j.

The easiest way to calculate the effect of a beamsplitter on single photons is to use the Heisenberg representation for which the quantum states are stationary but the operators evolve in time. In this formalism we can write the output modes as

$$\begin{pmatrix} \hat{a}_2 \\ \hat{a}_3 \end{pmatrix} = \hat{U} \begin{pmatrix} \hat{a}_0 \\ \hat{a}_1 \end{pmatrix} \tag{14.2}$$

Inverting (14.2) gives

$$\hat{a}_0 = t'^* \hat{a}_2 + r'^* \hat{a}_3$$
$$\hat{a}_1 = r^* \hat{a}_2 + t^* \hat{a}_3.$$

Suppose a photon is incident on port 0 and vacuum is incident on port 1. The input state is $|\psi\rangle_{\text{in}} = \hat{a}_0^\dagger |\text{vac}\rangle$ where $|\text{vac}\rangle$ represents vacuum in both input modes. The output state is

$$|\psi\rangle_{\text{out}} = t'|10\rangle_{23} + r'|01\rangle_{23}.$$

The notation $|ij\rangle_{23}$ represents i photons in mode 2 and j photons in mode 3. A 50:50 beamsplitter has $|t'| = |r'| = 1/\sqrt{2}$ so, apart from a beamsplitter dependent phase, the output is

$$|\psi\rangle_{\text{out}} = \frac{|10\rangle_{23} + |01\rangle_{23}}{\sqrt{2}}.$$

As expected the photon will be found at one of the two output ports with probability one half. In addition the output is in an entangled state and cannot be written as a product of the photon being in different modes.

A more interesting case occurs when one photon is present at both input ports. The input state is $|\psi\rangle_{\text{in}} = \hat{a}_0^\dagger \hat{a}_1^\dagger |\text{vac}\rangle$ and the output state is

$$|\psi\rangle_{\text{out}} = \hat{a}_0^\dagger \hat{a}_1^\dagger |\text{vac}\rangle = \sqrt{2}t'r|20\rangle_{23} + \sqrt{2}r't|02\rangle_{23} + (t't + r'r)|11\rangle_{23}.$$

It can be shown to follow from the Stokes relations that for any lossless 50:50 beamsplitter $t't + r'r = 0$ so the output is

$$|\psi\rangle_{\text{out}} = \frac{|20\rangle_{23} + e^{i\phi}|02\rangle_{23}}{\sqrt{2}}$$

where the phase ϕ depends on the beamsplitter construction. This is the celebrated Hong–Ou–Mandel effect [9] which says that both photons will always emerge from the same port. As we will see this effect is crucial for constructing a photonic two-qubit gate.

14.2 Probabilistic CNOT Gate

Using linear interference and single photon measurements we can construct an optical circuit that implements a CNOT gate with finite probability. Success of the circuit is heralded by detection of auxiliary photons. To set the stage for explaining how the CNOT gate works let's define two possible encoding schemes: polarization and dual-rail. For polarization encoding we may use horizontal and vertical polarization states and define $|0\rangle = |H\rangle, |1\rangle = |V\rangle$. An arbitrary state is

$$|\psi\rangle = \cos(\theta)|0\rangle + \sin(\theta)e^{i\phi}|1\rangle$$

which is the usual Bloch sphere representation of a pure state in terms of polar and azimuthal angles. Using optical waveplates we can construct arbitrary unitary transformations

$$\mathbf{U} = \begin{pmatrix} \cos(\theta) & ie^{-i\phi}\sin(\theta) \\ ie^{i\phi}\sin(\theta) & \cos(\theta) \end{pmatrix}.$$

An arbitrary state can be prepared starting with $|0\rangle$ and using

$$|\psi\rangle = \mathbf{U}|0\rangle = \begin{pmatrix} \cos(\theta) & ie^{-i\phi}\sin(\theta) \\ ie^{i\phi}\sin(\theta) & \cos(\theta) \end{pmatrix}\begin{pmatrix} 1 \\ 0 \end{pmatrix} = \cos(\theta)|0\rangle + \sin(\theta)e^{i\phi}|1\rangle.$$

For dual-rail encoding the qubit basis states correspond to which path the photon propagates on

$$- - -- \Longrightarrow$$

$$|0\rangle_{\text{dr}} \equiv$$

$$- - -- \longrightarrow$$

$$- - -- \longrightarrow$$

$$|1\rangle_{\text{dr}} \equiv$$

$$- - -- \Longrightarrow$$

Polarization encoding can be converted to dual-rail encoding using a beam splitter and a phase shifter. For example to prepare the state $|\psi\rangle$ we use the arrangement of Fig. 14.3. Propagating backwards through the same circuit transforms a dual-rail encoded qubit back into the polarization basis.

We will proceed to show how a CNOT gate can be constructed provided we have available a nonlinear sign gate. The nonlinear sign gate flips the sign of the two-photon part of the wavefunction. Suppose we have a state

$$|\psi\rangle = c_0|0\rangle + c_1|1\rangle + c_2|2\rangle$$

where $|i\rangle$ represents a state with i photons. The nonlinear sign gate acting on this state gives

$$U_{NS}|\psi\rangle = c_0|0\rangle + c_1|1\rangle - c_2|2\rangle.$$

This gate can be implemented probabilistically using the circuit in Fig. 14.4, as

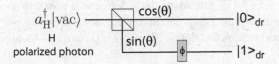

Fig. 14.3 Converting from polarization to dual-rail encoding to prepare the state $|\psi\rangle = \cos(\theta)|0\rangle + \sin(\theta)e^{i\phi}|1\rangle$ using a beamsplitter and a phase shift ϕ

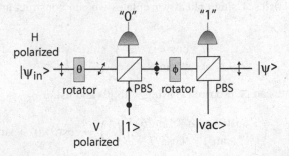

Fig. 14.4 Circuit for implementing a nonlinear sign gate [10]. The rotators rotate the polarization through angles θ and ϕ respectively. PBS = polarizing beam splitter that transmits horizontal polarization and reflects vertical polarization

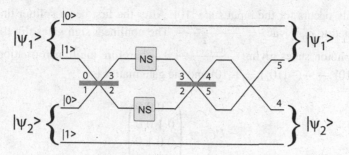

Fig. 14.5 C_Z circuit acting on dual-rail encoded qubits using nonlinear sign gates

well as many other possible circuits [3]. The circuit has rotators that are set to angles θ and ϕ satisfying

$$\frac{\cos(\theta)\cos(2\phi)}{\cos(\phi)} = 1, \quad \cos^2(\theta)\left[1 - 3\sin^2(\phi)\right] = -1$$

giving $\theta \simeq 150.49°$, $\phi \simeq 61.58°$. An auxiliary photon is injected into the open port of the first beam splitter and vacuum is presented to the second beam splitter. The output of the upper port of both beam splitters is detected. When the first detector observes no photon and the second detector observes a single photon this circuit implements U_{NS}. The probability of this happening, which depends on the input state, is at most 22.7%.

Given the U_{NS} gate we can construct a C_Z gate acting on dual-rail encoded qubits as shown in Fig. 14.5. To understand how this circuit works consider first the interior two rails. We will assume the beamsplitters have scattering matrices

$$U = \frac{1}{\sqrt{2}}\begin{pmatrix} 1 & i \\ i & 1 \end{pmatrix}$$

so $t = t' = 1/\sqrt{2}$ and $r = r' = i/\sqrt{2}$. Ignoring for the moment the U_{NS} gates and multiplying through we find $\begin{pmatrix} \hat{a}_5 \\ \hat{a}_4 \end{pmatrix} = i \begin{pmatrix} \hat{a}_0 \\ \hat{a}_1 \end{pmatrix}$. Therefore, still ignoring the U_{NS} gates, the two-qubit input states transform according to

$$U_{\text{linear}} = \begin{pmatrix} i & 0 & 0 & 0 \\ 0 & 1 & 0 & 0 \\ 0 & 0 & -1 & 0 \\ 0 & 0 & 0 & i \end{pmatrix}$$

which does not produce any entanglement.

To construct a two-qubit gate we include U_{NS} in each of the interior rails. This only changes the situation when there are two photons in the interior rails, which only occurs for the input state $|10\rangle$. After the first beam splitter this state is transformed to $\hat{a}_0^\dagger \hat{a}_1^\dagger |\text{vac}\rangle = \frac{|02\rangle_{23} + i|20\rangle_{23}}{\sqrt{2}}$. The nonlinear sign gates flip the sign of the two-photon states giving $-\frac{|02\rangle_{23} + i|20\rangle_{23}}{\sqrt{2}}$. Therefore the transformation for this state is $|10\rangle \rightarrow -(-|10\rangle) = +|10\rangle$ and the gate matrix is

$$U_{C_Z} = \begin{pmatrix} i & 0 & 0 & 0 \\ 0 & 1 & 0 & 0 \\ 0 & 0 & 1 & 0 \\ 0 & 0 & 0 & i \end{pmatrix}$$

which is a phase gate capable of producing maximal entanglement. This can be converted into a more standard form by adding single qubit phase shifts at the outputs.

To summarize we have shown how to construct a C_Z gate, which can be converted into a CNOT gate with single qubit operations, using a linear interferometer and two nonlinear sign gates that act probabilistically. Each nonlinear sign gate requires one ancilla photon so two ancillas are needed for the CNOT gate as well as photon number resolving detectors. Most significantly the CNOT gate only works when both nonlinear sign gates work and the probability of that occurring is about 5% on average. Although other designs have been invented that increase this probability, it can be shown that even with the use of additional independent ancilla photons a probabilistic linear optics CNOT gate cannot be constructed that has a success probability greater than $2/27 \simeq 7.4\%$ [11].

14.3   Gate Teleportation

Even accepting the circuitry requirements for linear optical gates it is not obvious how to implement a long calculation if the two-qubit gate at each step acts probabilistically. The advantage of quantum computing is that certain problems which require exponential resources on a classical machine can be solved on a quantum machine with resources that scale polynomially in the problem size. If the success probability of each gate is $P < 1$ then the probability of successfully executing N gates is P^N which is exponentially small for large N. Thus, there is no quantum advantage for a system based on probabilistic gates.

This difficulty led early researchers to believe that single photon based quantum computing was not scalable. However, an approach that is in principle scalable was designed in 2001 based on gate teleportation [7]. Just as quantum states can be deterministically teleported using entanglement as a resource the action of a quantum gate can be teleported using entanglement as a resource. This is more demanding than state teleportation and requires more entanglement. The idea is that even though a gate is probabilistic, we can deterministically teleport it using a certain entangled resource state. Preparation of the resource state is probabilistic, but we may keep trying until we succeed, and then teleport the gate into the computational circuit. The circuit for teleportation of a CNOT gate is shown in Fig. 14.6.

The required resources are four extra qubits, preparation of state $|r\rangle$ which requires two Hadamard gates and three CNOT gates, two measurements in the Bell basis, classical logic, two X gates and two Z gates. Provided we have long coherence memory to store $|c\rangle$ and $|t\rangle$ while $|r\rangle$ is being prepared, or to prepare and store many copies of $|r\rangle$ as a resource reservoir, it is possible to design a scalable architecture.

To understand how gate teleportation works consider the circuit of Fig. 14.7 which teleports the input states $|c\rangle$ and $|t\rangle$. For each state two ancillas and a measurement in the Bell basis, followed by classically controlled X and Z gates, provides teleportation. The operations on the interior qubit lines can be rearranged as shown in Fig. 14.8 in a way that prepares a resource state which enables teleportation of the two-qubit state $CNOT_{ct}$. The resource state is

$$|r\rangle = \frac{|0000\rangle + |0011\rangle + |1110\rangle + |1101\rangle}{2}.$$

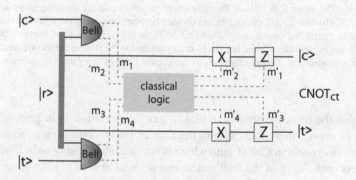

Fig. 14.6 Circuit for gate teleportation using a four-qubit resource state $|r\rangle$. The Bell measurement results $m_1 - m_4$ are sent over classical channels (dashed lines) to a logic circuit that generates signals $m'_1 - m'_4$ which control X and Z gates

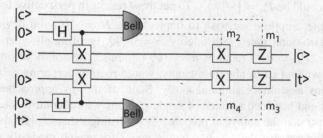

Fig. 14.7 Circuit for teleportation of qubits $|c\rangle$ and $|t\rangle$

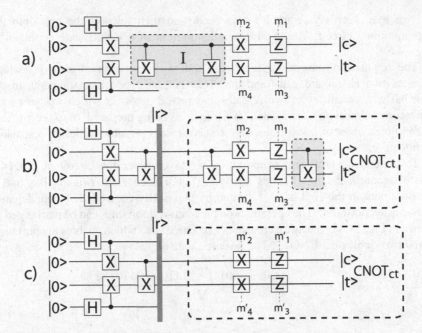

Fig. 14.8 (a) The inner four rails of the state teleportation circuit are unchanged by adding I = CNOT CNOT. (b) The CNOT operations are divided between preparing the resource state $|r\rangle$ and the classically controlled operations. Adding a CNOT on the right gives the output $\mathrm{CNOT_{ct}}$. (c) The dashed box is equivalent to the box in (b) minus the extra CNOT operations provided $m'_1 - m'_4$ are chosen appropriately. This corresponds to the central lines of the gate teleportation circuit in Fig. 14.6

Provided the resource state is available gate teleportation is in principle deterministic. However, there is a catch in the case of linear optics since the Bell basis measurements require a CNOT gate which is not available. Using only linear optics the success probability of the Bell measurements is at best $1/2$ [12].

This limitation was solved by KLM who showed that by using n photons in a multi-mode approach the success probability of qubit teleportation could be increased to $\frac{n}{n+1}$ so the success probability for gate teleportation, which involves two-qubits, could be $P_n = \left(\frac{n}{n+1}\right)^2$. To put these results in perspective let's consider a toy example. Suppose we wish to implement a $N = 10$ qubit device that can implement two-qubit gates with fidelity $\mathcal{F} = 0.99$. Ignoring all other errors, and setting $\mathcal{F} = P_n = 0.99$ requires $n = 199$ photons. The ability to implement N such gates, without extra photon memory, would require preparing $\sim 10 \times 200 = 2000$ photons essentially simultaneously. State of the art approaches [5] using multiplexed and heralded sources of single photons with a controllable delay line have demonstrated the ability to prepare ~ 30 indistinguishable photons at rates of about $1/s$. We see that the resource requirements for even a modestly sized linear optics quantum computer are far beyond present capabilities.

The KLM architecture based on nonlinear sign gates and teleportation is not the only possible approach. A CNOT success probability of 25% can be achieved using two entangled ancilla photons in a relatively simple architecture [13]. There are also alternatives to circuit model computing. In the one-way or measurement based approach [14, 15] a large cluster state that is highly entangled is prepared as a universal resource after which the computation proceeds using only measurements and single qubit gates. This has the advantage that once the cluster state is prepared, which can be done off-line in a heralded fashion, no additional two-qubit gates are required. Removing the requirement of two-qubit gate operation is a significant advantage although the requirement for simultaneous preparation of a large number of photons is unchanged. Furthermore, the applicability of error correction to the one-way model requires three dimensional connectivity which is non-trivial to implement [16].

Improvements that reduce resource requirements are the subject of active research. Nevertheless there is such a large gap between present capabilities and the resources needed for large scale linear optics quantum computing that the viability of this approach for reaching beyond classical capabilities remains to be demonstrated.

14.4 Nonlinear Optical Approaches

Another approach to optical quantum computing is based on implementing direct interactions between photons in nonlinear media. This can be based on qubits encoded in many-photon coherent states [17] or on using nonlinearity at the level of single photons to enable a deterministic entangling gate.

14.4.1 Single Photon Nonlinearities

A cross Kerr medium with a cubic nonlinearity can in principle be used to implement a deterministic phase gate. Consider the dual-rail encoding of Fig. 14.9.

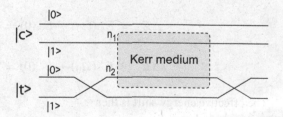

Fig. 14.9 A medium with a cross Kerr nonlinearity can be used to implement a phase gate acting on single photon qubits

The $|1\rangle$ rail of each qubit is sent through the medium which imparts a nonlinear phase ϕ according to the interaction

$$U_{\text{Kerr}} = e^{i\hat{\mathcal{H}}_{\text{Kerr}} t / \hbar} \tag{14.3}$$

where the cross Kerr Hamiltonian is $\hat{\mathcal{H}}_{\text{Kerr}} = \phi \hat{n}_1 \hat{n}_2$ with $\hat{n}_j = \hat{a}_j^\dagger \hat{a}_j$ the photon number operator of mode j. This gives a transformation of the basis states $|00\rangle \rightarrow |00\rangle$, $|01\rangle \rightarrow |01\rangle$, $|10\rangle \rightarrow |10\rangle$, $|11\rangle \rightarrow e^{i\phi}|11\rangle$ and for $\phi = \pi$ we recover a C_Z gate. The availability of such a deterministic phase gate would greatly reduce the resource requirements of linear optics quantum computing and help pave the way towards scalability.

The challenge is in finding a medium that has a strong nonlinearity at the level of single photons. The basic physics of the Kerr interaction is captured by a simple model of a two-level atom interacting with a quantized field [18]. Let the atom have levels $|g\rangle$, $|e\rangle$ and transition frequency ω_a. The photon has frequency ω that is detuned from the atomic transition by $\Delta = \omega - \omega_a$. The atom-field coupling strength for a single photon is Ω_0. The Jaynes–Cummings Hamiltonian describing the interaction is

$$\hat{\mathcal{H}} = \frac{\hbar \omega_a}{2} \hat{\sigma}_z + \hbar \omega \hat{a}^\dagger \hat{a} + \frac{\hbar \Omega_0}{2} (\hat{a}\hat{\sigma}_+ + \hat{a}^\dagger \hat{\sigma}_-).$$

The first and second terms are the atomic and field energies, and the third term is the field-atom interaction with $\hat{\sigma}_+ = |e\rangle\langle g|$, $\hat{\sigma}_- = \hat{\sigma}_+^\dagger$. The eigenergies of the so-called dressed states are $E_\pm(n) = (n + 1/2)\hbar\omega \pm \frac{\hbar}{2}\sqrt{\Delta^2 + \Omega_n^2}$ with $\Omega_n = \Omega_0\sqrt{n+1}$. The corresponding eigenstates are

$$|n, +\rangle = \cos(\phi_n/2)|e, n\rangle + \sin(\phi_n/2)|g, n+1\rangle$$
$$|n, -\rangle = -\sin(\phi_n/2)|e, n\rangle + \cos(\phi_n/2)|g, n+1\rangle$$

with $\phi_n = \tan^{-1}(\Omega_n/\Delta)$.

In order to minimize photon loss and establish a purely dispersive interaction we should choose a large detuning $|\Delta| \gg \Omega_n$ such that $E_\pm(n) \simeq (n + 1/2)\hbar\omega \pm \left(\frac{\hbar\Delta}{2} + \frac{\Omega_n^2}{4\Delta}\right)$. In this limit of large detuning $|n, +\rangle \simeq |e, n\rangle$, $|n, -\rangle \simeq |g, n+1\rangle$ and

$$\Delta E_+ = E_+(\Omega_0) - E_+(0) = \hbar \frac{\Omega_0^2}{4\Delta}(n+1)$$

$$\Delta E_- = E_-(\Omega_0) - E_-(0) = -\hbar \frac{\Omega_0^2}{4\Delta} n.$$

The effective energy shift is then

$$\Delta E = \hbar\chi \left(\Delta E_+ |e\rangle\langle e| + \Delta E_- |g\rangle\langle g|\right) = \hbar\chi \left((\hat{a}^\dagger \hat{a} + 1)|e\rangle\langle e| - \hat{a}^\dagger \hat{a}|g\rangle\langle g|\right)$$

with the interaction strength $\chi = \Omega_0^2/4\Delta$. When the atomic medium is in the ground state, which will be the case for large detuning even in the presence of a field, $\Delta E = -\hbar\chi\hat{a}^\dagger\hat{a}$. This gives a Kerr type interaction $U_{\text{Kerr}} = e^{-i\frac{\Delta E}{\hbar}t} = e^{i\chi\hat{a}^\dagger\hat{a}t}$, with t the interaction time corresponding to photon propagation through the medium.

To implement a phase gate we need a cross Kerr interaction as in Eq. (14.3) that provides a phase shift which depends on the product of the photon numbers in two modes. This can be implemented using atomic media with more than two levels [19]. Each of the photon modes is arranged to be resonant with a different atomic transition, for example by taking advantage of polarization selection rules. While the details are more complicated the result is analogous to that derived here for a scalar Kerr medium.

Experiments demonstrating a cross Kerr phase shift for single photons were first performed at optical frequencies using cold atoms as the nonlinear medium [20] and at microwave frequencies using Rydberg atoms [21]. The exceptionally large nonlinearity of superconducting devices at microwave frequencies has made it possible to reliably engineer single photon Kerr and cross Kerr nonlinearities for superconducting quantum computing, a topic which goes under the heading of circuit Quantum Electrodynamics (cQED) [22], and which we will return to in Chap. 15. The great challenge in the context of linear optical quantum computing is achieving a full π of phase shift without loss. Recent work has succeeded in demonstrating such large phase shifts, albeit with accompanying large absorption, although this remains a topic of active research [23, 24].

14.4.2 Continuous Variable Encoding

An alternative paradigm for optical quantum computing uses qubits encoded in states containing a large number of photons. There are some obvious potential advantages to this approach. If the qubit state contains many photons a small amount of absorption resulting in loss of a few photons will impart only a small error to the state. It is also easier to measure and prepare many photon states compared to the difficulty of working with single photons.

The most natural many photon states that we have available are the coherent states $|\alpha\rangle$ which are eigenstates of the annihilation operator $\hat{a}|\alpha\rangle = \alpha|\alpha\rangle$. Coherent states have a representation in terms of number states $|\alpha\rangle = e^{-\frac{1}{2}|\alpha|^2}\sum_{n=0}^{\infty}\frac{\alpha^n}{\sqrt{n!}}|n\rangle$. The output of a stable, monochromatic laser is to a good approximation a coherent state. Unfortunately coherent states with different amplitudes are not orthogonal, but have a scalar product

$$\langle\beta|\alpha\rangle = e^{\alpha\beta^* - \frac{1}{2}(|\alpha|^2 + |\beta|^2)}.$$

States with different α, β are not orthogonal and cannot directly be used as a qubit basis, but are almost orthogonal when $|\alpha - \beta| \gg 0$.

A convenient choice for qubit encoding turns out to be the superposition of coherent states with opposite sign amplitudes $|\alpha_{e/o}\rangle \sim |\alpha\rangle \pm |-\alpha\rangle$. These states are superpositions of states with only an even or odd number of photons and are orthogonal $\langle \alpha_o | \alpha_e \rangle = 0$ so they can be used for qubit encoding.

A distinct advantage of the continuous variable encoding compared to single photons is that Bell state measurements, gate teleportation, and therefore a CNOT gate can be implemented deterministically [25] which is in contrast to the 50% success rate limit for single photon Bell measurements [12]. The resource needed for high gate fidelity is a large optical nonlinearity so that highly squeezed optical states can be prepared. The experimental challenges are not unrelated to those that must be overcome for deterministic single photon quantum computing. Nevertheless, the infinite dimensional Hilbert space can potentially be used for error correcting redundant encoding [17] though it remains to be seen how far the continuous variable approach can be pushed.

14.5   Problems

1. A probabilistic C_Z gate can be implemented for single photons using an interferometer and a nonlinear sign (NS) gate.

   (a) Find the probability of success of the NS gate when each of the four computational basis states is input to the C_Z gate. Account for the probability of a correct detection at both the "0" detector and the "1" detector in the nonlinear sign gate. Using your results find the probability of success of the C_Z gate for a random two-qubit input state.

   (b) Assuming a perfect NS gate (which does not actually exist) the C_Z success probability would still depend on how well the interferometer functions. Assume the two input photons are in spatial modes $\mathcal{E}_a = c_a e^{-x^2/w_a^2}$, $\mathcal{E}_b = c_b e^{-x^2/w_b^2}$ (we are using a one-dimensional description). Here a, b label the two qubits, $\mathcal{E}_{a,b}$ are the amplitudes of the electric field, $w_{a,b}$ are beam widths and $c_{a,b}$ are normalization constants. For the four computational basis states find the probability of the C_Z gate giving a correct output state assuming a perfect deterministic NS gate and $w_a \neq w_b$. Express your answer in terms of $q = w_b/w_a$ and find how much q may differ from unity for the gate to succeed with probability $P > 0.9999$.

   Hint: When $q \neq 1$ there is imperfect mode matching of the interfering photons. In this situation there will be a small amplitude, say η, for each photon to leave the beamsplitter in a different port. We can estimate η from the expression $1 - \eta^2 = $ normalized overlap integral of the photon modes.

2. A general 2-qubit phase gate has the matrix representation

$$U_\phi = \begin{pmatrix} 1 & 0 & 0 & 0 \\ 0 & e^{i\phi_2} & 0 & 0 \\ 0 & 0 & e^{i\phi_3} & 0 \\ 0 & 0 & 0 & e^{i\phi_4} \end{pmatrix}$$

where we have assumed a global phase such that $\phi_1 = 0$. We wish to entangle the qubit states $|\psi_1\rangle = (|0\rangle + |1\rangle)/\sqrt{2}$ and $|\psi_2\rangle = (|0\rangle + |1\rangle)/\sqrt{2}$ using U_ϕ. Find a condition on the phases ϕ_2, ϕ_3, ϕ_4 such that the gate produces entanglement.

3. We analyzed a quantum circuit that implements teleportation of a CNOT gate using the auxiliary state $|r\rangle = (|0000\rangle + ||0011\rangle + |1110\rangle + |1101\rangle)/2$. The circuit includes classical logic that takes four binary values as inputs, m_1, m_2, m_3, m_4 and provides four values m'_1, m'_2, m'_3, m'_4 as outputs. Work out the truth table that maps the 16 possible input strings $m_1 m_2 m_3 m_4$ to 16 possible output strings $m'_1 m'_2 m'_3 m'_4$ for CNOT teleportation.

References

1. J. Yin, Y. Cao, Y.-H. Li, S.-K. Liao, L. Zhang, J.-G. Ren, W.-Q. Cai, W.-Y. Liu, B. Li, H. Dai, G.-B. Li, Q.-M. Lu, Y.-H. Gong, Y. Xu, S.-L. Li, F.-Z. Li, Y.-Y. Yin, Z.-Q. Jiang, M. Li, J.-J. Jia, G. Ren, D. He, Y.-L. Zhou, X.-X. Zhang, N. Wang, X. Chang, Z.-C. Zhu, N.-L. Liu, Y.-A. Chen, C.-Y. Lu, R. Shu, C.-Z. Peng, J.-Y. Wang, J.-W. Pan, Satellite-based entanglement distribution over 1200 kilometers. Science **356**, 1140 (2017)
2. S.-K. Liao, W.-Q. Cai, J. Handsteiner, B. Liu, J. Yin, L. Zhang, D. Rauch, M. Fink, J.-G. Ren, W.-Y. Liu, Y. Li, Q. Shen, Y. Cao, F.-Z. Li, J.-F. Wang, Y.-M. Huang, L. Deng, T. Xi, L. Ma, T. Hu, L. Li, N.-L. Liu, F. Koidl, P. Wang, Y.-A. Chen, X.-B. Wang, M. Steindorfer, G. Kirchner, C.-Y. Lu, R. Shu, R. Ursin, T. Scheidl, C.-Z. Peng, J.-Y. Wang, A. Zeilinger, J.-W. Pan, Satellite-relayed intercontinental quantum network. Phys. Rev. Lett. **120**, 030501 (2018)
3. P. Kok, W.J. Munro, K. Nemoto, T.C. Ralph, J.P. Dowling, G.J. Milburn, Linear optical quantum computing with photonic qubits. Rev. Mod. Phys. **135**, 2007 (2007)
4. H. Wang, J. Qin, X. Ding, M.-C. Chen, S. Chen, X. You, Y.-M. He, X. Jiang, L. You, Z. Wang, C. Schneider, J. J. Renema, S. Höfling, C.-Y. Lu, J.-W. Pan, Boson sampling with 20 input photons and a 60-mode interferometer in a 10^{14}-dimensional Hilbert space. Phys. Rev. Lett. **123**, 250503 (2019)
5. F. Kaneda, P.G. Kwiat, High-efficiency single-photon generation via large-scale active time multiplexing. Sci. Adv. **5**, eaaw8586 (2019)
6. F. Marsili, V.B. Verma, J.A. Stern, S. Harrington, A.E. Lita, T. Gerrits, I. Vayshenker, B. Baek, M.D. Shaw, R.P. Mirin, S.W. Nam, Detecting single infrared photons with 93% system efficiency. Nat. Photon. **7**, 210 (2013)
7. E. Knill, R. Laflamme, G.J. Milburn, A scheme for efficient quantum computation with linear optics. Nature **409**, 46 (2001)
8. G.G. Stokes, On the perfect blackness of the central spot in Newton's rings and on the verification of Fresnel's formulae for the intensities of reflected and refracted rays. Camb. Dublin Math. J. **4**, 1 (1849)

9. C.K. Hong, Z.Y. Ou, L. Mandel, Measurement of subpicosecond time intervals between two photons by interference. Phys. Rev. Lett. **59**, 2044–2046 (1987)

10. T. Rudolph, J.W. Pan, A simple gate for linear optics quantum computing (2001). arXiv:quant-ph/0108056

11. D.B. Uskov, L. Kaplan, A.M. Smith, S.D. Huver, J.P. Dowling, Maximal success probabilities of linear-optical quantum gates. Phys. Rev. A **79**, 042326 (2009)

12. J. Calsamiglia, N. Lütkenhaus, Maximum efficiency of a linear-optical Bell-state analyzer. Appl. Phys. B **72**, 67 (2001)

13. J. Zeuner, A.N. Sharma, M. Tillmann, R. Heilmann, M. Gräfe, A. Moqanaki, A. Szameit, P. Walther, Integrated-optics heralded controlled-NOT gate for polarization-encoded qubits. npj Qu. Inf. **4**, 13 (2018)

14. R. Raussendorf, H.J. Briegel, A one-way quantum computer. Phys. Rev. Lett. **86**, 5188 (2001)

15. R. Raussendorf, D.E. Browne, H.J. Briegel, Measurement-based quantum computation on cluster states. Phys. Rev. A **68**, 022312 (2003)

16. R. Raussendorf, J. Harrington, K. Goyal, A fault-tolerant one-way quantum computer. Ann. Phys. **321**, 2242 (2006)

17. D. Gottesman, A. Kitaev, J. Preskill, Encoding a qubit in an oscillator. Phys. Rev. A **64**, 012310 (2001)

18. E.T. Jaynes, F.W. Cummings, Comparison of quantum and semiclassical radiation theories with application to the beam maser. Proc. IEEE **51**, 89 (1963)

19. H. Schmidt, A. Imamoğlu, Giant Kerr nonlinearities obtained by electromagnetically induced transparency. Opt. Lett. **21**, 1936 (1996)

20. Q.A. Turchette, C.J. Hood, W. Lange, H. Mabuchi, H.J. Kimble, Measurement of conditional phase shifts for quantum logic. Phys. Rev. Lett. **75**, 4710–4713 (1995)

21. A. Rauschenbeutel, G. Nogues, S. Osnaghi, P. Bertet, M. Brune, J.M. Raimond, S. Haroche, Coherent operation of a tunable quantum phase gate in cavity QED. Phys. Rev. Lett. **83**, 5166 (1999)

22. A. Blais, R.-S. Huang, A. Wallraff, S.M. Girvin, R.J. Schoelkopf, Cavity quantum electrodynamics for superconducting electrical circuits: an architecture for quantum computation. Phys. Rev. A **69**, 062320 (2004)

23. J.D. Thompson, T.L. Nicholson, Q.-Y. Liang, S.H. Cantu, A.V. Venkatramani, S. Choi, I.A. Fedorov, D. Viscor, T. Pohl, M.D. Lukin, V. Vuletić, Symmetry-protected collisions between strongly interacting photons. Nature **542**, 206 (2017)

24. D. Tiarks, S. Schmidt-Eberle, T. Stolz, G. Rempe, S. Dürr, A photon–photon quantum gate based on Rydberg interactions. Nat. Phys. **15**, 124 (2018)

25. S. Takeda, A. Furusawa, Toward large-scale fault-tolerant universal photonic quantum computing. APL Photon. **4**, 060902 (2019)

Chapter 15
Solid State Qubits

In this chapter we will study qubits encoded in superconducting circuits and semi-conductor quantum dots. In contrast to the qubits discussed so far these approaches do not involve optics or lasers and are therefore closer to the technologies that are widely used today for information processing. On the other hand maintaining and controlling quantum coherence in a solid state material is difficult at room temperatures so the approaches to be discussed require cooling of the circuit elements to mK temperatures. Representative examples of modern solid state devices are shown in Fig. 15.1.

15.1 Superconducting Qubits: Overview

Superconducting qubits are fundamentally different than the atomic and optical approaches we have discussed so far. Instead of encoding a qubit in the quantum state of a single particle, or a few correlated particles in the case of an atom or ion, the qubit is encoded in the mesoscopic collective state of a large number of electrons. Mesoscopic refers to an entity that is neither microscopic nor macroscopic but straddles the region in between. A mesoscopic object may still contain a large number of particles, but is small enough that quantum fluctuations away from mean values are significant.

There has been remarkable progress in superconducting circuits for quantum information to the point where the first demonstrations of digital quantum circuits with high fidelity gates running on processors with more than 50 qubits were made with superconducting qubits [2]. Many believe that superconductors represent the leading path towards a practical quantum computer and a great deal of commercial investment is currently directed at superconducting devices.

We will start with a review of how to quantize an electrical LC circuit serving as an analog to a mechanical harmonic oscillator. To be directly useful for qubit

© Springer Nature Switzerland AG 2021
J. A. Bergou et al., *Quantum Information Processing*, Graduate Texts in Physics,
https://doi.org/10.1007/978-3-030-75436-5_15

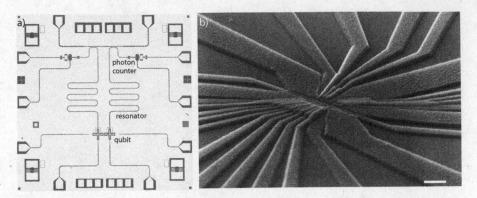

Fig. 15.1 Solid state qubit devices. (**a**) A superconducting chip with two qubits, resonators, and microwave photon counters [1]. The die is 6.25 mm on a side. Image courtesy of Robert McDermott. (**b**) A gate-defined quadruple quantum dot, capable of hosting two double-dot qubits or four single-spin qubits. The scale bar is 500 nm. Image courtesy of Samuel Neyens and Mark Eriksson

encoding anharmonicity is needed and this is provided by a nonlinear inductor based on the Josephson effect. We will then show how different types of superconducting qubits can be designed around superconducting elements and continue with a description of how the DiVincenzo criteria are implemented in a superconducting processor. There are a great range of approaches that have been explored in the superconducting qubit community. We will only cover a small fraction of the results to date and refer to review articles for more extensive coverage [3, 4].

15.2 Circuit Quantization

We are familiar with basic linear circuit elements: the resistor R, the capacitor C and the inductor L. The relations between charge Q, magnetic flux Φ, current I, and voltage V for these elements are

$$R = \frac{V}{I}, \quad C = \frac{Q}{V}, \quad L = \frac{\Phi}{I}. \tag{15.1}$$

Charge and current on a capacitor are related by

$$I = \frac{dQ}{dt} = C\frac{dV}{dt}$$

while magnetic flux and voltage in an inductor are related by

$$V = -\int d\boldsymbol{\ell} \cdot \mathbf{E} = \frac{d\Phi}{dt} = L\frac{dI}{dt}.$$

Note that correct definition of signs in the above relations can be ambiguous and is best done with reference to a circuit diagram. The power dissipated in a resistor is $P = IV = \frac{V^2}{R} = RI^2$ and the stored energy in a capacitor is $U_C = \frac{1}{2}CV^2$ and in an inductor $U_L = \frac{1}{2}LI^2$. The resistor is dissipative and heat is generated by a current. In contrast ideal capacitors and inductors are lossless. Real components are predominantly of one type but contain small effective contributions of other types of elements.

An LC circuit containing an ideal inductor and capacitor connected in series will oscillate at frequency ω. The circuit relations are

$$V = L\frac{dI}{dt}$$

and

$$-\frac{dV}{dt} = \frac{I}{C} = -L\frac{d^2I}{dt^2}$$

so

$$\frac{d^2I}{dt^2} = -\frac{1}{LC}I.$$

The solution is $I = I_0 \cos(\omega t)$ with $\omega = \frac{1}{\sqrt{LC}}$.

Some typical values of small elements are $C = 0.3\,\text{pF}$, $L = 3\,\text{nH}$ giving $\omega = 3.3 \times 10^{10}\,\text{rad/s}$, or in cycles per second $f = \frac{\omega}{2\pi} = 5.3\,\text{GHz}$. The dimensions of C and L with these values are less than 1 mm. On the other hand the wavelength of the oscillation is $\lambda = \frac{c}{f} \simeq 5.7\,\text{cm}$. We see that $\lambda \gg$ the size of the circuit elements. This situation is referred to as a lumped element circuit.

The magnetic flux $\Phi = LI_0 \cos(\omega t)$ is a collective degree of freedom. In a mechanical analogy it corresponds to the position of a mass in a mechanical oscillator. The charge Q corresponds to the momentum, and Φ and Q are conjugate coordinates. This analogy follows from the Hamiltonian for the circuit. For a mechanical oscillator the energy or Hamiltonian is

$$H = \frac{p^2}{2m} + \frac{\kappa x^2}{2}$$

with m the mass, p the momentum, and κ the spring constant. For the LC circuit the stored energy is

$$H = \frac{1}{2}CV^2 + \frac{1}{2}LI^2 = \frac{1}{2}\frac{Q^2}{C} + \frac{1}{2}\frac{\Phi^2}{L}.$$

Both Hamiltonians are quadratic in a pair of conjugate variables. We can therefore quantize the electronic oscillator by analogy with quantization of a mechanical

Table 15.1 Mechanical and electronic oscillators and their quantized analogs. The commutators follow from the harmonic oscillator relations $[\hat{a}^\dagger, \hat{a}] = 1$

Mechanical		Electronic	
Position	x	Flux	Φ
Momentum	p	Charge	Q
Mass	m	Capacitance	C
Spring constant	κ	Inverse inductance	$1/L$
Frequency	$\omega = \sqrt{\kappa/m}$		$\omega = 1/\sqrt{LC}$
Quantization			
Hamiltonian	$H = \frac{p^2}{2m} + \frac{\kappa x^2}{2}$		$H = \frac{Q^2}{2C} + \frac{\Phi^2}{2L}$
Coordinate	$x \to \hat{x}$		$\Phi \to \hat{\Phi}$
	$= \sqrt{\frac{\hbar}{2m\omega}}(\hat{a} + \hat{a}^\dagger)$		$= \sqrt{\frac{\hbar\sqrt{L}}{2\sqrt{C}}}(\hat{a} + \hat{a}^\dagger)$
Momentum	$p \to \hat{p} = -i\hbar\frac{\partial}{\partial x}$		$Q \to \hat{Q} = -i\hbar\frac{\partial}{\partial \Phi}$
	$= i\sqrt{\frac{m\hbar\omega}{2}}(\hat{a}^\dagger - \hat{a})$		$= i\sqrt{\frac{\hbar\sqrt{C}}{2\sqrt{L}}}(\hat{a}^\dagger - \hat{a})$
Commutator	$[\hat{x}, \hat{p}] = i\hbar$		$[\hat{\Phi}, \hat{Q}] = i\hbar$

oscillator. The analogs between the mechanical and LC oscillators are listed in Table 15.1.

The energy levels of the LC oscillator are quantized, but they are equidistant so this is not directly useful as a qubit. If we were to encode a qubit in the lowest two levels, a resonant field would also couple the second level to the third level, and so on, leading to leakage out of the computational basis. While it is interesting to note that there is a way to encode a qubit in a harmonic oscillator [5], other challenges arise such as the need to prepare highly nonclassical states to perform gate operations.

The approach that is instead followed is to introduce a nonlinear element so that the quantized energy levels are no longer equally spaced. In this way the two lowest levels can be independently addressed and used as a qubit. The nonlinear element that is used is the Josephson tunnel junction which effectively provides a nonlinear inductance as shown in Fig. 15.2.

Additionally, the superconducting character of the circuit eliminates the ohmic dissipation present in room temperature wires which prevents the observation of quantized levels in ordinary circuits. We note that the condition on the dissipation that could be tolerated and still observe quantized levels can be estimated from the requirement that the coherence time of the circuit is greater than the inverse oscillation frequency. If all losses including dissipation and radiation are lumped into an effective resistance R an excited harmonic oscillator state will decay in a time that scales as $1/R$. Setting $\omega\tau > 1$ implies an upper limit on R to resolve quantized levels.

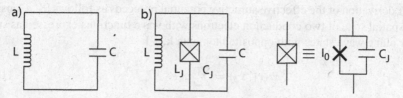

Fig. 15.2 Linear and nonlinear oscillator circuits. (**a**) Linear LC oscillator. (**b**) Nonlinear circuit with a Josephson element. The box with a cross represents a Josephson tunnel junction with nonlinear inductance L_J and capacitance C_J which can be reduced to a nonlinear (phase dependent) current source I_0 represented by a cross in parallel with C_J

The superconducting aspect of the circuit makes observation of quantized energy levels possible. There is also a fundamental difference, compared to a classical circuit, in that the classical variables of charge and current are now represented by quantum wavefunctions and can be in superposition states. Thus the charge may be positive and negative at the same time and the current may be in a superposition of flowing in opposite directions even though a mesoscopic number of electrons are involved. Thus the quantum degrees of freedom have a *collective* character.

15.3 Superconductivity

In order to understand how Josephson junction qubits operate we should first learn something about superconductivity. In any normal metal electrons, which carry current, scatter off impurities and are repelled by other electrons all of which have the same electronic charge. The result is a finite resistivity and dissipation of energy in the current carrying material. Dissipation is undesirable for qubit devices and can be avoided at low temperatures where many materials become superconducting and carry current while exhibiting zero resistivity. A microscopic theory of superconductivity was first developed by Bardeen et al. [6]. Here we give a simplified argument for how electrons, even though they electrically repel each other, may exhibit a weak attraction inside a material.

The physical picture is that a conduction electron exerts an attractive force on a positive ion in the valence band, giving a small charge displacement. A second conduction electron is attracted by the displaced ion leading to an effective attraction between electrons. The effect is weak so in order for the attractive force to dominate over thermal fluctuations the material must be sufficiently cold. The attractive force leads to electron pairing into so-called Cooper pairs [7]. The paired electrons form an integer spin boson, with many Cooper pairs contributing to a supercurrent with the elementary charge of the current carriers being $2e$.

A derivation of the effective attractive potential proceeds as follows [8]. Consider the spatial state of two conduction electrons with wave functions expressed as sums over plane wave states with opposite momenta $\mathbf{k}$, $-\mathbf{k}$

$$\psi(\mathbf{r}_1, \mathbf{r}_2) = \sum_{\mathbf{k}} a_{\mathbf{k}} e^{\iota \mathbf{k} \cdot (\mathbf{r}_1 - \mathbf{r}_2)} \tag{15.2}$$

The total wavefunction includes the spin state $\chi(1, 2)$ and the requirement of antisymmetry implies that $\Psi(1, 2)$ is a sum of $\cos[\mathbf{k} \cdot (\mathbf{r}_1 - \mathbf{r}_2)]$ terms times the spin singlet state or a sum of $\sin[\mathbf{k} \cdot (\mathbf{r}_1 - \mathbf{r}_2)]$ terms times one of the three spin triplet states. In the case of an attractive interaction the spin singlet state will have a lower energy since the cos dependence has a higher probability for the electrons to be closer together. We therefore write

$$\Psi(\mathbf{r}_1 - \mathbf{r}_2) = \sum_{\mathbf{k}} a_{\mathbf{k}} \cos[\mathbf{k} \cdot (\mathbf{r}_1 - \mathbf{r}_2)] \chi_S(1, 2) \tag{15.3}$$

with $\chi_S(1, 2)$ the singlet state.

The time independent Schrödinger equation for this state takes the form

$$(E - 2\epsilon_k) a_{\mathbf{k}} = \sum_{\mathbf{k}'} V_{\mathbf{k}\mathbf{k}'} a_{\mathbf{k}'} \tag{15.4}$$

with the single particle energy at momentum k, $\epsilon_k = \frac{\hbar^2 k^2}{2m}$ and $V_{\mathbf{k}\mathbf{k}'} = \frac{1}{V_{sc}} \int d^3 r \, V(\mathbf{r}) e^{\iota(\mathbf{k}' - \mathbf{k}) \cdot \mathbf{r}}$ the interaction energy between momentum eigenstates normalized by the volume V_{sc}.

We then make a zero temperature approximation which says that below the Fermi wavenumber, $k < k_F$, all states are filled so $a_k = 0$. Additionally above a critical wavenumber, $k > k_c$, the coupling to high momentum fast moving electrons is weak. We therefore set $V_{\mathbf{k},\mathbf{k}'} = 0$ for $k' < k_F$ and $V_{\mathbf{k}\mathbf{k}'} = 0$ for $k' > k_c$ as shown in Fig. 15.3. We also approximate $V_{\mathbf{k}\mathbf{k}'}$ by a constant $-V$, with $V > 0$ for $k_F < k, k' < k_c$. With these approximations and replacing the sum by an integral over the density of states $N(\epsilon)$, $\sum_{\mathbf{k}} \to \int d\epsilon \, N(\epsilon) \simeq N_F \int d\epsilon$, where N_F is the density of states at

Fig. 15.3 Momentum dependent interaction term

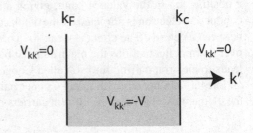

the Fermi energy $E_F = \frac{\hbar^2}{2m}\left(3\pi^2\rho\right)^{2/3}$ (ρ is the electron density), we find

$$e^{2/N_F V} = \frac{2E_F + 2E_c - E}{2E_F - E}.$$

Here E_c is the energy at momentum k_c. In 3D for non-interacting electrons the density of states $\sim E^{1/2}$ and assuming low energy so that $N_F V \ll 1$ we arrive at

$$E \simeq 2E_F - 2E_c e^{-2/N_F V}.$$

The second term on the right hand side is negative so $E < 2E_F$. E is the energy of a pair of electrons so the result implies a bound state with negative energy relative to the Fermi surface even though the electrons have $k > k_F$.

The paired electrons are integer spin bosons and we can have as many as we want in the same state, including the ground state of the conduction band. Since they are all in the same state they do not scatter off each other which results in a supercurrent of Cooper pairs that has zero resistance. A rigorous calculation is more complicated [6] but this toy calculation shows a plausible explanation.

15.4   Flux Quantization

We introduce a collective wavefunction for the supercurrent of Cooper pairs

$$\psi(\mathbf{r}, t) = \sqrt{n(\mathbf{r}, t)}\, e^{i\theta(\mathbf{r}, t)}$$

with $n = |\psi|^2$ the charge density and θ the phase.

The electric current is

$$\mathbf{J} = (2e)\int d^3 r\, \psi^* \mathbf{v} \psi.$$

In the presence of a vector potential the canonical momentum is $\mathbf{p} = m\mathbf{v} + \frac{q}{c}\mathbf{A}$. Using $\mathbf{p} = -i\hbar\nabla$ we find for the velocity $\mathbf{v} = -i\frac{\hbar}{m}\nabla - \frac{q}{mc}\mathbf{A}$ where $q = 2e$ is the charge of a Cooper pair. Assuming uniform charge density, vector potential and phase gradient we find

$$\mathbf{J} = \frac{2en}{m}\left(\hbar\nabla\theta - \frac{2e}{c}\mathbf{A}\right). \tag{15.5}$$

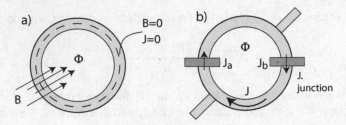

Fig. 15.4 Superconducting ring circuits. (**a**) Superconducting ring in an applied magnetic field. (**b**) Superconducting quantum interference device (SQUID) with two tunnel junctions embedded in the ring

It then follows from the Maxwell equations[1] that the magnetic field satisfies

$$\nabla^2 \mathbf{B} = \frac{1}{\lambda^2} \mathbf{B}$$

with $\lambda^2 = \frac{c}{4\pi} \frac{mc}{4ne^2}$. This is known as the London equation. Solving in a one-dimensional greometry shows that the magnetic field inside a superconductor vanishes exponentially with a penetration length λ. Taking $n \sim 10^{28}\,\mathrm{m}^{-3}$ for the charge density gives $\lambda \sim 25\,\mathrm{nm}$. Vanishing of the magnetic field inside a superconductor is known as the Meissner effect.

Consider a superconducting ring as in Fig. 15.4a with zero current and magnetic field inside the material of the ring. From Eq. (15.5) zero current implies

$$\nabla\theta = \frac{2e}{\hbar c}\mathbf{A}$$

where Φ is the magnetic flux inside the ring. Integrating the phase around the dashed contour we find

$$\oint d\mathbf{l} \cdot \nabla\theta = \frac{2e}{\hbar c} \oint d\mathbf{l} \cdot \mathbf{A}$$

$$= \frac{2e}{\hbar c} \oint d\mathbf{a} \cdot \nabla \times \mathbf{A}$$

$$= \frac{2e}{\hbar c} \oint d\mathbf{a} \cdot \mathbf{B}$$

$$= \frac{2e}{\hbar c} \Phi. \tag{15.6}$$

---

[1]$\nabla \times \mathbf{J} \sim \nabla \times \mathbf{A} = \mathbf{B}$, and $\nabla \times \mathbf{B} \sim \mathbf{J}$ so $\nabla \times \nabla \times \mathbf{B} = -\nabla^2 \mathbf{B} \sim \mathbf{B}$.

Requiring that the wavefunction be single valued implies that the change in phase after one loop around the ring is a multiple of 2π or

$$\Phi = \Phi_0 p$$

with p an integer and $\Phi_0^{(\text{cgs})} = \frac{hc}{2e}$ $\left(\Phi_0^{(\text{SI})} = \frac{h}{2e}\right)$ the quantum of magnetic flux. Note the charge is not e but $2e$ which corresponds to a Cooper pair.

15.5 Josephson Junction

The Josephson junction can be modeled as a thin piece of insulating material separating two superconductors as shown in Fig. 15.5. Assume there are wavefunctions ψ_1, ψ_2 in the two regions. We model the junction by

$$i\hbar\frac{\partial \psi_1}{\partial t} = U_1\psi_1 + K\psi_2,$$

$$i\hbar\frac{\partial \psi_2}{\partial t} = U_2\psi_2 + K\psi_1.$$

$U_{1,2}$ are the energies on each side and K is a coefficient that depends on the junction.

Suppose we connect the two sides with a voltage V in series so $U_1 - U_2 = qV$ where the charge is $q = 2e$ corresponding to a Cooper pair. Referencing the energies to potential zero we have $U_{1,2} = \pm qV/2$ and

$$i\hbar\frac{\partial \psi_1}{\partial t} = \frac{qV}{2}\psi_1 + K\psi_2,$$

$$i\hbar\frac{\partial \psi_2}{\partial t} = -\frac{qV}{2}\psi_2 + K\psi_1.$$

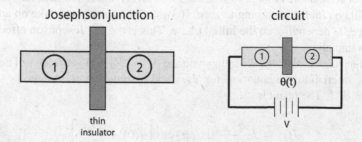

Fig. 15.5 A Josephson junction consists of two superconducting leads separated by a thin insulating region through which charges can tunnel. When the junction is placed in a circuit with a voltage source a phase dependent current flows

Then put $\psi_1 = \sqrt{n_1}e^{\iota\theta_1}$, $\psi_2 = \sqrt{n_2}e^{\iota\theta_2}$ to get

$$\frac{dn_1}{dt} = -\frac{dn_2}{dt} = \frac{2K}{\hbar}\sqrt{n_1 n_2}\sin(\theta_2 - \theta_1),$$

$$\frac{d\theta_1}{dt} = -\frac{K}{\hbar}\sqrt{\frac{n_2}{n_1}}\cos(\theta_2 - \theta_1) - \frac{qV}{2\hbar},$$

$$\frac{d\theta_2}{dt} = -\frac{K}{\hbar}\sqrt{\frac{n_1}{n_2}}\cos(\theta_2 - \theta_1) + \frac{qV}{2\hbar}.$$

The first equation shows there is a current due to a flow of charge carriers across the junction

$$J = \frac{dn_1}{dt} = J_0\sin(\theta)$$

where $J_0 = \frac{2K}{\hbar}n$, we have set $n = n_1 = n_2$ for identical superconductors, and $\theta = \theta_2 - \theta_1$. Even though there is a current we can set $n_1 = n_2$ since the voltage source maintains a constant potential across the junction. Note though that the current depends on the phase difference but not on V. In turn the phase difference does depend on V according to

$$\frac{d\theta}{dt} = \frac{qV}{\hbar}.$$

The equations of motion for a Josephson junction with an applied voltage V are therefore

$$J(t) = J_0\sin(\theta(t)), \quad \theta(t) = \theta(0) + \frac{q}{\hbar}\int_0^t dt'\, V(t').$$

Now suppose we apply a constant voltage $V(t) = V_0$. Then $\theta(t)$ will change rapidly for reasonable voltage values since $\hbar$ is a very small number and the current will oscillate rapidly, averaging to zero. If $V_0 = 0$ the current can take on any value between $\pm J_0$ depending on the initial phase. This is the DC Josephson effect which is due to tunneling.

We can also apply an oscillating voltage $V = V_0 + V_{ac}\cos(\omega t)$. Then solve for $\theta(t)$, insert into the equation for $J(t)$ and assume the ac voltage is small so $qV_{ac}/\hbar\omega \ll 1$. The result is

$$J(t) \approx J_0\frac{qV_{ac}}{\hbar\omega}\sin(\omega t)\cos\left[\theta(0) + \frac{qV_0}{\hbar}t\right].$$

For general values of V_0, ω the current averages to zero. However for the special case of $\omega = qV_0/\hbar$ we get a nonzero average current. This is the ac Josephson effect.

The Josephson junction can be modeled as a nonlinear inductor. To see this write the current and phase equations as $I = qJ = I_0 \sin(\theta)$ and $V = \frac{\hbar}{q}\frac{d\theta}{dt} = \frac{\Phi_0}{2\pi}\frac{d\theta}{dt}$ where $\Phi_0 = h/2e$ is one flux quantum. The derivative of the current flow is therefore

$$\frac{dI}{dt} = I_0 \cos(\theta)\frac{d\theta}{dt} = I_0 \cos(\theta)\frac{2\pi}{\Phi_0}V.$$

Comparing to the equation for an inductor $\frac{dI}{dt} = V/L$ we identify the Josephson inductance as

$$L_J = \frac{\Phi_0}{2\pi I_0 \cos(\theta)}.$$

Thus we have a nonlinear, phase dependent inductance that is infinite for $\theta \to (2p+1)\pi/2$ and is negative for $\pi/2 < \theta < 3\pi/2$. At zero phase $L_{J0} = \Phi_0/(2\pi I_0)$ and I_0 is defined as the critical current.

It will be convenient to re-express the phase in terms of the magnetic flux across the junction. We have

$$\theta(t) = \frac{q}{\hbar}\int_{-\infty}^{t} dt'\, V(t') = \frac{2\pi}{\Phi_0}\int_{-\infty}^{t} dt'\, V(t').$$

For a loop of wire with a time varying magnetic flux passing through it the voltage generated can be expressed as (Faraday's law) $V = \frac{d\Phi_{\text{loop}}}{dt}$ so $\Phi_{\text{loop}}(t) = \int_{-\infty}^{t} dt'\, V(t')$. We can generalize this to any electrical element with two leads and define $\Phi(t) = \int_{-\infty}^{t} dt'\, V(t')$ so that $\theta(t) = 2\pi\Phi(t)/\Phi_0$ and the Josephson current relation in terms of the flux is

$$I(t) = I_0 \sin\left[2\pi\,\Phi(t)/\Phi_0\right].$$

A very useful circuit construction is shown in Fig. 15.4b. Known as a superconducting quantum interference device or SQUID it consists of a superconducting ring with two Josephson junctions. The total circulating current is $J = J_a + J_b$ and for two identical junctions $J_a = J_0 \sin(\theta_a)$, $J_b = J_0 \sin(\theta_b)$. Since the voltage difference around the loop must vanish $\theta_{a,b} = \theta(0) \pm \frac{q}{\hbar}\int_0^t dt'\, V(t')$. We can therefore write

$$J = J_0\left[\sin\left[\theta(0) + \frac{q}{\hbar}\int_0^t dt'\, V(t')\right] + \sin\left[\theta(0) - \frac{q}{\hbar}\int_0^t dt'\, V(t')\right]\right]$$

$$= 2J_0 \sin[\theta(0)]\cos\left[\frac{q}{\hbar}\int_0^t dt'\, V(t')\right].$$

However $\frac{q}{\hbar} \int_0^t dt' V(t') = \frac{\theta_a - \theta_b}{2} = \pi \Phi/\Phi_0$ where Φ is the control flux (see Eq. (15.6)). Therefore the current is

$$J = 2J_0 \sin[\theta(0)] \cos(\pi \Phi/\Phi_0) = -2J_0 \sin[\theta(0)] \sin[\pi(\Phi - \Phi_0/2)/\Phi_0]$$
$$(15.7)$$

which depends on the flux through the junction in units of the flux quantum. This circuit element will be useful for flux control of qubits.

15.6   Qubit Designs

There are a variety of ways in which linear and nonlinear elements can be combined to create superconducting qubits. The design trade offs involve minimizing sensitivity to charge or flux noise in the environment while maintaining adequate anharmonicity. The three basic qubit types are the charge qubit, also called a Cooper pair box, the flux qubit based on an RF-SQUID, and the phase qubit based on a current biased junction. These are shown in Fig. 15.6. It should be noted that the names charge, flux, and phase qubit do not imply that the qubits are encoded in states of definite charge, flux, or phase. In general the qubit states are superpositions of these variables. The nomenclature refers to the fact that in order to control the qubit dynamics we couple to the charge, flux, or phase.

15.6.1   Charge Qubit

The charge qubit in the number, phase basis is described by the Hamiltonian

$$H = E_C(\hat{n} - n_g)^2 - E_J \cos(\hat{\theta}). \qquad (15.8)$$

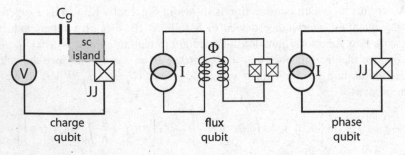

Fig. 15.6   Primary tupes of superconducting qubits: charge, flux, and phase

Here $E_C = \frac{(2e)^2}{2(C_J + C_g)}$ is the electrostatic charging energy of the superconducting island formed by the gate capacitor and the tunnel junction, $\hat{n}$ is the charge number operator and $n_g = C_g V / (2e)$ is the offset charge on the superconducting island formed by one side of the gate capacitor and the tunnel junction. The last term is the Josephson energy with $E_J = I_0 \Phi_0 / 2\pi$. To see this we recognize that the work done on the junction by the current is

$$U = \int_0^t dt' \, I V(t') = \int_0^t dt' \, I \frac{\Phi_0}{2\pi} \frac{d\theta}{dt'}$$

$$= \frac{\Phi_0}{2\pi} \int_0^\theta d\theta \, I_0 \sin(\theta) = \frac{\Phi_0 I_0}{2\pi} (1 - \cos(\theta)).$$

Neglecting the constant term gives the Josephson energy. This was the first superconducting qubit that was demonstrated [9] and is operated in the limit of $E_C \gg E_J$. Choosing V so that $n_g = 1/2$ the charge states $n = 0$ and $n = 1$ are degenerate. The degeneracy is broken by E_J.

It can be shown that the effective qubit Hamiltonian (15.8) acting on the two lowest charge eigenstates can be written as

$$H = E_C(n_g - 1/2)\sigma_z - \frac{E_J}{2}\sigma_x. \tag{15.9}$$

The operators are expressed in the basis $|+\rangle, |-\rangle$ of $n = 1, 0$. Eigenstates $|0\rangle, |1\rangle$ which can serve as a qubit basis are

$$|0\rangle = \cos(\chi/2)|+\rangle + \sin(\chi/2)|-\rangle$$
$$|1\rangle = -\sin(\chi/2)|+\rangle + \cos(\chi/2)|-\rangle$$

with eigenvalues $U_{0,1} = \pm U$, $U = \sqrt{E_C^2(n_g - 1/2)^2 + E_J^2/4}$, $\cos(\chi) = E_C(n_g - 1/2)/U$, $\sin(\chi) = -(E_J/2)/U$. When the charging energy dominates, $E_C \gg E_J$, $|0\rangle \approx |+\rangle$, $|1\rangle \approx |-\rangle$ which are states of definite charge. At $n_g = 1/2$ the eigenstates are equal superpositions of the charge states and there is an avoided crossing of width E_J. Choosing V so $n_g = 1/2$ the energy spacing is first order insensitive to n_g which minimizes the sensitivity to noise.

Also the phase degree of freedom can be controlled if we modify the circuit by replacing the Josephson junction with two junctions in a SQUID geometry as shown in Fig. 15.4b. Using Eq. (15.7) for the current the qubit Hamiltonian becomes

$$H = E_C(\hat{n} - n_g)^2 + 2E_J \sin[\theta(0)] \cos[\pi(\Phi - \Phi_0)/\Phi_0]. \tag{15.10}$$

The current, and therefore the Josephson energy, can be changed between the maximum and minimum values by changing the control flux by an amount Φ_0.

Returning to the charge qubit, to perform one-qubit gates we can bias the qubit at a point with $n_g \neq 1/2$ that is not close to the degeneracy point. The eigenstates are then approximately states of definite charge $|+\rangle$ and $|-\rangle$. One qubit Z gates are applied by changing n_g by a small amount δn_g for a short time δt which imparts a differential phase $\phi = E_c \delta n_g \delta t$. One qubit X gates are performed by rapidly switching to the degeneracy point $n_g = 1/2$ waiting a time δt and then switching back. This gives the operation $U = e^{i \frac{E_J \delta t}{2\hbar} \sigma_x}$. Combining these operations we can perform arbitrary one-qubit gates.

Alternatively if the junction is replaced by a flux controllable SQUID we can use flux pulses to tune E_J. If we set V so $n_g = 1/2$ and $\Phi = \Phi_0/2$ the Hamiltonian vanishes and the qubit state will not evolve in time. Changing V or Φ in a controlled manner will then selectively generate rotations about z or x so arbitrary single qubit gates can be performed.

The quantum state of the charge qubit can be measured in several ways. One possibility is to couple a charge trap to the qubit and use a single electron transistor (SET) to readout the charge. When the qubit is operated away from $n_g = 1/2$ the basis states are nearly pure charge states so a measurement of the charge reveals the qubit state. To perform two-qubit gates coupling between the qubits is needed. This was originally done using a capacitor between two qubits fabricated on the same chip very close to each other [10].

15.6.2   Flux Qubit

The next type of qubit is the flux qubit based on an RF SQUID [11]. In this device a transformer is used for control instead of a gate capacitor. It can be shown that the Hamiltonian is

$$H = \frac{q^2}{2C_J} + \frac{\phi^2}{2L} - E_J \cos\left[2\pi(\phi - \Phi_{\text{ext}})/\Phi_0\right].$$

Here ϕ is the voltage across the inductor and Φ_{ext} is the externally supplied flux. The device is biased with a flux $\Phi_0/2$ which results in two potential wells with a barrier in between. The situation is akin to that of the ammonia molecule. The lowest energy states are the symmetric and antisymmetric combinations of the wavefunctions localized in each well. The energy separation of these states is determined by the strength of the tunnel coupling between the potential wells which is exponentially sensitive to the height of the barrier. The splitting can be written as

$$E_S \sim \sqrt{E_B E_{C_J}}\, e^{-\xi \sqrt{E_B/E_{C_J}}}.$$

Here E_B is the barrier height, E_{C_J} is the charging energy of the junction capacitance, and ξ has to be found numerically from modeling of device parameters. The flux qubit is operated with $E_J \gg E_C$, the opposite limit of the charge qubit.

The control parameter is the external flux which is modulated by sending a current through the transformer. This type of qubit has been demonstrated but has not reached a level of performance comparable to other approaches [12].

15.6.3 Phase Qubit

The phase qubit is the third basic type [13, 14]. The junction is biased with a current source. The resulting potential is anharmonic and can be shown to be of the form

$$U(\theta) = \Phi_0 \left[-I_0 \cos(\theta) - I\theta \right]$$

with θ the phase across the junction. The combination of the linear and sinusoidal term gives a so-called washboard potential. The qubit states are the lowest two energy levels in the anharmonic potential. Readout of the qubit state is accomplished by adjusting the dc bias of the circuit which lowers the barrier and allows the $|1\rangle$ state (or higher states) to tunnel out thereby creating a measurable current.

Initial devices had short coherence times well below $1\,\mu s$. Improvements in the device design and extensive studies to identify alternative materials with fewer defects increased the coherence time to the μs scale. These improved phase qubits were the first superconducting qubits to demonstrate high fidelity Bell states with $\mathcal{F} = 0.87$ after correction for measurement errors [15]. Nevertheless it has not been possible to further improve the coherence properties and phase qubits are not currently a prime contender for scaling up superconducting quantum circuits.

15.7 Transmon

The three basic types of charge, flux, and phase qubit operate with different ratios of the charging and Josephson energies. The charge qubit has $E_J \ll E_C$. The flux qubit has $E_J > E_C$ and the phase qubit has $E_J \gg E_{C_J}$. For all types a primary challenge has been increasing the coherence time. A large number of different qubit designs have been proposed and demonstrated: fluxonium, quantronium, Xmon, gatemon, transmon. The transmon in particular has demonstrated that much longer coherence times and high gate fidelity are possible. These devices form the basis of the current era of superconducting processors with more than 50 qubits.

The transmon is essentially a charge qubit based on a Cooper pair box but with a much larger ratio of E_J/E_C typically 10–100, whereas the charge qubit has $E_J \ll E_C$. Increasing this ratio has two important effects. The dependence of the qubit energy levels on n_g is exponentially reduced with the ratio E_J/E_C. The result is that sensitivity to charge noise and to the external voltage V that controls the gate charge n_g is reduced resulting in much longer coherence times.

At the same time the anharmonicity of the potential is decreased, which is undesirable. However the anharmonicity only decreases as a power law in E_J/E_C which is tolerable. With the reduced anharmonicity and lack of sensitivity to V the qubit is more difficult to control. The use of phase control with the RF-SQUID type approach described in the section on charge qubits, together with developments in pulse design have enabled high fidelity gates despite the smaller anharmonicity.

The Hamiltonian of the charge qubit is

$$H = E_C(\hat{n} - n_g)^2 - E_J \cos(\hat{\theta}). \quad (15.8)$$

In the phase representation the number operator is $\hat{n} = -i\frac{\partial}{\partial\theta}$ so the time independent Schrödinger equation is

$$\left[E_C \left(-i\frac{\partial}{\partial\theta} - n_g \right)^2 - E_J \cos(\theta) \right] \psi(\theta) = E\psi(\theta).$$

Introducing the function $g(\phi) = e^{-i2n_g\phi}\psi(2\phi)$ [16] we obtain

$$\frac{d^2 g(\phi)}{d\phi^2} + \left[\frac{E}{(E_C/4)} + \frac{E_J}{(E_C/4)} \cos(2\phi) \right] g(\phi) = 0.$$

This is a Mathieu equation that can be solved exactly for the eigenfunctions and eigenvalues.

As the ratio E_J/E_C is increased two things happen. First, the eigenenergies become more and more insensitive to n_g which implies a reduced sensitivity to the control voltage V. Second, the anharmonicity is reduced which makes the circuit less suitable for qubit encoding. Detailed analysis [16] shows that eigenstate k has an energy dispersion proportional to $\epsilon_k \cos(2\pi n_g)$ with

$$\epsilon_k \simeq (-1)^k E_C \frac{2^{4k+3}}{k!} \sqrt{\frac{2}{\pi}} \left(\frac{2E_J}{E_C} \right)^{\frac{k}{2}+\frac{3}{4}} e^{-\sqrt{2E_J/E_C}}$$

which decreases exponentially with the ratio E_J/E_C. At the same time the relative anharmonicity of levels 2 and 1 compared to levels 1 and 0 is

$$\alpha = \frac{E_{21} - E_{10}}{E_{10}} \simeq -\frac{1}{\sqrt{32E_J/E_C}}.$$

The anharmonicity must be sufficient to selectively address the $0 \leftrightarrow 1$ transition with a pulse that is short compared to the T_1, T_2 coherence times yet satisfies $\omega_{10}|\alpha| \gg 1/t$. Putting in numbers for $t = 10\,\text{ns}$ we find the condition $E_J/E_C \ll 10^4$. Indeed the energy ratio can be very large while retaining sufficient anharmonicity to address the qubit transition with minimal coupling to the next transition. The expression for the energy dispersion then implies an exponentially small sensitivity to n_g. Transmon devices are typically operated with $E_J/E_C \sim$ 10–50.

15.8 Coupling Qubits to Resonators

Circuit architectures based on coupling of qubits to microwave resonators are known as circuit-Quantum Electrodynamics (cQED). This name is in analogy to Cavity Quantum Electrodynamics (CQED) which originated in studies of the quantum mechanical coupling between single atoms and single photons in optical resonators. The same type of physics can be accessed with superconducting qubits coupled to microwave frequency resonators, which instead of being built with three dimensional mirrors are based on planar waveguides that act as resonators. An example is shown in Fig. 15.1a. Compared to experiments with single atoms, including highly excited Rydberg atoms that have large transition matrix elements, the qubit-microwave coupling strength in cQED is exceptionally large which makes possible a broad range of quantum experiments.

The basic model describing linear coupling of a two-level system (the transmon) to a single mode oscillator is the quantum Rabi model

$$\hat{\mathcal{H}} = \frac{1}{2}E\hat{\sigma}_z + \hbar\omega_r\left(\frac{1}{2} + \hat{a}^\dagger\hat{a}\right) + \hbar g\hat{\sigma}_x(\hat{a} + \hat{a}^\dagger).$$

The first term is the qubit energy $E = E_1 - E_0$, the second term is the energy in the resonator mode at frequency ω_r, and the third term is the resonator—qubit coupling with coupling constant g. This model includes rotating and counter-rotating terms which can be separated out by introducing the qubit raising and lowering operators $\hat{\sigma}_\pm = (\hat{\sigma}_x \pm i\hat{\sigma}_y)/2$. This gives

$$\hat{\mathcal{H}} = \frac{1}{2}E\hat{\sigma}_z + \hbar\omega_r\left(\frac{1}{2} + \hat{a}^\dagger\hat{a}\right) + \hbar g(\hat{\sigma}_+\hat{a} + \sigma_-\hat{a}^\dagger) + \hbar g(\hat{\sigma}_+\hat{a}^\dagger + \sigma_-\hat{a}).$$

The last term is the counter-rotating term we met before when discussing atom-field coupling. It corresponds to raising the qubit state while emitting a photon or lowering the qubit state while annihilating a photon. Neglecting this term we are left with the energy conserving qubit-field interaction which corresponds to making the rotating wave approximation. The terms describing the energy of the qubit and the field and the energy conserving interaction comprise the Jaynes–Cummings model.

The behavior of the coupled qubit-resonator system depends on five parameters: the qubit—field coupling g, the qubit energy E, the resonator frequency ω_r, the resonator decay rate or line width κ, and the qubit decay rate $\gamma = 1/T_1$. The most important regime for qubit experiments is the strong coupling limit of

$$\gamma, \kappa \ll g \ll E, \omega_r$$

together with validity of the rotating wave approximation when $|\hbar\omega_r - E| = |\Delta| \ll \omega_r$. These quantities can be combined into a single cooperativity parameter

$$C = \frac{g^2}{\kappa\gamma}.$$

When $C \gg 1$ the single photon—single qubit coupling rate is large compared to the resonator and qubit decoherence rates which makes coherent quantum dynamics possible. Parameters for realistic devices are $g/2\pi \sim 10\,\text{MHz}$, resonator quality factors $Q \sim 10^7$ at $\omega_r/2\pi \sim 10\,\text{GHz}$, so $\kappa/2\pi = \omega_r/2\pi Q \sim 1\,\text{kHz}$, and $T_1 = 100\,\mu\text{s}$ giving $\gamma/2\pi = 1.6\,\text{kHz}$. With these values we find $C > 10^7$. Such large cooperativity is unheard of in experiments with real atoms where $C \lesssim 200$ has been the limit.

Neglecting the zero point energy of the field we have the Hamiltonian

$$\hat{\mathcal{H}}_{\text{JC}} = \frac{1}{2}E\hat{\sigma}_z + \hbar\omega_r\hat{a}^\dagger\hat{a} + \hbar g(\hat{\sigma}_+\hat{a} + \sigma_-\hat{a}^\dagger).$$

In the limit of large detuning $|\Delta| \gg g$ we get the approximate Hamiltonian

$$\hat{\mathcal{H}} = \frac{1}{2}\left(E + \frac{\hbar g^2}{\Delta}\right)\hat{\sigma}_z + \left(\hbar\omega_r + \frac{\hbar g^2}{\Delta}\sigma_z\right)\hat{a}^\dagger\hat{a}.$$

This is essentially the same approximation as in the discussion of the Kerr nonlinearity for single photon phase gates in Sect. 14.4. In this limit the qubit energy is Stark shifted by g^2/Δ and the field energy is shifted depending on the state of the qubit. By recording the phase of the oscillator field this can be used to make a non-destructive measurement of the qubit state. This type of qubit-resonator coupling was first demonstrated in Ref. [17].

Single qubit gates can be implemented using the techniques already discussed for the charge qubit. Two-qubit gates can be performed by direct capacitive coupling of the qubit charges, by coupling via a Josephson junction circuit, or by coupling via the resonator. In this last case the effective Hamiltonian for identical qubits and identical couplings takes the form [18, 19]

$$\hat{\mathcal{H}}_{2q} = \frac{1}{2}\left(E + \frac{\hbar g^2}{\Delta}\right)(\hat{\sigma}_{z,1} + \hat{\sigma}_{z,2}) + \left[\hbar\omega_r + \frac{\hbar g^2}{\Delta}(\hat{\sigma}_{z,1} + \hat{\sigma}_{z,2})\right]\hat{a}^\dagger\hat{a}$$

$$+ \frac{\hbar g^2}{\Delta}\left(\hat{\sigma}_{+,1}\hat{\sigma}_{-,2} + \hat{\sigma}_{-,1}\hat{\sigma}_{+,2}\right). \tag{15.11}$$

The first two terms are the additive Stark shifts of the qubits and the resonator by dispersive coupling. The last term provides a direct qubit-qubit interaction due to exchange of virtual photons across the resonator. This allows for qubit coupling without physical proximity as well as the capability to couple multiple qubits, all connected via a shared resonator bus.

When the qubits are sufficiently far detuned from the resonator their interaction is weak. By changing the qubit tuning the resonator mediated interaction can be turned on and used to implement a two-qubit gate. The last term in Eq. (15.11) swaps the state of the qubits. Turning on the interaction for a time $g^2 t / \Delta = \pi/2$ gives a swap operation up to a factor of i, also called the iSWAP gate. Combining two iSWAP gates or two $\sqrt{\text{iSWAP}}$ gates with single qubit operations it is possible to synthesize a CNOT gate. Therefore the resonator coupling can be used to implement a universal set of quantum gates.

15.9 Protected Qubits

Scaling quantum computing to large numbers of qubits and arbitrarily deep circuits remains an outstanding challenge. In large part because of the huge overheads associated with quantum error correction. With gate errors at the 10^{-3}–10^{-4} levels that appear within reach of optimized hardware and error correcting surface codes with thresholds at the 10^{-2} level there is a very substantial resource requirement in terms of qubit count to suppress the logical error rate. Long computations that provide strong quantum advantage may require only a few hundred protected logical qubits, but $>10^5$ physical qubits.

A possible alternative to error correcting codes is to design hardware that is intrinsically protected against decoherence, thereby reducing the overhead of logical error correction. One direction for this research has been in engineering materials that support exotic non-Abelian anyons which afford topological protection [20]. Another approach that is closer to present experimental capabilities is based on new types of superconducting qubits.

One such approach is the "$0 - \pi$" qubit [21, 22]. This qubit is based on a circuit with energy that is a function of the phase θ between the two leads. There are two nearly degenerate ground states with $\theta = 0$ and π. The energy splitting is exponentially small as a function of external parameters and stable with respect to weak local perturbations so it is expected to be resistant to decoherence due to local noise.

It can be shown that the circuit in Fig. 15.7 has energy

$$E = f(\theta_2 + \theta_4 - \theta_1 - \theta_3) + \mathcal{O}\left(e^{-\frac{1}{8}\sqrt{\frac{L}{C}}}\right)$$

where $f(\theta)$ is a 2π periodic function. Provided $L \gg C$ the additional correction is exponentially small. To make a qubit the circuit is twisted and reconnected so θ_2

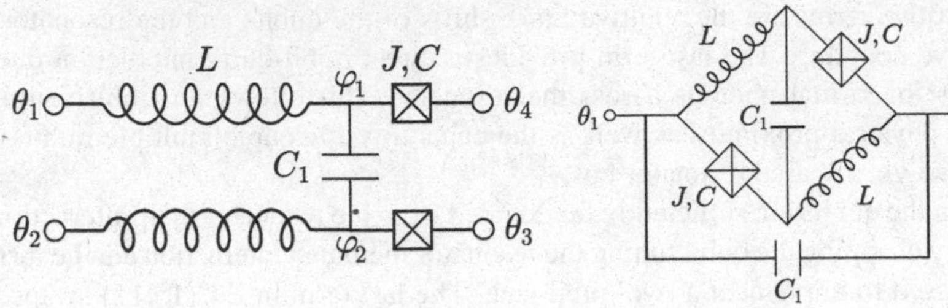

Fig. 15.7 Circuit of "$0 - \pi$" qubit from [21]

and θ_4 are the same and θ_1 and θ_3 are the same. The energy is then

$$E = f(2(\theta_2 - \theta_1)) + \mathcal{O}\left(e^{-\frac{1}{8}\sqrt{\frac{L}{C}}}\right)$$

which is periodic in the phase difference across the circuit with period π. The smallness of the correction term depends on a large ratio L/C. One way to achieve this is with a chain of N Josephson junctions. The inductance of the chain will scale with N and the capacitance as $1/N$ so $\sqrt{L/C} \sim N$. Ratios of $\sqrt{L/C} \approx 20$ have been demonstrated with this approach. Detailed analysis including gate protocols and error estimates have been given in [21]. Estimates of decoherence times suggest that 20 ms, which would be an order of magnitude improvement over current devices, may be possible with available devices [23].

15.10 Outlook

The cQED architecture with transmon type qubits connected via microwave resonators forms the basis for the 50 qubit scale processors that have been implemented by Google and IBM. In the Google device [2] the qubits are each connected via resonators to four nearest neighbors on a square grid. The qubits are tuned into resonance and the interaction is additionally controlled using tunable couplers. In the IBM device the qubit-resonator coupling is fixed and the qubit frequencies are controlled to implement gate operations. Current efforts are aimed at further scaling of the parameters needed for quantum computing including coherence time, gate fidelity and qubit count, as well as development of scalable methods for qubit measurement and control.

There is also an interesting complementary approach that switches the roles of superconducting elements and microwaves. The cQED architecture uses superconducting circuits as the qubits and virtual microwave excitation of resonators for coupling. Instead we might use microwave photons as the qubits and Josephson junctions to effectively implement photon-photon coupling.

Using 3D resonators with larger volume to surface ratio than is possible with planar waveguides very long coherence times exceeding 1 ms, about 10× longer than has been possible for Josephson junction qubits, have been achieved in this architecture. The 3D resonators support multiple resonant frequencies and can therefore store multiple qubits. With this approach qubits stored as photons have been deterministically entangled [24] a feat that has not been achieved with linear optics approaches. It remains to be seen how well this architecture will scale to large qubit number.

15.11 Quantum Dot Qubits: Overview

Advances in information processing that permeate the modern world are built on semiconductor electronic devices. Modern computer processors produced in sophisticated facilities pack billions of logic elements into a single chip with remarkable levels of engineered complexity. In light of our capability to fabricate highly complex semiconductor electronics it is natural to ask if we can implement qubits in semiconductor devices, and leverage established engineering methodologies to build scalable quantum processors.

In principle single electrons with spin 1/2 are ideal qubits. The spin degree of freedom may be used to encode a qubit with states $|0\rangle = |\uparrow\rangle, |1\rangle = |\downarrow\rangle$ corresponding to spin projection $m = \pm 1/2$. In a magnetic field B directed along the quantization direction the states have energies $U_{0/1} = \pm\mu_B B$, with μ_B the Bohr magneton. In a large, but not extreme, field of 0.1 T the states are separated by 2.8 GHz, a value comparable to what we have encountered with atomic or superconducting qubits.

The challenge is how to trap, control, and measure single electron qubits in order to satisfy the DiVincenzo criteria for computation [25]. The hyperfine qubits discussed in Chap. 13 are encoded in coupled states of electronic and nuclear degrees of freedom with the neutral atoms or ions essentially providing a localized potential that binds the electrons. Implementations require substantial optical and laser technologies that are difficult to integrate at the scale that has been achieved with semiconductor electronics. An alternative approach to implementing electronic qubits that does not require lasers is based on preparing an array of single electrons floating on a film of liquid He [26]. Another idea is to use a Penning trap for electrons, instead of ions [27]. While intriguing, neither of these ideas have been developed particularly far.

The semiconductor approach to single electron qubits that has been extensively developed stems from the ideas of Loss and DiVincenzo [28] who proposed a scalable architecture based on quantum dots. The basic architecture is illustrated in Fig. 15.8. A semiconductor heterostructure is patterned with electrodes on the top surface. Voltages applied to the electrodes define potential wells that trap single electrons. By appropriate adjustment of the voltage settings it is possible to prepare stable charge states at the single or few electron level. The trapping potential

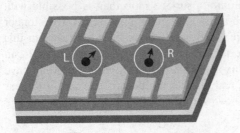

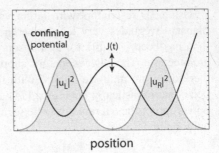

position

Fig. 15.8 Quantum dot qubits in a semiconductor heterostructure. Left: Two quantum dot qubits in top-gated potential wells. Voltages applied to the pads are used to define the confining potential. Typical scale of devices is a dot separation of 100 nm. Right: When the barrier is high the electrons are localized in individual wells. Adjusting the barrier height implements gates by modifying the wavefunction overlap and exchange coupling

has orbital bound states akin to those found in an atom, so this structure can be thought of as an artificial atom. Initialization, measurement, and gate operations are implemented with a combination of DC and AC voltages and fields applied to the device, as we proceed to discuss in more detail.

15.12 State Preparation and Measurement

Electrons are loaded into the top gated heterostructure by careful adjustment of gate voltages until the desired states with single electron occupancy are attained. The procedures to achieve this are delicate, but are now well established [29]. After loading, electrons can be trapped essentially indefinitely, provided the apparatus is well isolated from stray charges.

A static magnetic field is applied to define a quantization axis for the spins. When the Zeeman energy is large compared to $k_B T$ the electron will eventually relax to the lower energy spin down state. A 2.8 GHz qubit frequency corresponds to $T = 135$ mK so dilution refrigerators at mK temperatures are required, in part for initialization, but also to ensure long coherence times in the solid state host. One way to evade the low temperature requirement is to work at large field strengths. Qubit operation at a high field strength near 1.4 T corresponding to a qubit frequency ~40 GHz has been demonstrated in a 1.5 K cryostat [30].

Relaxation to the lowest energy spin state proceeds at a rate determined by local fields and scattering from material imperfections. The time for this to occur may be inconveniently long in which case initialization can be accelerated by active coupling to interface states that cause enhanced spin-orbit coupling and faster relaxation.

Measurement of the spin state of a single electron can be performed using spin dependent tunneling. The basic idea is to use the difference in Zeeman energy of the spin states in a magnetic field to allow $|\uparrow\rangle$ say, to tunnel out of the well while $|\downarrow\rangle$ is confined. A quantum point contact (QPC), essentially a narrow constriction between two conducting regions, is located next to the quantum dot. The QPC is electrostatically coupled to the charge state of the dot which modifies the tunneling current through the QPC. In this way the presence or absence of a charge, which reveals the spin state, can be used to switch a nA current that is detectable. Initial experiments [31] had relatively low measurement fidelity ~65%, but refinements have been developed that have pushed single shot measurement fidelities to above 99%. The spin dependent measurement can also be operated in reverse as an alternative method of initialization [32]. Thus single electron tunneling from a charge reservoir that has been prepared in a known spin state can be used to simultaneously load and initialize the quantum dot.

This type of spin-dependent tunneling is intrinsically a destructive measurement since electrons in one of the spin states are removed. For practical error correction it is desirable to make non-destructive measurements so the qubit can be reused. This has been accomplished by entangling the qubit to be measured with an ancilla, and then measuring the state of the ancilla which enables a projective measurement of the qubit state [33]. The method still requires replacement of the ancilla that is measured destructively, but does provide additional architectural flexibility in the design of circuits, as well as the ability to repeatedly observe the dynamics of a single electron.

Alternatively it is possible to use spin-dependent tunneling of a qubit encoded in the singlet and triplet states of two electrons to make a non-destructive measurement [34]. Due to the Pauli exclusion principal the triplet state of two electrons in a single well must involve occupation of an excited orbital whereas in the singlet state both electrons can be in the ground orbital. The difference in tunneling rates for excited and ground state orbitals makes it possible to arrange for the triplet state to tunnel out of the dot, and the change in charge to be detected by a QPC, followed by tunneling back to the original state, whereas the singlet state does not tunnel.

15.13 One-Qubit Gates

A large dc magnetic field B_0 defines a quantization axis for qubits encoded in the spin state. One-qubit gates are most directly implemented with a resonant microwave field at frequency $\omega_q = 2\mu_B B_0/\hbar$. The magnetic field couples to the spin leading to the same Rabi equations as we have seen earlier for atomic qubits. Working with a magnetic field $\mathbf{B} = B_0 \mathbf{e}_z + B_x \cos(\omega_q t)\mathbf{e}_x$, B_0 sets the qubit frequency and B_x drives rotations about the $\mathbf{e}_x$ axis. Making the usual rotating wave approximation we find a Rabi frequency of $\Omega = \mu_B B_x/\hbar$. In order to achieve fast rotations with moderate microwave power it is necessary to use near field coupling to concentrate the energy in an area much less than a square wavelength. This was first achieved in a GaAs double quantum dot structure with Rabi frequencies near 10 MHz demonstrated [35].

It is also possible to resonantly couple to an oscillating electric field in materials such as GaAs that have large spin-orbit coupling [36]. The electric field causes the electron position to oscillate and the spin-orbit coupling results in an effective magnetic field that couples to the spin. In systems without spin-orbit coupling the same mechanism can effectively be achieved by applying a magnetic field gradient. Essentially the same type of magnetic gradient approach has been used with trapped ion qubits to implement gates without relying on laser pulses [37]. Using coupling to a microwave frequency electric field, Rabi frequencies in excess of 100 MHz and gate fidelities $\sim$95% have been demonstrated on a three-electron hybrid qubit [38].

15.14 Two-Qubit Gates

Two-qubit gates are implemented by raising and lowering the electrostatic barrier between the wells. When the barrier is high tunneling is suppressed and the qubit states do not evolve. When the barrier is lowered tunneling is allowed and the spins experience a Heisenberg interaction

$$\hat{\mathcal{H}}(t) = J(t)\hat{s}_1 \cdot \hat{s}_2. \tag{15.12}$$

Here $J(t)$ is a time dependent coupling constant that depends on the control voltages and barrier parameters, $\hat{s}$ is the spin operator and 1, 2 label the qubits. The interaction of Eq. (15.12) accurately describes interacting quantum dots provided that the higher lying orbital states of the dot are not occupied. This will be the case if the level spacing within the dot is large compared to the thermal energy, $\delta U \gg k_B T$ and the gate time t_{gate} satisfies $t_{\text{gate}} \gg \hbar/\delta U$.

This form of interaction is not due to the direct coupling of the electron spins $U_{\text{dd}} \sim \mu_B^2/d^3$ with d the electron separation, which is negligible for typical experimental parameters. Rather it is an exchange interaction that arises from the Coulomb coupling of the electrons, analogous to that which takes place in any two-electron atom. Electrons obey Fermi-Dirac statistics so the quantum state $\Psi(1, 2)$ of two-electrons including spatial and spin degrees of freedom must be antisymmetric upon particle exchange. The lowest energy state has one electron in each well to minimize the Coulomb repulsion. These states are in general a superposition of symmetric and antisymmetric spatial states denoted $\psi_s(1, 2)$, $\psi_{\text{as}}(1, 2)$ and symmetric and antisymmetric spin triplet and singlet states denoted $\chi_s(1, 2)$, $\chi_{\text{as}}(1, 2)$. A general two-electron state in the low energy sector can then be written as

$$\Psi(1, 2) = c_1 \psi_s(1, 2)\chi_{\text{as}}(1, 2) + c_2 \psi_{\text{as}}(1, 2)\chi_s(1, 2)$$

with c_1, c_2 complex coefficients and $\psi_s(1, 2) = \frac{u_L(\mathbf{r}_1)u_R(\mathbf{r}_2)+u_R(\mathbf{r}_1)u_L(\mathbf{r}_2)}{\sqrt{2}}$, $\psi_{\text{as}}(1, 2) = \frac{u_L(\mathbf{r}_1)u_R(\mathbf{r}_2)-u_R(\mathbf{r}_1)u_L(\mathbf{r}_2)}{\sqrt{2}}$ where u_L, u_R are wavefunctions localized in the left and right wells.

The energy eigenstates including the Coulomb interaction as a perturbation are found from solving $\hat{\mathcal{H}}\Psi(1,2) = E\Psi(1,2)$ with $\hat{\mathcal{H}} = \hat{\mathcal{H}}_0 + \hat{\mathcal{H}}_C$ where $\hat{\mathcal{H}}_0$ is due to the external potential and $\hat{\mathcal{H}}_C = \frac{q^2}{|\mathbf{r}_1 - \mathbf{r}_2|}$ is the Coulomb interaction with $q^2 = \frac{e^2}{4\pi\epsilon}$, ϵ is the local dielectric constant in the host. Since the Hamiltonian is independent of the spin state we need consider only the spatial states. The symmetric and antisymmetric spatial states are degenerate with respect to the external potential which leaves the perturbative term $\hat{\mathcal{H}}_C\psi(1,2) = E_C\psi(1,2)$. In the basis of symmetric and antisymmetric states this can be written as

$$\begin{pmatrix} \hat{\mathcal{H}}_{11} & \hat{\mathcal{H}}_{12} \\ \hat{\mathcal{H}}_{21} & \hat{\mathcal{H}}_{22} \end{pmatrix} \begin{pmatrix} c_1 \\ c_2 \end{pmatrix} = E_C \begin{pmatrix} c_1 \\ c_2 \end{pmatrix}.$$

The components of the Hamiltonian can be written in the form

$$\hat{\mathcal{H}}_{11} = D + K,$$
$$\hat{\mathcal{H}}_{22} = D - K,$$
$$\hat{\mathcal{H}}_{12} = \hat{\mathcal{H}}_{21} = 0$$

where the direct and exchange integrals are

$$D = q^2 \int \int d\mathbf{r}_1 d\mathbf{r}_2 \frac{|u_L(\mathbf{r}_1)|^2 |u_R(\mathbf{r}_2)|^2}{|\mathbf{r}_1 - \mathbf{r}_2|},$$

$$K = q^2 \int \int d\mathbf{r}_1 d\mathbf{r}_2 \frac{u_L^*(\mathbf{r}_1) u_R(\mathbf{r}_1) u_L(\mathbf{r}_2) u_R^*(\mathbf{r}_2)}{|\mathbf{r}_1 - \mathbf{r}_2|},$$

and $K = K^*$. We see that the symmetric and antisymmetric states are eigenstates of the Coulomb interaction with an energy splitting of $2K$.

The two electron eigenstates are therefore $\Psi_S = \psi_s(1,2)\chi_{as}(1,2)$ and $\Psi_T = \psi_{as}(1,2)\chi_s(1,2)$ where S, T stand for singlet and triplet and the eigenvalues are $E_{S,T} = D \pm K$. The symmetric spatial state has higher energy since the electrons have higher probability density near $|\mathbf{r}_1 - \mathbf{r}_2| = 0$ than the antisymmetric state does, and therefore stronger Coulomb repulsion. Thus the Coulomb interaction combined with the symmetry imposed by fermionic statistics leads to an effective spin dependent or exchange interaction. The eigenvalues can be written in terms of the spin operators as

$$E_{S,T} = D - \frac{K}{2} - 2K\langle \mathbf{s}_1 \cdot \mathbf{s}_2 \rangle$$

since $\langle \mathbf{s}_1 \cdot \mathbf{s}_2 \rangle = -3/4, 1/4$ for the singlet and triplet states. Neglecting the spin independent offset we have an effective exchange Hamiltonian given by $\hat{\mathcal{H}}_{ex} = -2K\mathbf{s}_1 \cdot \mathbf{s}_2$ which recovers Eq. (15.12) with $J(t) = -2K(t)$. The exchange coefficient K acquires a time dependence by changing the barrier height which affects the overlap of the wavefunctions in the integral.

The protocol for a two-qubit gate proceeds as follows. The barrier voltage is initially set high so that $J(0)$ is negligible. The voltage is then reduced to increase $J(t)$ for a desired time, and then returned to the high value. The time evolution operator is

$$\mathbf{U}(t) = e^{\iota \int_0^t dt'\, \hat{\mathcal{H}}_{ex}(t')/\hbar} = e^{\iota 2 \int_0^t dt'\, K(t') \mathbf{s}_1 \cdot \mathbf{s}_2/\hbar}.$$

Choosing the voltages and gate time such that $\int_0^t dt'\, 2K(t')/\hbar = \pi$ we obtain the two-qubit interaction

$$\mathbf{U} = e^{\iota \frac{\pi}{4}\sigma_1 \cdot \sigma_2} = e^{\iota \pi/4} \begin{pmatrix} 1 & 0 & 0 & 0 \\ 0 & 0 & 1 & 0 \\ 0 & 1 & 0 & 0 \\ 0 & 0 & 0 & 1 \end{pmatrix} \tag{15.13}$$

which is a SWAP gate in the spin basis $\{\uparrow\uparrow, \uparrow\downarrow, \downarrow\uparrow, \downarrow\downarrow\}$ up to a global phase factor. If we instead set $\int_0^t dt'\, 2K(t')/\hbar = \pi/2$ we get $\mathbf{U} = \sqrt{\text{SWAP}}$ up to a global phase of $\pi/8$. The $\sqrt{\text{SWAP}}$ gate can be converted into a standard C_Z gate with the sequence

$$C_Z = e^{-\iota 3\pi/4} e^{\iota \frac{\pi}{4} Z_1} e^{-\iota \frac{\pi}{4} Z_2} \sqrt{\text{SWAP}} e^{\iota \frac{\pi}{2} Z_1} \sqrt{\text{SWAP}}. \tag{15.14}$$

It should be noted that the two electrons allow for four spin states: the singlet state $|S\rangle = \frac{|\uparrow\downarrow\rangle - |\downarrow\uparrow\rangle}{\sqrt{2}}$, and the triplet states $|T_0\rangle = \frac{|\uparrow\downarrow\rangle + |\downarrow\uparrow\rangle}{\sqrt{2}}$, $|T_1\rangle = |\uparrow\uparrow\rangle$, and $|T_{-1}\rangle = |\downarrow\downarrow\rangle$. The bias field B_0 shifts $|T_1\rangle, |T_{-1}\rangle$ away from $|S\rangle, |T_0\rangle$ which have no Zeeman shift, and it is only these states that are resonantly coupled by the exchange interaction. The time needed for a C_Z gate depends on the magnitude of the exchange coupling K and on the time needed for the one-qubit rotations. The exchange based coupling underlying the $\sqrt{\text{SWAP}}$ gate, as described here, was demonstrated in 2005 [39] in GaAs quantum dots. Entanglement of two-electron qubits on a sub-microsecond time scale was achieved a few years later with single electron qubits [40] and with qubits encoded in the $|S\rangle$ and $|T_0\rangle$ states of two electrons [41]. In the case of $|S\rangle - |T_0\rangle$ encoding the gate operation relies on the presence of both exchange and capacitive coupling of the charges [42].

While the exchange based $\sqrt{\text{SWAP}}$ gate has long been the natural approach for entangling quantum dot qubits [28] it is also possible to implement a C_Z gate more directly by combining the exchange interaction with resonant driving between spin states in the presence of a magnetic field gradient [43]. The spin states $|\uparrow\downarrow\rangle, |\downarrow\uparrow\rangle$ that are resonantly coupled by the exchange mechanism acquire an energy splitting proportional to the field gradient. This enables a controllable spin dependent phase shift giving an interaction of the form $\mathbf{U} = \text{diag}\{1, e^{\iota\phi_1}, e^{\iota\phi_2}, 1\}$, where the phases are proportional to the interaction time. Choosing $\phi_1 + \phi_2 = \pi$ gives a C_Z gate up to single qubit phases. This type of gate was demonstrated in [44] and used to demonstrate a programmable two-qubit processor [45].

15.15 Variations on a Theme

The Loss and DiVincenzo proposal of qubits encoded in single electrons has provided a rich source of inspiration for extensions to the original ideas that have led to performance in small systems approaching that needed for scalable operation. A major challenge has been the need to increase T_1 and T_2 coherence times. Spin qubits experience spin-flips and dephasing mediated by spin-phonon coupling, spin-orbit coupling, and hyperfine coupling to nuclear spins in the host. Measurements in GaAs/AlGaAs heterostructures showed spin flip times $T_1 < 1\,\mathrm{ms}$ [31], which is long compared to gate times, but the inhomogeneous dephasing time was observed to be $T_2^* \sim 10\,\mathrm{ns}$ [39], orders of magnitude shorter. Using dynamical decoupling sequences this has been increased to $T_2 \sim 200\,\mu s$ [46].

Longer coherence times are possible in materials that have reduced spin-phonon, and spin-orbit coupling as well as zero nuclear spin. This has motivated the development of spin qubits in Si and SiGe materials [47]. Compared to III–V semiconductors Si has much smaller spin-phonon and spin-orbit coupling. Natural Si is 95.3% ^{28}Si and ^{30}Si which have zero nuclear spin, and 4.7% ^{29}Si with $I = 1/2$, and therefore much reduced hyperfine dephasing. The situation can be improved even further using isotopically purified Si. Also natural Ge which has 7.7% ^{73}Ge with $I = 9/2$ and all other isotopes with zero spin is a favorable host material. Even working with unenriched Ge in a SiGe heterostructure nuclear spin effects are suppressed since a small fraction of the electron density resides in the Si/Ge barriers above and below the wells. In natural Si $T_2^* \sim 1\,\mu s$ and with purified ^{28}Si, $T_2^* = 120\,\mu s$ has been demonstrated in a heterostructure fabricated with silicon metal oxide semiconductor (MOS) technology [48]. With dynamical decoupling techniques $T_2 \sim 30\,\mathrm{ms}$ has been achieved.

Although charge noise does not directly couple to the spin state, two-qubit gate operations rely on the exchange mechanism which involves electron tunneling that is sensitive to the local charge environment. In order to mitigate the impact on gate performance, as well as the desire to provide better local control, and reduced dependence on resonant fields, encodings based on more than one electron have been developed. As has already been mentioned spin qubits can be encoded in two electrons using the $|S\rangle$ and $|T_0\rangle$ spin states as a computational basis. These singlet-triplet qubits can be advantageous compared to single electron qubits for non-destructive measurements [34]. With encodings based on three electrons, each in a separate dot, it is possible to implement a universal gate set with only electrostatic control of the exchange interaction [49], thereby removing the need for resonant driving which simplifies the control infrastructure. A related concept that allows for simpler gate implementations, better coherence properties, and very fast operations at the 100 ps time scale is the three electron hybrid spin-charge qubit with two electrons in one well and one electron in the other well [50]. For completeness we note that optical excitations of quantum dots, excitons, provide yet another control modality for qubits encoded in electrons. One- and two-qubit gate operations have been demonstrated [51], but the nsec scale coherence times present a challenge for scalability.

In addition to the range of possible implementations with electron qubits in gate defined potentials there has been extensive work on donor based qubits. The original idea of this type stems from a proposal of Kane [52] to use qubits encoded in nuclear spin states that can have exceptionally long coherence times. Donor electrons are coupled to the nuclear spin through the hyperfine interaction, and nuclear-nuclear two-qubit interactions are mediated by electron-electron exchange coupling as in gate defined qubits. To have a strong nuclear-electron hyperfine coupling a positively charged donor ion that attracts a high electron density near the nucleus is desirable. At the same time the electron wavefunction should be sufficiently delocalized to enable coupling to neighboring nuclei that are not too close. These requirements can be satisfied using shallow-level donors for which electron wavefunctions extend up to a few 100 Å away from the nucleus.

A suitable shallow donor is ^{31}P in Si. The ^{31}P nucleus has $I = 1/2$ making it ideal for qubit encoding. Using electrostatic gates to distort the electronic wavefunction the nucleus-donor electron hyperfine coupling can be turned on and off, and with additional gates the exchange interaction between donor electrons attached to separate nuclei can be used to implement a two-qubit exchange gate, as in [28]. A requirement in this approach is that donor implantation is controlled at sub nm scales. Although extremely challenging, substantial progress has been made including the demonstration of coherent control of the nuclear and electron spins and a nuclear coherence time of 30 s in an isotopically purified Si device with a single ^{31}P donor [53].

15.16   Outlook

Semiconductor quantum dot based approaches to quantum computing have advanced to the point where all of the DiVincenzo criteria have been demonstrated with fidelities approaching those needed for scalability. While the motivation for a semiconductor based approach lies in part in the hope that the remarkable scalability of classical semiconductor electronics can be transferred to the quantum domain, that scalability remains to be achieved in practice. The current state of the art includes arrays with four [54] and nine [55] quantum dots.

One of the challenges is the small 100 nm feature scale of quantum dot devices as seen in Fig. 15.1b. While the small footprint is in principle an advantage for scalability, and 100 nm is also large compared to the sub 10 nm scale of state of the art semiconductor chips, performing quantum control with low noise at that scale presents new challenges for integration of the necessary leads at each dot. Designs based on interleaving small arrays of dots with control electronics, and coupling channels between arrays, are being developed [56]. These architectures rely on long range coupling that is possible based on cQED approaches utilizing coplanar microwave waveguides analogous to those that have been well developed for superconducting qubits [57].

15.17   Problems

1. Consider a linear LC circuit at room temperature. Quantized energy levels are not observed. This may in part be due to the finite resistance and energy dissipation of the wires. Model the dissipation as a resistor R in series with L and C.

   (a) Estimate an upper bound on the value of $R = R_{max}$ such that discrete harmonic oscillator energy levels are observable. Express R_{max} in terms of the characteristic impedance of the circuit $Z = \sqrt{L/C}$. In this part ignore any other loss mechanisms and thermal effects.
   (b) Find numerical values for R_{max} and $\omega/2\pi$ using $C = 0.3\,pF$ and $L = 3.\,nH$.
   (c) Is R_{max} a value that could be achieved in a room temperature circuit with normal wires? If the answer is yes, then why do you think quantized energy levels are not normally observed at room temperature?
   (d) Let $R = 0$ and estimate the temperature T for which discrete energy levels can in principle be observed using the parameters of part (b).

2. Fill in the steps leading to Eq. (15.4).
3. Show that Eq. (15.9) follows from (15.8) when we restrict the states to the number basis $|0\rangle$, $|1\rangle$. Hint: Although a unique Hermitian phase operator does not exist in quantum mechanics the operator $\cos(\hat{\phi})$ can be given a definite meaning using the polar decompositions of the creation and annihilation operators: $\hat{a} = e^{i\hat{\phi}}\hat{n}^{1/2}$, $\hat{a}^{\dagger} = \hat{n}^{1/2}e^{-i\hat{\phi}}$.
4. Consider a superconducting qubit prepared in the state $|\psi\rangle = c_0|0\rangle + c_1|1\rangle$. The qubit state is mapped to a microwave photon excitation of a resonator such that $|0\rangle \to |0\rangle_v, |1\rangle \to |1\rangle_v$ where $|n\rangle_v$ represents an n photon microwave excitation. At finite temperature the resonator is also thermally excited according to $n_{th} = \frac{1}{e^{\hbar\omega/k_BT}-1}$ where ω is the angular photon frequency. The thermal state of the resonator can be described by a density matrix ρ_{th}.

   A full solution of this problem requires solving the master equation for the dynamics of the coupled system of qubit+resonator+thermal environment. Here we will take a simpler approach to estimate the fidelity of the resonator state, without solving the full problem. You can assume for the calculations that the temperature is low enough such that we can truncate the thermal density matrix to zero and one photon excitations.

(a) It can be argued that a good approximation to the density matrix of the thermal resonator after qubit mapping is

$$\rho = \frac{1}{1+e^{-\beta}}\begin{pmatrix}|c_0|^2 & c_0 c_1^* \\ c_0^* c_1 & |c_1|^2 + e^{-\beta}\end{pmatrix}, \qquad (15.15)$$

with $\beta = \hbar\omega/k_B T$. The density matrix has been truncated to zero and one photon excitations and is therefore a 2×2 matrix. Show that $\lim_{T\to 0}\rho = \rho_{id}$ with ρ_{id} the density matrix we would obtain after mapping if there were no thermal excitations. In addition show that $\lim_{c_1\to 0}\rho = \rho_{th}$ where ρ_{th} is the thermal density matrix, truncated to zero and one photons.

(b) Find the microwave state preparation fidelity $\mathcal{F} = \left(\mathrm{Tr}\left[\sqrt{\sqrt{\rho}\,\rho_{id}\sqrt{\rho}}\right]\right)^2$ as a function of T, ω and c_0, c_1. Assuming $c_0 = c_1 = 1/\sqrt{2}$ and $\omega = 2\pi \times 5\,\mathrm{GHz}$ what is the maximum value of T for $\mathcal{F} = 0.99$? You may want to compare your answer with that from a full master equation calculation which can be found in Phys. Rev. A **89**, 010301(R) (2014), Fig. 4.

(c) The approximation in part (b) is good at low temperatures but is too crude at higher temperatures so it overestimates the fidelity. A better approximation would be to use

$$\rho = \frac{1}{1+\frac{1}{e^{\beta}-1}}\begin{pmatrix}|c_0|^2 & c_0 c_1^* \\ c_0^* c_1 & |c_1|^2 + \frac{1}{e^{\beta}-1}\end{pmatrix}.$$

Repeat part (b) with this expression for ρ.

(d) Now consider two superconducting qubits prepared in the entangled singlet state $|\psi\rangle = \frac{|01\rangle - |10\rangle}{\sqrt{2}}$. The qubits are mapped to microwave photon excitations of two resonators. Find the maximum temperature T for which the microwave excitations of the resonators are entangled. What is the entanglement fidelity, using the coherence criterion (or some other criterion of your choice) at the temperature found in parts (b) and (c).

5. The exchange interaction can be used to implement a two-qubit gate for quantum dots. In addition to exchange, electrons have a magnetic dipole interaction. Find the strength of the magnetic dipole interaction for two electrons in the state $|1/2, 1/2\rangle_z$ that are separated by 100 nm in the x direction (perpendicular to the quantization axis). Express your answer in Hz.

6. Consider a simplified model of a two-site quantum dot device with sites separated by d along the x axis. The trapping potential is $V = V_0\left[e^{-(x-d/2)^2/w^2}e^{-(x+d/2)^2/w^2}\right]$. Assume the effective electron mass is the bare mass, $V_0 = -k_B T$ with $T = 10\,\mathrm{K}$, $d = 100\,\mathrm{nm}$, and $w = 20\,\mathrm{nm}$.

(a) In the tight binding approximation $x_0 \ll d, w$ find x_0, ω_0 for the given parameters with x_0, ω_0 the harmonic oscillator length and vibrational frequency of the electron wavefunction.

(b) Calculate the direct and exchange coefficients D, K.

7. Verify that the expression in Eq. (15.14) for a C_Z gate is correct.

References

1. A. Opremcak, I.V. Pechenezhskiy, C. Howington, B.G. Christensen, M.A. Beck, E. Leonard, Jr., J. Suttle, C. Wilen, K.N. Nesterov, G.J. Ribeill, T. Thorbeck, F. Schlenker, M.G. Vavilov, B.L.T. Plourde, R. McDermott, Measurement of a superconducting qubit with a microwave photon counter. Science **361**, 1239 (2018)
2. F. Arute et. al., Quantum supremacy using a programmable superconducting processor. Nature **574**, 505 (2019)
3. G. Wendin, Quantum information processing with superconducting circuits: a review. Rep. Prog. Phys. **80**, 106001 (2017)
4. P. Krantz, M. Kjaergaard, F. Yan, T. Orlando, S. Gustavsson, W.D. Oliver, A quantum engineer's guide to superconducting qubits. Appl. Phys. Rev. **6**, 021318 (2019)
5. D. Gottesman, A. Kitaev, J. Preskill, Encoding a qubit in an oscillator. Phys. Rev. A **64**, 012310 (2001)
6. J. Bardeen, L.N. Cooper, J.R. Schrieffer, Theory of superconductivity. Phys. Rev. **108**, 1175 (1957)
7. L.N. Cooper, Bound electron pairs in a degenerate Fermi gas. Phys. Rev. **104**, 1189 (1956)
8. M. Tinkham, *Introduction to Superconductivity*, 2nd edn. (McGraw-Hill, New York, 2010)
9. Y. Nakamura, Y.A. Pashkin, J.S. Tsai, Coherent control of macroscopic quantum states in a single-Cooper-pair box. Nature **398**, 786 (1999)
10. T. Yamamoto, Y.A. Pashkin, O. Astafiev, Y. Nakamura, J.S. Tsai, Demonstration of conditional gate operation using superconducting charge qubits. Nature (London) **425**, 941 (2003)
11. J.E. Mooij, T.P. Orlando, L. Levitov, L. Tian, C.H. van der Wal, S. Lloyd, Josephson persistent-current qubit. Science **285**, 1036 (1999)
12. I. Chiorescu, Y. Nakamura, C. Harmans, J. Mooij, Coherent quantum dynamics of a superconducting flux qubit. Science **299**, 1869 (2003)
13. R. Ramos, M. Gubrud, A. Berkley, J. Anderson, C. Lobb, F. Wellstood, Design for effective thermalization of junctions for quantum coherence. IEEE Trans. Appl. Supercond. **11**, 998 (2001)
14. J.M. Martinis, S. Nam, J. Aumentado, C. Urbina, Rabi oscillations in a large Josephson junction qubit. Phys,. Rev. Lett. **89**, 117 (2002)
15. M. Steffen, M. Ansmann, R.C. Bialczak, N. Katz, E. Lucero, R. McDermott, M. Neeley, E.M. Weig, A.N. Cleland, J.M. Martinis, Measurement of the entanglement of two superconducting qubits via state tomography. Science **313**, 1423–1425 (2006)
16. J. Koch, T.M. Yu, J. Gambetta, A.A. Houck, D.I. Schuster, J. Majer, A. Blais, M.H. Devoret, S.M. Girvin, R.J. Schoelkopf, Charge-insensitive qubit design derived from the Cooper pair box. Phys. Rev. A **76**, 042319 (2007)
17. A. Wallraff, D.I. Schuster, A. Blais, L. Frunzio, R.S. Huang, J. Majer, S. Kumar, S.M. Girvin, R.J. Schoelkopf, Strong coupling of a single photon to a superconducting qubit using circuit quantum electrodynamics. Nature **431**, 162 (2004)
18. A. Blais, R.-S. Huang, A. Wallraff, S.M. Girvin, R.J. Schoelkopf, Cavity quantum electrodynamics for superconducting electrical circuits: an architecture for quantum computation. Phys. Rev. A **69**, 062320 (2004)

19. A. Blais, J. Gambetta, A. Wallraff, D.I. Schuster, S.M. Girvin, M.H. Devoret, R.J. Schoelkopf, Quantum-information processing with circuit quantum electrodynamics. Phys. Rev. A **75**, 032329 (2007)
20. C. Nayak, S.H. Simon, A. Stern, M. Freedman, S. Das Sarma, Non-Abelian anyons and topological quantum computation. Rev. Mod. Phys. **80**, 1083–1159 (2008)
21. P. Brooks, A. Kitaev, J. Preskill, Protected gates for superconducting qubits. Phys. Rev. A **87**, 052306 (2013)
22. J.M. Dempster, B. Fu, D.G. Ferguson, D.I. Schuster, J. Koch, Understanding degenerate ground states of a protected quantum circuit in the presence of disorder. Phys. Rev. B **90**, 094518 (2014)
23. P. Groszkowski, A. DiPaolo, A.L. Grimsmo, A. Blais, D.I. Schuster, A.A. Houck, J. Koch, Coherence properties of the $0 - \pi$ qubit. New J. Phys. **20**, 043053 (2018)
24. Y.Y. Gao, B.J. Lester, K.S. Chou, L. Frunzio, M.H. Devoret, L. Jiang, S.M. Girvin, R.J. Schoelkopf, Entanglement of bosonic modes through an engineered exchange interaction. Nature **566**, 509 (2019)
25. D. DiVincenzo, The physical implementation of quantum computers. Fort. Phys. **48**, 771 (2000)
26. P.M. Platzman, M.I. Dykman, Quantum computing with electrons floating on liquid helium. Science **284**, 1967 (1999)
27. L. Lamata, D. Porras, J.I. Cirac, J. Goldman, G. Gabrielse, Towards electron-electron entanglement in Penning traps. Phys. Rev. A **81**, 022301 (2010)
28. D. Loss, D.P. DiVincenzo, Quantum computation with quantum dots. Phys. Rev. A **57**, 120 (1998)
29. R. Hanson, L.P. Kouwenhoven, J.R. Petta, S. Tarucha, L.M.K. Vandersypen, Spins in few-electron quantum dots. Rev. Mod. Phys. **79**, 1217–1265 (2007)
30. C.H. Yang, R.C.C. Leon, J.C.C. Hwang, A. Saraiva, T. Tanttu, W. Huang, J.C. Lemyre, K.W. Chan, K.Y. Tan, F.E. Hudson, K.M. Itoh, A. Morello, M. Pioro-Ladrière, A. Laucht, A.S. Dzurak, Operation of a silicon quantum processor unit cell above one kelvin. Nature **580**, 350 (2020)
31. J.M. Elzerman, R. Hanson, L.H.W. van Beveren, B. Witkamp, L.M.K. Vandersypen, L.P. Kouwenhoven, Single-shot read-out of an individual electron spin in a quantum dot. Nature **430**, 431 (2004)
32. M. Friesen, C. Tahan, R. Joynt, M.A. Eriksson, Spin readout and initialization in a semiconductor quantum dot. Phys. Rev. Lett. **92**, 037901 (2004)
33. T. Nakajima, A. Noiri, J. Yoneda, M.R. Delbecq, P. Stano, T. Otsuka, K. Takeda, S. Amaha, G. Allison, K. Kawasaki, A. Ludwig, A.D. Wieck, D. Loss, S. Tarucha, Quantum non-demolition measurement of an electron spin qubit. Nat. Nanotech. **14**, 555 (2019)
34. T. Meunier, I.T. Vink, L.H. Willems van Beveren, F.H.L. Koppens, H.P. Tranitz, W. Wegscheider, L.P. Kouwenhoven, L.M.K. Vandersypen, Nondestructive measurement of electron spins in a quantum dot. Phys. Rev. B **74**, 195303 (2006)
35. F.H.L. Koppens, C. Buizert, K.J. Tielrooij, I.T. Vink, K.C. Nowack, T. Meunier, L.P. Kouwenhoven, L.M.K. Vandersypen, Driven coherent oscillations of a single electron spin in a quantum dot. Nature **442**, 766 (2006)
36. K.C. Nowack, F.H.L. Koppens, Y.V. Nazarov, L.M.K. Vandersypen, Coherent control of a single electron spin with electric fields. Science **318**, 1430 (2007)
37. C. Ospelkaus, U. Warring, Y. Colombe, K.R. Brown, J.M. Amini, D. Leibfried, D.J. Wineland, Microwave quantum logic gates for trapped ions. Nature **476**, 181 (2011)
38. D. Kim, D.R. Ward, C.B. Simmons, D.E. Savage, M.G. Lagally, M. Friesen, S.N. Coppersmith, M.A. Eriksson, High-fidelity resonant gating of a silicon-based quantum dot hybrid qubit. NPJ Quant. Inf. **1**, 15004 (2015)
39. J. Petta, A. Johnson, J. Taylor, E. Laird, A. Yacoby, M. Lukin, C. Marcus, M.P. Hanson, A. Gossard, Coherent manipulation of coupled electron spins in semiconductor quantum dots. Science **309**, 2180 (2005)

40. K.C. Nowack, M. Shafiei, M. Laforest, G.E.D.K. Prawiroatmodjo, L.R. Schreiber, C. Reichl, W. Wegscheider, L.M.K. Vandersypen, Single-shot correlations and two-qubit gate of solid-state spins. Science **333**, 1269 (2011)

41. M.D. Shulman, O.E. Dial, S.P. Harvey, H. Bluhm, V. Umansky, A. Yacoby, Demonstration of entanglement of electrostatically coupled singlet-triplet qubits. Science **336**, 202 (2012)

42. J.M. Taylor, H.A. Engel, W. Dür, A. Yacoby, C.M. Marcus, P. Zoller, M.D. Lukin, Fault-tolerant architecture for quantum computation using electrically controlled semiconductor spins. Nat. Phys. **1**, 177 (2005)

43. T. Meunier, V.E. Calado, L.M.K. Vandersypen, Efficient controlled-phase gate for single-spin qubits in quantum dots. Phys. Rev. B **83**, 121403 (2011)

44. M. Veldhorst, C. Yang, J. Hwang, W. Huang, J. Dehollain, J. Muhonen, S. Simmons, A. Laucht, F. Hudson, K. Itoh, A. Morello, A. Dzurak, A two-qubit logic gate in silicon. Nature **526**, 410 (2015)

45. T.F. Watson, S.G.J. Philips, E. Kawakami, D.R. Ward, P. Scarlino, M. Veldhorst, D.E. Savage, M.G. Lagally, M. Friesen, S.N. Coppersmith, M.A. Eriksson, L.M.K. Vandersypen, A programmable two-qubit quantum processor in silicon. Nature **555**, 633 (2018)

46. H. Bluhm, S. Foletti, I. Neder, M. Rudner, D. Mahalu, V. Umansky, A. Yacoby, Dephasing time of GaAs electron-spin qubits coupled to a nuclear bath exceeding $200\,\mu s$. Nat. Phys. **7**, 109 (2011)

47. F.A. Zwanenburg, A.S. Dzurak, A. Morello, M.Y. Simmons, L.C.L. Hollenberg, G. Klimeck, S. Rogge, S.N. Coppersmith, M.A. Eriksson, Silicon quantum electronics. Rev. Mod. Phys. **85**, 961–1019 (2013)

48. M. Veldhorst, J.C.C. Hwang, C.H. Yang, A.W. Leenstra, B. deRonde, J.P. Dehollain, J.T. Muhonen, F.E. Hudson, K.M. Itoh, A. Morello, A.S. Dzurak, An addressable quantum dot qubit with fault-tolerant control-fidelity. Nat. Nanotech. **9**, 981 (2014)

49. D.P. DiVincenzo, D. Bacon, J. Kempe, G. Burkard, K.B. Whaley, Universal quantum computation with the exchange interaction. Nature **408**, 339 (2000)

50. D. Kim, Z. Shi, C.B. Simmons, D.R. Ward, J.R. Prance, T.S. Koh, J.K. Gamble, D.E. Savage, M.G. Lagally, M. Friesen, S.N. Coppersmith, M.A. Eriksson, Quantum control and process tomography of a semiconductor quantum dot hybrid qubit. Nature **511**, 70 (2014)

51. D. Kim, S.G. Carter, A. Greilich, A.S. Bracker, D. Gammon, Ultrafast optical control of entanglement between two quantum-dot spins. Nat. Phys. **7**, 223 (2011)

52. B.E. Kane, A silicon-based nuclear spin quantum computer. Nature **393**, 133–137 (1998)

53. J.T. Muhonen, J.P. Dehollain, A. Laucht, F.E. Hudson, R. Kalra, T. Sekiguchi, K.M. Itoh, D.N. Jamieson, J.C. McCallum, A.S. Dzurak, A. Morello, Storing quantum information for 30 seconds in a nanoelectronic device. Nat. Nanotech. **9**, 986 (2014)

54. S.F. Neyens, E. MacQuarrie, J. Dodson, J. Corrigan, N. Holman, B. Thorgrimsson, M. Palma, T. McJunkin, L. Edge, M. Friesen, S. Coppersmith, M. Eriksson, Measurements of capacitive coupling within a quadruple-quantum-dot array. Phys. Rev. Appl. **12**, 064049 (2019)

55. D.M. Zajac, T.M. Hazard, X. Mi, E. Nielsen, J.R. Petta, Scalable gate architecture for a one-dimensional array of semiconductor spin qubits. Phys. Rev. Appl. **6**, 054013 (2016)

56. L.M.K. Vandersypen, H. Bluhm, J.S. Clarke, A.S. Dzurak, R. Ishihara, A. Morello, D.J. Reilly, L.R. Schreiber, M. Veldhorst, Interfacing spin qubits in quantum dots and donors—hot, dense, and coherent. npj Qu. Inf. **3**, 34 (2017)

57. T. Frey, P.J. Leek, M. Beck, A. Blais, T. Ihn, K. Ensslin, A. Wallraff, Dipole coupling of a double quantum dot to a microwave resonator. Phys. Rev. Lett. **108**, 046807 (2012)

Index

© Springer Nature Switzerland AG 2021
J. A. Bergou et al., *Quantum Information Processing*, Graduate Texts in Physics,
https://doi.org/10.1007/978-3-030-75436-5

Printed in the United States
by Baker & Taylor Publisher Services

Printed in the United States
by Baker & Taylor Publisher Services